INTERNATIONAL CENTRE FOR MECHANICAL SCIENCES

COURSES AND LECTURES - No. 433

SELECTED TOPICS IN BOUNDARY INTEGRAL FORMULATIONS FOR SOLIDS AND FLUIDS

EDITED BY

VLADIMIR KOMPIŠ
UNIVERSITY OF ŽILINA

This volume contains 91 illustrations

Printed in Italy
SPIN 10890847

In order to make this volume available as economically and as rapidly as possible the authors' typescripts have been reproduced in their original forms. This method unfortunately has its typographical limitations but it is hoped that they in no way distract the reader.

ISBN 3-211-83693-4 Springer-Verlag Wien New York

DATE DUE

TH

CISM COURSES AND LECTURES

The series presents lecture notes, monographs, edited works and proceedings in the field of Mechanics, Engineering, Computer Science and Applied Mathematics.
Purpose of the series is to make known in the international scientific and technical community results obtained in some of the activities organized by CISM, the International Centre for Mechanical Sciences.

PREFACE

This book contains the complete lectures presented by L. Gaul, V. Kompiš, S. Mukherjee and G. Szeidl at the Advanced School on "Selected Topics in Boundary Integral Formulations for Solids and Fluids" in CISM Udine, during July 9-13, 2001. The lectures given by G. Novati are only partly contained in the book, because they were partially shared with the topics given by his coworker G. Maier in another course, and will be published in the book for that course. Also, the topics concerning design sensitivity for materially non-linear problems, application of BEM to metal forming, thermal aspects of machining problems, conduction-convection, damage in grinding applications and a hybrid micro-macro BE formulation, given by A. Chandra, have already been published in the book "A. Chandra and S. Mukherjee: Boundary Element Methods in Manufacturing, Oxford University Press, 1997", and are not contained in this book.

The aim of the course was to present some current non-classical boundary integral formulations: non-singular Boundary Integral Equation (BIE) formulations, reciprocity and hybrid multi-domain FEM/BEM, the Boundary Contour and Boundary Node Methods, dual reciprocity BEM and their applications in linear and non-linear Computational Mechanics.

Chapter 1 of the book is an introduction to non-linear continuum mechanics.

The problems of large deformation of solids using Trefftz functions (i.e. functions satisfying the governing equations for corresponding linear problem) for boundary integral formulation with non-singular kernels are shown in the Chapter 2. The total Lagrange, multi-domain (MD) formulation for the sub-domain solution results in a formulation compatible with classical FEM. Fast convergence is documented even for large rotations and strains in examples.

Chapter 3 contains some smoothing procedures for the gradient fields (stresses) for multi-domain formulations for both linear and non-linear problems. It is shown that the same rate of convergence in these fields can be obtained as that obtained in primary field.

Errors in MD Trefftz formulations and modeling of local effects are presented in the Chapter 4.

In Chapter 5, BIE formulations using the Trefftz reciprocity principle are introduced for some CFD problems described by the Poisson, Laplace, Helmholtz, or Navier-Stokes equations.

For the solution of fluid-structure interaction problems, symmetric Hybrid BEM are derived in Chapter 6 for the acoustic field. Multi-field variational principles lead to Hybrid Displacement and Hybrid Stress formulations of BEM. Calculated sound field examples in frequency and time domain indicate the accuracy and efficiency of the methods.

The noise radiation from rolling tires is studied numerically by using a sequential FEM – Hybrid BEM approach in Chapter 7. An Arbitrary Lagrangian-Eulerian description is used to solve the non-linear contact problem for steady state rolling. The superimposed vibratory response is calculated from an eigenvalue problem in the deformed state, providing the Neumann data for the acoustic field.

Numerical solutions of the equations of piezoelectricity by domain and boundary discretization methods, FEM and BEM, respectively are described in Chapter 8. Excellent agreement of FEM and BEM results as well as the superiority of the BEM for the calculation of elastic stresses and the electric field are documented.

In Chapter 9 transient viscoelastodynamic BE formulations are shown to be a powerful tool for the computation of dynamic structural response in frequency and time domain. Fundamental solutions are used as weighting functions in the BIE, which fulfill the Sommerfeld radiation condition, ensuring the energy radiation into an infinite domain. The transient response of viscoelastic continua such as soil is obtained by inverse Fourier transform in each time step, or alternatively by the Convolution Quadrature Method. A numerical study of wave propagation problems in a 3D viscoelastic waveguide and a halfspace are presented.

The BE approach in Chapter 10 avoids singular boundary integrals for problems of elastostatics and elastodynamics by placing the load points at a proper, error minimizing distance outside of the problem boundary. A weighted superposition of static fundamental solutions is used for the field approximation in the domain, whereas the displacement and stress fields on the boundary are interpolated by polynomial shape functions. Separation of time and space dependency provides a boundary formulation with time independent mass and stiffness matrices such that efficient numerical solutions can be obtained.

Chapter 11 presents the Boundary Contour Method (BCM) for 3D linear elasticity, together with shape sensitivity analysis, shape optimization, and error analysis and adaptivity with the BCM. Some numerical examples are given.

The (meshfree) Boundary Node Method (BNM) for 3D linear elasticity, with surface approximants, is introduced in Chapter 12. Error estimation and adaptivity, together with numerical examples, are presented here for this method.

A formulation of the symmetric Galerkin BEM for 3D elastic fracture mechanics problems and some relevant computational aspects are presented in Chapter 13. The method is employed for the evaluation of stress intensity factors and for modeling the fatigue crack growth. Two examples solving the fatigue growth of surface breaking cracks are given.

Chapter 14 is devoted to a BC formulation in the dual system of plane elasticity. The governing equations of the problem are given in terms of stress functions of order one. It is shown that the integrand of the direct boundary integral equation is divergence free in the dual formulation as well. The shape functions are quadratic. Numerical examples are also presented.

We, the lecturers, really enjoyed having such a knowledgeable and responsive group of young people in this course. Also, support from the administrative staff of CISM Udine during the course was excellent, and is greatly appreciated.

I am glad that I could give together such a good team of lecturers.

Vlado Kompiš

CONTENTS

Chapter 1
Introduction to nonlinear continuum mechanics

V. Kompiš

Department of Mechanical Engineering, University of Žilina,
Veľký Diel, 010 26, Žilina, Slovak Republic
e-mail: kompis@fstroj.utc.sk

Abstract. In this chapter we will briefly give the basic introduction to nonlinear continuum mechanics. The reader can find a more detail description of the problems in [Belytschko, T, Liu W. K, and Moran, B. (2000)], [Okrouhlik, M. (ed.). (1995)] ,[Altenbach, J, Altenbach, H. (1994)]. The most general analysis case is the one in which the material is subjected to large displacements and large strains. The basic problem in a general nonlinear analysis is to find the state of equilibrium of a body corresponding to the applied loads.Continuum mechanics is concerned with models of solids and fluids in which properties and response can be characterised by smooth function of spatial variables. The objective of continuum mechanics is to provide models for macroscopic behaviour of fluids, solids and structures. Three aspects of description of governing equations are classified: 1. the mesh description, 2. the kinetic description determined by the choice of the stress tensor and the form of the momentum equation and 3. The kinematic description determined by the choice of the strain measure.

1 Description of motion.

Spatial (Eulerian) coordinates are, denoted by **x**, specify the location of points in space. *Material (Lagrangian) coordinates*, denoted by **X**, specify a material point: each material point has a unique material coordinates. The domain of a body in the initial state is denoted by Ω_0 and it is called *initial*, or *reference configuration*. It is often considered to be an *undeformed configuration*. The domain of the current configuration of the body is denoted by Ω. This will often be called the *deformed configuration*.
The *motion* or *deformation* of a body is described by a function

$$\mathbf{x} = \phi(\mathbf{X}, t) \tag{1}$$

with material coordinate **X** and time coordinate t as the independent variables.
The displacement **u** of a material point is

$$\mathbf{u}(\mathbf{X}, t) = \phi(\mathbf{X}, t) - \mathbf{X} \tag{2}$$

and it is a difference between its current and original (undeformed) positions.

The differences between Lagrangian and Eulerian meshes are most clearly seen in the behaviour of the nodes. If the mesh is Eulerian, the coordinates of nodes are fixed, i.e. they are coincident with spatial points. If the mesh is Lagrangian, the Lagrangian (material) coordinates of nodes are time invariant, i.e. the nodes are coincident with material points.

The vectors are defined by their components and corresponding unit vectors $\mathbf{e}_i$ of a rectangular Cartesian coordinate system, e.g.

$$\mathbf{X} = X_i \mathbf{e}_i \tag{3}$$

with the summation after repeated indices

The *velocity*

$$\mathbf{v}(\mathbf{X},t) = \frac{\partial \phi(\mathbf{X},t)}{\partial t} = \frac{\partial \mathbf{u}(\mathbf{X},t)}{\partial t} = \dot{\mathbf{u}} \tag{4}$$

is the rate of change of the position vector for a material point, i.e. the time derivative with $\mathbf{X}$ held constant (called material time derivative). The *material time derivatives* are also called *total derivatives.* Note that $\mathbf{X}$ is independent of time.

Similarly, acceleration is given by

$$\mathbf{a}(\mathbf{X},t) = \frac{\partial \mathbf{v}(\mathbf{X},t)}{\partial t} = \frac{\partial^2 \mathbf{u}(\mathbf{X},t)}{\partial t} = \ddot{\mathbf{u}} \tag{5}$$

When the velocity, v(x, t), is expressed in terms of the spatial coordinates and the time, i.e. in an Eulerian description, the material time derivative is obtained by:

$$\frac{Dv_i(\mathbf{x},t)}{Dt} = \frac{\partial v_i(\mathbf{x},t)}{\partial t} + \frac{\partial v_i(\mathbf{x},t)}{\partial x_j}\frac{\partial \phi_j(\mathbf{x},t)}{\partial t} = \frac{\partial v_i}{\partial t} + \frac{\partial v_i}{\partial x_j} v_j \tag{6}$$

which can be written in tensor notation as

$$\frac{D\mathbf{v}(\mathbf{x},t)}{Dt} = \frac{\partial \mathbf{v}(\mathbf{x},t)}{\partial t} + \mathbf{v}.\nabla\mathbf{v} = \frac{\partial \mathbf{v}}{\partial \mathbf{t}} + \mathbf{v}.grad\mathbf{v} \tag{7}$$

An important variable in the description of deformation is the *deformation gradient*

$$F_{ij} = \frac{\partial \phi_i}{\partial X_j} = \frac{\partial x_i}{\partial X_j} = x_{i,j} \qquad \text{or} \qquad \mathbf{F} = \frac{\partial \phi}{\partial \mathbf{X}} = (\nabla_0 \mathbf{\Phi})^T \tag{8}$$

The determinant of $\mathbf{F}$ is denoted by J and called *Jacobian determinant,* or simply *Jacobian*

$$J = \det(\mathbf{F}) \tag{9}$$

The Jacobian can be used to relate integrals in the current and reference configurations by

$$\int_{\Omega} f(\mathbf{x},t)d\Omega = \int_{\Omega_0} f(\phi(\mathbf{X},t),t)Jd\Omega_0 \tag{10}$$

The mapping of a differential element $d\mathbf{X}$ into $d\mathbf{x}$ is given by

$$d\mathbf{x} = \mathbf{F}.d\mathbf{X} \text{ or } dx_i = x_{i,j}dX_j = F_{ij}dX_j \tag{11}$$

The deformation gradient is a second order tensor and can be decomposed into orthogonal (rotational) and symmetric (stretch) tensors by

$$\mathbf{F} = \mathbf{R}.\mathbf{U} = \mathbf{V}.\mathbf{R} \tag{12}$$

where $\mathbf{R}$ is the rotation tensor and $\mathbf{U}$ and $\mathbf{V}$ are the right and left stretch tensors, respectively.

Decomposition of the deformation gradient tensor into the rotation and stretch tensors is known as the polar decomposition. The tensors $\mathbf{U}$ and $\mathbf{R}$ can be found from

$$\begin{aligned} \mathbf{U} &= (\mathbf{F}^T.\mathbf{F})^{1/2} \\ \mathbf{R} &= \mathbf{F}.\mathbf{U}^{-1} \end{aligned} \tag{12a}$$

2 Strain measures.

The difference in the squares of the final length ds and original length dS of the differential element is given by

$$\begin{aligned} ds^2 - dS^2 &= dx_k dx_k - dX_k dX_k \\ &= dx_k dx_k - dX_i dX_j \delta_{ij} \\ &= x_{k,i}x_{k,j} dX_i dX_j - dX_i dX_j \delta_{ij} \\ &= (x_{k,i}x_{k,j} - \delta_{ij}) dX_i dX_j \\ &= 2E_{ij} dX_i dX_j \end{aligned} \tag{13}$$

where δ_{ij} is the Kronecker symbol and E_{ij} is the *Green-Lagrange strain tensor*

$$E_{ij} = \frac{1}{2}\left(x_{k,i}x_{k,j} - \delta_{ij}\right) = \frac{1}{2}\left(F_{ki}F_{kj} - \delta_{ij}\right) \tag{14}$$

or in tensor notation

$$\mathbf{E} = \frac{1}{2}\left(\mathbf{F}^T.\mathbf{F} - \mathbf{I}\right) \tag{15}$$

where **I** is the identity tensor.

Using (1.1) and (1.2) in index notation

$$x_i = X_i + u_i \tag{16}$$

we can express

$$x_{k,i} = \delta_{ki} + u_{k,i} \tag{17}$$

and

$$\begin{aligned} x_{k,i}x_{k,j} &= \delta_{ki}\delta_{kj} + \delta_{ki}u_{k,j} + \delta_{kj}u_{k,i} + u_{k,i}u_{k,j} \\ &= \delta_{ij} + u_{i,j} + u_{j,i} + u_{k,i}u_{k,j} \end{aligned} \tag{18}$$

which results in

$$E_{ij} = \frac{1}{2}\left(u_{i,j} + u_{j,i} + u_{k,i}u_{k,j}\right) \tag{19}$$

The last equation in the tensor notation is

$$\mathbf{E} = \frac{1}{2}\left((\nabla_0\mathbf{u})^T + \nabla_0\mathbf{u} + \nabla_0\mathbf{u}.(\nabla_0\mathbf{u})^T\right) \tag{20}$$

We can show that in rigid body motion ($\mathbf{F} = \mathbf{R}$)

$$\mathbf{E} = \frac{1}{2}\left(\mathbf{R}^T.\mathbf{R} - \mathbf{I}\right) = \frac{1}{2}\left(\mathbf{I} - \mathbf{I}\right) = 0 \tag{21}$$

This demonstrates that the Green-Lagrange strain will vanish in any rigid body motion and similarly, that it is independent of rotations, as it can be seen by substituting **F = R . U.** It is an important requirement of a strain measure.

The second kinematic measure to be considered here is the *rate-of-deformation* **D,** which is also called the *velocity strain*. It is a rate measure of deformation. We first define the *velocity gradient* **L** by

$$\mathbf{L} = \frac{\partial \mathbf{v}}{\partial \mathbf{x}} = (\nabla \mathbf{v})^T = (grad\mathbf{v})^T \quad \text{or} \quad L_{ij} = \frac{\partial v_i}{\partial x_j} \tag{22}$$

The velocity gradient tensor can be decomposed into symmetric and skew-symmetric parts

$$\mathbf{L} = \mathbf{D} + \mathbf{W} \tag{23}$$

where

$$\mathbf{D} = \frac{1}{2}\left(\mathbf{L} + \mathbf{L}^T\right) \quad \text{or} \quad D_{ij} = \frac{1}{2}\left(\frac{\partial v_i}{\partial x_j} + \frac{\partial v_j}{\partial x_i}\right) \tag{24}$$

$$\mathbf{W} = \frac{1}{2}\left(\mathbf{L} - \mathbf{L}^T\right) \quad \text{or} \quad W_{ij} = \frac{1}{2}\left(\frac{\partial v_i}{\partial x_j} - \frac{\partial v_j}{\partial x_i}\right) \tag{25}$$

The rate-of-deformation is a measure of the rate of change of the square of the length of the differential element:

$$\frac{\partial}{\partial t}\left(ds^2\right) = 2d\mathbf{x}.\mathbf{D}.d\mathbf{x} \tag{26}$$

In the absence of deformation, the spin tensor (the skew-symmetric tensor **W**) and angular velocity tensor are equal: $\mathbf{W} = \mathbf{\Omega}$. When the body undergoes deformation in addition to rotation, the spin tensor generally differs from the angular velocity tensor.

It can be shown [Belytschko, T, Liu W. K, and Moran, B. (2000)] that the velocity gradient can be expressed in terms of the deformation gradient as

$$\mathbf{L} = \dot{\mathbf{F}}.\mathbf{F}^{-1} \quad \text{or} \quad L_{ij} = \dot{F}_{ik}\, F^{-1}_{kj} \tag{27}$$

and the rate-of-deformation tensor can be related to the rate of the Green-Lagrange strain tensor by

$$\mathbf{D} = \mathbf{F}^{-T}.\dot{\mathbf{E}}.\mathbf{F}^{-1} \quad \text{or} \quad D_{ij} = F^{-T}_{ik}\, \dot{E}_{kl}\, F^{-1}_{ij} \tag{28}$$

The two measures are two ways of viewing the same process: the rate of Green-Lagrange strain expresses in the reference configuration what the rate-of-deformation expresses in the current configuration. However, the properties of the two forms are somewhat different. The integral of the Green-Lagrange strain rate in time is path independent, whereas the integral of the rate-of-deformation is not path independent.

3 Stress measures.

The stresses are defined as follows [Belytschko, T, Liu W. K, and Moran, B. (2000)]:

$$\mathbf{n}.\boldsymbol{\sigma}\, d\Gamma = d\mathbf{f} = \mathbf{t} d\Gamma \tag{29}$$

$$\mathbf{n}_0 \mathbf{P} d\Gamma_0 = d\mathbf{f} = \mathbf{t}_0 d\Gamma_0 \tag{30}$$

$$\mathbf{n}_0.\mathbf{S} d\Gamma_0 = \mathbf{F}^{-1}.d\mathbf{f} = \mathbf{F}^{-1}.\mathbf{t}_0 d\Gamma_0 \tag{31}$$

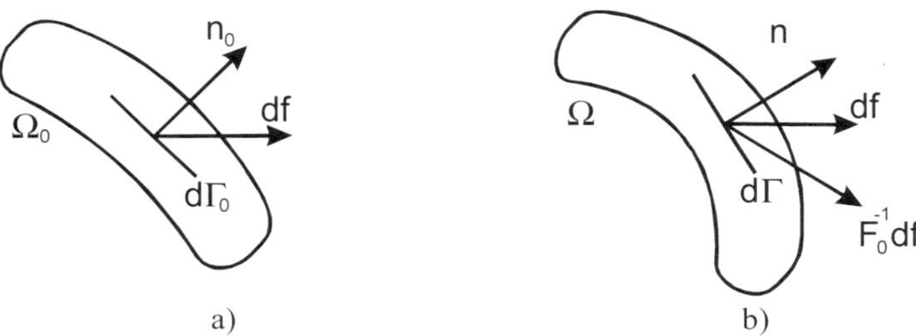

Figure 1: a) Reference cofiguration
b) Curient configuration

where $d\Gamma$, $d\mathbf{f}$, $\mathbf{t}$ and $\mathbf{n}$ denote the elementary surface, force and traction acting on the surface and the surface normal, respectively. $\mathbf{P}$, $\mathbf{S}$ and $\boldsymbol{\sigma}$ are the nominal (or the first Piola-Kirchhoff), the second Piola-Kirchhoff and Cauchy stress tensors, respectively. The index 0 denotes the undeformed configuration. The Cauchy stress involves normal and the traction to the current surface and is often called the physical stress or true stress. Also the trace of the Cauchy stress,

$$\frac{1}{3} trace(\boldsymbol{\sigma}) = \frac{1}{3}\sigma_{ii} = -p \tag{32}$$

gives the true pressure p commonly used in fluid mechanics.

The relations between stress tensors are obtained from (29-30) with Nanson's relation

$$n_i d\Gamma = J n_j^0 F_{ji}^{-1} d\Gamma_0 \quad \text{or} \quad n d\Gamma = J n_0.F^{-1} d\Gamma_0 \tag{33}$$

So, for example the Cauchy stress is related to the second Piola-Kirchhoff stress by

$$\sigma_{ij} = J^{-1} F_{ik} S_{kl} F_{lj}^{T} \tag{34}$$

Conservation equations. Many of the laws of physics can be expressed in the form of statement that some physical quantity is preserved.
Mass conservation requires that the mass of any material domain be constant. It can be expressed as

$$\frac{D\rho}{Dt} + \rho div(\mathbf{v}) = 0 \quad \text{or} \quad \frac{D\rho}{Dt} + \rho v_{i,i} = 0 \quad \text{or} \quad \dot{\rho} + \rho v_{i,i} = 0 \tag{35}$$

This equation is often called the *continuity equation.* If the definition of a material time derivative (as it was done in (1.6)) is invoked, this equation can be written in the form

$$\frac{\partial \rho}{\partial t} + \rho_{,i} v_i + \rho v_{i,i} = \frac{\partial \rho}{\partial t} + (\rho v_i)_{,i} = 0 \quad \text{or} \quad \frac{\partial \rho}{\partial t} + \nabla.(\rho \mathbf{v}) = 0 \tag{36}$$

This is called the conservative form of the mass conservation equation and is often preferred in *computational fluid dynamics* (CFD).
For Lagrangian descriptions, this equation can be written in the form

$$\rho(\mathbf{X},t) J(\mathbf{X},t) = \rho_0(\mathbf{X}) \quad \text{or} \quad \rho J = \rho_0 \tag{37}$$

The equation of *linear momentum conservation* is a key equation in nonlinear numerical models (FEM, BEM, etc.) and is often called the balance of momentum principle. Using the Cauchy stress, this equation has the form

$$\rho \frac{D\mathbf{v}}{Dt} = \nabla.\boldsymbol{\sigma} + \rho \mathbf{b} \equiv div\boldsymbol{\sigma} + \rho \mathbf{b} \quad \text{or} \quad \rho \frac{Dv_i}{Dt} = \frac{\partial \sigma_{ji}}{\partial x_j} + \rho b_i \tag{38}$$

where **b** is a force per unit mass.
This form of the *momentum equation* is applicable to both Lagrangian and Eulerian descriptions. In a Lagrangian description, the momentum equation is

$$\rho(\mathbf{X},t) \frac{\partial \mathbf{v}(\mathbf{X},t)}{\partial t} = div\boldsymbol{\sigma}\left(\phi^{-1}(\mathbf{x},t),t\right) + \rho(\mathbf{X},t)\mathbf{b}(\mathbf{X},t) \tag{39}$$

Note that the stress must be expressed as a function of the Eulerian coordinates so that the spatial divergence of the stress field can be evaluated.

In an Eulerian description, the momentum equation becomes

$$\rho(\mathbf{x},t)\left(\frac{\partial\mathbf{v}(\mathbf{x},t)}{\partial t}+\left(\mathbf{v}(\mathbf{x},t).grad\mathbf{v}(\mathbf{x},t)\right)\right)=div\boldsymbol{\sigma}(\mathbf{x},t)+\rho(\mathbf{x},t)\mathbf{b}(\mathbf{x},t)$$

$$\text{or}\qquad \rho\left(\frac{\partial v_i}{\partial t}+v_{i,j}v_j\right)=\frac{\partial\sigma_{ij}}{\partial x_j}+\rho b_i \tag{40}$$

In the mathematical description of material behaviour, the response of the material is characterised by a constitutive equation which gives the stress as a function of the deformation history of the body. For many materials can be considered as *Kirchhoff materials*, by which the relation between stress and strain is described as

$$S_{ij}=C_{ijkl}E_{kl} \tag{41}$$

where $\mathbf{C}$ is the fourth order tensor of elastic moduli. If the material is isotropic, this tensor is defined by two material (Lamé) constants λ and μ

$$C_{ijkl}=\lambda\delta_{ij}\delta_{kl}+\mu\left(\delta_{ik}\delta_{jl}+\delta_{il}\delta_{jk}\right) \tag{42}$$

Other types of materials are described in [Belytschko, T, Liu W. K, and Moran, B. (2000)] and in other books.

In many problems, the loads are applied slowly and the inertial forces are very small and can be dropped. The momentum equation is then called the *equilibrium equation*.

Taking the cross-product of each term of linear momentum principle with the position vector $\mathbf{x}$ (conservation of angular momentum) states that the Cauchy stress be a symmetric tensor. The principle of *conservation of energy* states that the rate of change of total energy is equal to the work done by the body forces and surface tractions plus the heat energy delivered by the flux and other sources of heat.

The motion of the continuous fluid medium is governed by the principles described above and they can be expressed in a fixed Cartesian coordinate reference frame (Eulerian mesh) in the forms for mass, momentum and energy conservation, *called general conservative Navier-Stokes equations* [Bathe, K. J. (1996)]:

$$\begin{aligned}&\frac{\partial \rho}{\partial t}+\nabla.(\rho\mathbf{v})=0\\&\frac{\partial(\rho\mathbf{v})}{\partial t}+\nabla.(\rho\mathbf{v}\mathbf{v}-\boldsymbol{\sigma})=\mathbf{b}\\&\frac{\partial(\rho E)}{\partial t}+\nabla.(\rho\mathbf{v}E-\boldsymbol{\sigma}.\mathbf{v}+\mathbf{q})=\mathbf{b}.\mathbf{v}+q^B\end{aligned} \tag{43}$$

where E is the specific total energy, $\boldsymbol{\tau}$ is the stress tensor and they are defined as

$$E=\frac{1}{2}\mathbf{v}.\mathbf{v}+e \tag{44}$$

$$\tau=(p-\lambda\nabla.\mathbf{v})\mathbf{I}+2\mu\mathbf{e} \tag{45}$$

$\mathbf{q}$ is heat flux and q^B is the specific rate of heat generation, e is the specific internal energy, p is pressure, λ and μ are the two coefficients of fluid viscosity and $\mathbf{e}$ is the velocity strain tensor

$$\mathbf{e}=\frac{1}{2}\left(\nabla\mathbf{v}+(\nabla\mathbf{v})^T\right) \tag{46}$$

The heat flux is assumed to obey the Fourier's law of heat conduction

$$\mathbf{q}=-k\nabla\theta \tag{47}$$

where θ is the temperature and k is the heat conductivity coefficient. The other quantities are identical with corresponding quantities of solid mechanics.

Based on the types of state equations, various forms of equations can be derived for describing incompressible, compressible, viscous, inviscid, etc. fluid flows.

References

Belytschko, T, Liu W. K, and Moran, B. (2000). *Nonlinear Finite Elements for Continua and Structures*, John Willey, Chichester.

Okrouhlik, M. (ed.). (1995). *Implementation of Nonlinear Continuum Mechanics in Finite Element Code, notes to lectures*, Inst. of Thermomech., Prague.

Altenbach, J, Altenbach, H. (1994). *Einführung in die Kontinuumsmechanik* (*Introduction to Continuum Mechanics* in German), Teubner, Stuttgart.

Bathe, K. J. (1996). *Finite Element Procedures*, Prentice Hall Englewood Cliffs, NJ.

Chapter 2
Finite Displacements in Reciprocity Based Multi-Domain BE/FE Formulations

Vladimir Kompiš, Pavol Novák, Marián Handrik

Faculty of Mechanical Engineering,
University of Žilina, Velký diel, 010 26 Žilina, Slovakia
e-mail:kompis@fstroj.utc.sk, paul@mppserv.utc.sk, handrik@mppserv.utc.sk.

Abstract. In this chapter, Trefftz (T-) functions are used for the development of Multi-Domain (MD) BEM/FEM based on the reciprocity relations. This reciprocity principles are well known from the Boundary Element formulations, however, using the Trefftz functions (polynomials, fundamental solutions with the source point defined outside the sub-domain, or other type of non-singular T-function) in the reciprocity relations instead of the fundamental solutions yields the non-singular integral equations for the evaluation of corresponding sub-domain relations. A weak form satisfaction of the equilibrium is used for the inter-domain connectivity relations. For linear problems, the element stiffness matrices are defined in the boundary integral equation form. In non-linear problems the total Lagrangian formulation leads to the evaluation of the boundary integrals over the original (related) sub-domain evaluated only once during the solution and to the domain integrals containing the non-linear terms. Considering the examples of simple tension, pure bending and tension of fully clamped rectangular 2D domain (2D stress/strain problems) for large strain-large rotation problems, the use of the initial stiffness, the Newton-Raphson procedure, and the incremental Newton- Raphson procedure is considered.

1 Introduction

The Trefftz (T-) functions [Trefftz, E. (1926)] are those which satisfy all the governing equations of the problem, i.e. the differential equations inside the domain. The polynomial T-functions can be found in a simple way for many important 2D and 3D solid and fluid mechanics problems by the methods of symbolic algebra or by numerical methods [Kompiš V. (1994), Kompiš, V., Jakubovičová, L., Konkol F.(1999)]. Also, the well-known fundamental solutions with the source point located outside the domain and other non-singular T-functions (harmonic functions, Bessel functions, etc.) can be used as the T-weight functions. For general problems, obtaining the solution satisfying all governing equations is difficult; only an approximate solution can be found. The simplest way to satisfy the boundary conditions is the collocation method, where the boundary conditions are satisfied in the discrete points of the boundary only [Kolodziej, J. A., Uscilowska A. (1997)]. Such an approximation does not guarantee the convergence of the solution in the multi-domain (MD) formulation. Further more, many other approaches can be found in which the inter-domain continuity and boundary condition satisfaction is enforced in the weak (integral) sense (e.g. weighted residual, variational form, integral least squares [Jirousek, J., Wróblewski A. (1997)]).

In the hybrid FEM formulations, the internal and boundary fields are chosen independently; the internal field variables are approximated by T-functions, and the boundary field enforces both the inter-domain continuity and the satisfaction of the boundary conditions in a weak sense [Jirousek, J., Wróblewski A. (1997)-Kompiš, V., Fraštia, L. (1997)]. In such formulations, the internal primary fields (displacements in hybrid-displacement FEM formulations) are incompatible between the elements, but usually the corresponding boundary fields are taken as representative of the solution.

In this Chapter, FE approximation using T-functions is shown in the reciprocity based MD-BEM/FEM formulations; these are well known from the BEM [Cruse, T. A. (1997),Cruse, T. A. (1974)]. If non-singular T-functions are used for the weight functions, difficulties encountered with numerical integration are not present [Lachat, J. C., Watson, J. O. (1976), Cheung, Y. K., Jin, W. G., Zienkiewicz, O. C. (1989)]. Numerical integration is a special procedure necessary for the integration of the singular integrals with weak, strong or hyper-singular kernels or the integrals with quasi-singularities. When the polynomial weight functions are used, then the complexity of the interpolation fields increases with the complexity of the solved problem; therefore, the MD formulation was introduced [Cheung, Y. K., Jin, W. G., Zienkiewicz, O. C. (1991), Kompiš, V., Oravec, J., Búry, J. (1999)]. If the fundamental solution with the source located outside the domain is used For the weight function, the resulting system of equations is worse conditioned than in classical BEM solutions. The MD formulation increases the conditioning and creates banded and sparse system of equations instead of fully populated in classical formulation.

The form of both hybrid and reciprocity based BEM/FEM can be more general than that by using other FEM formulations because the integration is executed over the element boundaries only for linear problems. Consequently, large elements can also be used for the regions with complicated fields of variables (singularities, large gradients, local effects, elements with holes, etc.).

Finding the T-functions for a general problem is not possible for non-linear problems. The non-linearity yields volume integrals, with non-linear functions in the integrand. In solid statics, a suitable form for the reciprocity based BEM/FEM is the total Lagrangian formulation. In this formulation, the reference configuration does not change during the solution process and the integration is performed only once. The volume integrals containing the non-linear terms update the right side of the discretized form of the solution, leading to the initial stiffness formulation that does not converge by large strains. Improvement is achieved using an updated (tangential) stiffness matrix obtained from the nodal displacements of the previous iteration step.

The stress (and also strain) field is discontinuous between the elements in most FEM formulations, and its rate of convergence is lower than the rate of convergence of the displacements. If the stress is interpolated using T-polynomial interpolation functions with the moving least squares procedure from the nodal displacements (and known static conditions in the points near the boundaries), much better accuracy can be obtained with same rate of convergence for both displacements and stresses.

Considering the examples of simple tension, pure bending, and tension of fully clamped rectangular plate (2D stress/strain problems) for large strain-large rotation problems, the use of the initial stiffness, the Newton-Raphson procedure, and the incremental Newton- Raphson procedure will be discussed.

2 Formulation for small strain, small rotation problems

When using the basic equations for time independent small strain plasticity, the current state of deformation does not depend on only the current loading but also on the complete history of loading. At any loading time the total strain tensor can be split into the elastic and plastic parts

$$\varepsilon_{ij} = \varepsilon_{ij}^{e} + \varepsilon_{ij}^{p} \tag{2.1}$$

The strain is derived from displacements by the kinematic relation

$$\varepsilon_{ij} = \frac{1}{2}\left(u_{i,j} + u_{j,i}\right) \tag{2.2}$$

The stress-strain relation for isotropic material is

$$\sigma_{ij} = C_{ijkl}(\varepsilon_{kl} - \varepsilon_{kl}^{p}) = 2G\left[\left(\varepsilon_{ij} - \varepsilon_{ij}^{p}\right) + \frac{\nu}{1-\nu}\left(\varepsilon_{kk} - \varepsilon_{kk}^{p}\right)\delta_{ij}\right] \tag{2.3}$$

where C_{ijkl} is the elasticity tensor, G and ν are shear modulus and Poisson's ratio, respectively. Einstein's notation is used, in which the index after the comma denotes the partial derivative in the corresponding direction and with summation after repeated indices. The equilibrium equations in stresses

$$\sigma_{ji,j} = -b_i \tag{2.4}$$

are in displacements given by the Lame-Navier equations

$$u_{i,jj} + \frac{1}{1-2\nu}u_{j,ji} = -\frac{b_i}{G} \tag{2.5}$$

with

$$b_i = \bar{b}_i - 2G\left[\varepsilon_{ij,j}^{p} + \frac{\nu}{1-2\nu}\varepsilon_{j,ji}^{p}\right] \tag{2.6}$$

where b_i is a pseudo-body force and the bar denotes prescribed value. The last relation contains the influence of the non-linear material behaviour.

The weak formulation of equilibrium (2.4) using the principle of weighted residuals can be written in the form

$$\int_{\Omega}\left(\sigma_{ji,j} + b_i\right)U_i d\Omega = 0 \tag{2.7}$$

We can choose the weight function U_i to be the T-displacement field of a linear-elastic reference problem of the same body with body force absent. Letters denoted with a star will denote the quantities corresponding to this field.

The tractions corresponding to the stress field σ_{ij} are given by

$$t_i = \sigma_{ij} n_j \tag{2.8}$$

where n_i denotes the outer normal to the boundary.

Applying the integration by parts and the Gauss theorem to the first part of equation (2.7) we obtain

$$\int_\Gamma t_i U_i d\Gamma - \int_\Omega \sigma_{ij} U_{j,i} d\Omega + \int_\Omega b_i U_i d\Omega = 0 \tag{2.9}$$

Using Hook's law and identity

$$\Sigma_{ij}\varepsilon_{ij} = \Sigma_{ij} u_{j,i} \tag{2.10}$$

we can write

$$\sigma_{ij} U_{j,i} = \Sigma_{ij} u_{j,i} - \Sigma_{ij}\varepsilon^p_{ij} \tag{2.11}$$

where Σ_{ij} denotes the stress state corresponding to the displacement field U_i. T-displacements and T-stresses satisfy according to the definition of the following homogeneous equilibrium equations

$$U_{i,jj} + \frac{1}{1-2\nu} U_{j,ji} = 0$$

$$\Sigma_{ji,j} = 0 \tag{2.12}$$

Then the eq. (2.9) can be written in the form

$$\int_\Omega \Sigma_{ij} u_{j,i} d\Omega = \int_\Gamma t_i U_i d\Gamma + \int_\Omega \Sigma_{ij}\varepsilon^p_{ij} d\Omega + \int_\Omega b_i u_i d\Omega \tag{2.13}$$

Again, using integration by parts and Gauss theorem applied to the left side of the last equation, together with the equilibrium condition of the reference problem, the generalized form of Betti's theorem is obtained

$$\int_\Gamma T_i u_i d\Gamma = \int_\Gamma t_i U_i d\Gamma + \int_\Omega \Sigma_{ij}\varepsilon^p_{ij} d\Omega + \int_\Omega b_i U_i d\Omega \tag{2.14}$$

Where T_i are T-tractions derived from T-stress Σ_{ij}. This equation expresses the reciprocity of works done by two systems of forces: the one denoted by letters without stars which is looked for, and the other, reference state (for which all, displacements, strains, stresses and tractions are known inside and on the domain boundaries), denoted by letters with stars.

In the classical BEM, the reference state defined by the Kelvin fundamental solution leads to the singular integral equation problem. Using T-polynomials (see Appendix A) for the reference state, all integrals are regular; however, with complex problems many different reference states are needed for a numerical solution, and complex and thus, inefficient high order T-polynomials are to be used. Splitting the whole domain into sub-domains, a MD formulation is obtained with sub-domain, or element matrices defined by the boundary integral equations (2.14) and their corresponding numerical resolution by the BEM. The displacements between the sub- domains are chosen to be compatible: the displacements on the element boundaries are common to the neighbour elements. However, the tractions will be incompatible between the elements; therefore, the inter-element equilibrium and natural boundary conditions will be satisfied in a weak (integral) sense as follows

$$\int_{\Gamma_t} \delta u_i \left(t_i - \bar{t}_i\right) d\Gamma + \int_{\Gamma_i} \delta u_i \left(t_i^A - t_i^B\right) d\Gamma = \int_{\Gamma_e} \delta u_i t_i d\Gamma - \int_{\Gamma_t} \delta u_i \bar{t}_i d\Gamma = 0 \qquad (2.15)$$

where Γ_i, Γ_t and Γ_e are the inter-element boundaries, the boundaries with prescribed tractions and element boundaries, respectively. The upper indices A and B denote the neighbouring elements.

For the discretization, the boundary displacements and tractions can be expressed by their values in nodal points (denoted by the dash) and by shape functions, N, as

$$u_i^e(\xi) = N_{Ju}(\xi)\hat{u}_{Ji}^e \qquad \text{or} \qquad \mathbf{u}^e = \mathbf{N}_u \hat{\mathbf{u}}^e \qquad (2.16)$$

$$t_i^e(\xi) = N_{Jt}(\xi)\hat{t}_{Ji}^e \qquad \text{or} \qquad \mathbf{t}^e = \mathbf{N}_t \hat{\mathbf{t}}^e \qquad (2.17)$$

where ξ is a local co-ordinate of an element boundary point and the capital letter index denotes the nodal point. The upper index e denotes the correspondence to the element.

Note that the tractions are discontinuous in the corner points, and thus a double node for tractions exists.

Then the equation (2.14)leads to the system of equations

$$T_{iIJ}\hat{u}_{iJ}^e = U_{iIK}\hat{t}_{iK}^e + f_I^{ep} + f_I^{eb} \qquad (2.18)$$

or, in the matrix form

$$\mathbf{T}\mathbf{u}^e = \mathbf{U}\mathbf{t}^e + \mathbf{f}^\mathbf{p} + \mathbf{f}^\mathbf{b} \qquad (2.19)$$

where $\mathbf{u}^e$ and $\mathbf{t}^e$ are vectors of element nodal displacements and tractions and

$$\begin{aligned} T_{iIJ} &= \int_{\Gamma_e} T_{iI}(x(\xi))N_{Ju}(\xi)d\Gamma = \sum_j \left(T_{iI}(x(\xi^{(j)}))N_{Ju}(\xi^{(j)})J(\xi^{(j)})w^{(j)}\right) \\ U_{iIK} &= \int_{\Gamma_e} U_{iI}(x(\xi))N_{Kt}(\xi)d\Gamma = \sum_j \left(U_{iI}(x(\xi^{(j)}))N_{Kt}(\xi^{(j)})J(\xi^{(j)})w^{(j)}\right) \\ f_I^{ep} &= \int_{\Omega_e} \Sigma_{ijI}(x(\xi))\varepsilon_{ij}^p(x(\xi))d\Omega = \sum_\alpha \left(\Sigma_{ijI}(x(\xi^{(\alpha)}))\varepsilon_{ij}^p(x(\xi^{(\alpha)}))J(\xi^{(\alpha)})w^{(\alpha)}\right) \\ f_I^{eb} &= \int_{\Omega_e} U_{iI}(x(\xi))b_i(x(\xi))d\Omega = \sum_\alpha \left(U_{iI}(x(\xi^{(\alpha)}))b_i(x(\xi^{(\alpha)}))J(\xi^{(\alpha)})w^{(\alpha)}\right) \end{aligned} \tag{2.20}$$

ξ_j and w_j are co-ordinates and weights in the Gauss quadrature formulas and J is the Jacobian. Lower case indices in these expressions correspond to the field (vector or tensor) components, the lower case index I corresponds to the I-th Trefftz function, and the lower case index J corresponds to the nodal value. In order to distinguish the integration over the volume of the element from that over the element boundaries, the volume Gauss points are denoted by α.

Eq. (2.15) is written in to the discrete form as

$$\sum_e \sum_j \sum_L N_{Ku}(\xi^{(j)})N_{Lt}(\xi^{(j)})J(\xi^{(j)})w^{(j)}\hat{t}_{iL} = \sum_e \sum_j N_{Ku}(\xi^{(j)})\bar{t}_i(\xi^{(j)})J(\xi^{(j)})w^{(j)} \tag{2.21}$$

or in the matrix form

$$\sum_e \mathbf{M}\mathbf{t}^e = \sum_e \mathbf{p^e} \tag{2.22}$$

with summation over all elements.

The lower case indices K and L in (2.21) correspond to the nodal displacements and tractions, respectively.

Setting for tractions from (2.22) into (2.19), the resulting system of discretized equations is obtained

$$\sum_e \mathbf{M}\mathbf{U}^{-1}\mathbf{T}\mathbf{u}^e = \sum_e \left(\mathbf{p^e} + \mathbf{M}\mathbf{U}^{-1}\left(\mathbf{f^b} + \mathbf{f^p}\right)\right) \tag{2.23}$$

or shortly

$$\mathbf{K}\mathbf{u} = \mathbf{p} \tag{2.24}$$

This is a system of linear equations in which the non-linear term (containing the plastic strains) is the second term of the right hand side of eq. (2.23).

Note that in order to obtain non-singular matrices, the vector of element nodal tractions must have as many, or more, independent components than the vector of element nodal displacements. Also, we must choose as many, or more, T-functions for each element than we have element nodal tractions.

Note also that in order to obtain non-singular integral equations in the sub-domain BEM relations we can use distributed Kelvin tractions (see Appendix B), or Boussinesq half plane, or half space solution for the reciprocal states of the body, when the source points will be located outside corresponding sub-domain. If the distance of the source points from the sub-domain boundaries is large enough, then also a comparable number of Gauss integration points to that using the T-polynomials suffices to ensure necessary accuracy of the solution. In our experiments the source points were chosen on a circle (on a sphere for 3D problems) with the diameter equal to double size of the domain and with the origin in its centre. Also a combination of polynomial and Kelvin functions can be used for this purpose.

3 The total Lagrangian formulation for finite deformation problems

In this section, the application of the MD BEM formulation to geometrically and physically non-linear problems in the total Lagrangian approach will be illustrated. The basic equations refer to the undeformed configuration of the body.

Let X_i denote the coordinate of a material particle X in the undeformed body. After the deformation, the co-ordinate of this particle will be x_i. The cartesian components F_{ij} of the deformation gradient are defined by

$$F_{ij}(X) = \frac{\partial x_i(X)}{\partial X_j} \tag{3.1}$$

Using the displacement u_i of a material particle X

$$u_i = x_i - X_i \tag{3.2}$$

leads to the alternative expression for the deformation gradient

$$F_{ij}(X) = \delta_{ij} + \frac{\partial u_i(X)}{\partial X_j} = \delta_{ij} + u_{i,j}(X) \tag{3.3}$$

Since the formulation used is presented in the undeformed configuration, partial derivatives denoted by $(.)_{,i}$ are taken with respect to the undeformed co-ordinates X_i. The deformation gradient can be used to define the Green strain tensor

$$E_{ij} = \frac{1}{2}\left(F_{ki}F_{kj} - \delta_{ij}\right) = \frac{1}{2}\left(u_{i,j} + u_{j,i}\right) + \frac{1}{2}u_{k,i}u_{k,j} \tag{3.4}$$

and the symmetric 2nd Piola-Kirchhoff stress tensor

$$S_{ij} = JF_{ik}^{-1}\sigma_{kl}F_{jl}^{-1} \qquad \text{with} \qquad J = \det(F_{ij}) \tag{3.5}$$

both referring to the undeformed configuration. σ_{ij} is the Cauchy stress tensor.

The equilibrium equation

$$\frac{\partial \sigma_{ij}}{\partial x_j} + b_i = 0 \tag{3.6}$$

can be transformed to the initial (undeformed) configuration

$$\left(S_{jk}F_{ik}\right)_{,j} + b_i^o = 0 \tag{3.7}$$

where b_i^0 and b_i denote the body force with respect to the initial and deformed configuration, respectively. The relation between the tractions, t_i^0, which measure the force per unit undeformed area $d\Gamma^0$, and the tractions, t_i, in the deformed configuration $d\Gamma$ is given by

$$t_i^0 = F_{ij}S_{kj}n_k^0 = t_i\frac{d\Gamma}{d\Gamma^0} = \sigma_{ki}n_k\frac{d\Gamma}{d\Gamma^0} \tag{3.8}$$

where n_i^0 and n_i is the outer surface normal in the initial and deformed configuration. By using the derivation of the reciprocity relations, a similar procedure as in the case of infinitesimal displacements is followed, starting from the equilibrium equation of the deformed body relative to the initial configuration (3.7). Again, the T- displacements will be taken as the weight function in the weak formulation of the balance

$$\int_\Omega \left[(S_{kl}F_{il})_{,k} + b_i^0\right] U_i d\Omega = 0 \tag{3.9}$$

The T-functions are taken in co-ordinates of the undeformed (initial) configuration. Due to the total Lagrangian approach, Ω is the domain of the undeformed body, we omit its upper index 0 (similarly we do for the surface Γ) and all derivatives are taken with respect to this configuration. Applying integration by parts and the Gauss' theorem to eq. (3.9) we obtain

$$\int_\Gamma t_i^o U_i d\Gamma - \int_\Omega (S_{kl}F_{il})U_{i,k} d\Omega + \int_\Omega b_i^0 U_i d\Omega = 0 \tag{3.10}$$

Substituting the displacement gradients for the deformation gradient from (3.3) into (3.10) results in

$$\int_\Gamma t_i^0 U_i d\Gamma + \int_\Omega b_i^0 U_i d\Omega - \int_\Omega S_{ji}U_{i,j} d\Omega - \int_\Omega (S_{jk}u_{i,k})U_{i,j} d\Omega = 0 \tag{3.11}$$

Again, the strain tensor can be split into the elastic and plastic parts

$$E_{ij} = E^e_{ij} + E^p_{ij} \tag{3.12}$$

and because of the linear dependence between the elastic part of the Green strain tensor and the 2nd Piola Kirchhoff stress tensor, the reciprocity relation can be found in the form

$$S_{ij}U_{j,i} = u_{j,i}\Sigma_{ij} + \frac{1}{2}u_{k,i}u_{k,j}\Sigma_{ij} + E^p_{ij}\Sigma_{ij} \tag{3.13}$$

Using this relation, the integration by parts, and the Gauss' theorem the eq. (3.11) can be written in the form

$$\int_\Gamma t^0_i U_i d\Gamma + \int_\Omega b^0_i U_i d\Omega - \int_\Gamma u_i T_i d\Gamma - \int_\Omega \frac{1}{2}u_{k,i}u_{k,j}\Sigma_{ij} d\Omega - \\ - \int_\Omega S_{jk}u_{i,k}U_{i,j} d\Omega - \int_\Omega E^p_{ij}\Sigma_{ij} d\Omega = 0 \tag{3.14}$$

If eq. (3.14) is applied for the computation of the relation between the boundary displacements $\mathbf{u}$ and the tractions $\mathbf{t^0}$ for each sub-domain (element), and the inter- domain traction continuity (2.15) is used to the weak satisfaction of equilibrium, the procedure described in the previous section can be used.

4 Linearization of resulting equations for large strain problems

For large strains we have to linearize the expressions in the integrands. For this purpose the displacements inside the element can be approximated from its nodal values using the shape functions as it is done by classical (displacement) FE formulations [Niu, Q., Shephard, M. S. (1993), Zienkiewicz, O. C., Taylor, R. L. (1991)]

$$u_i = N_J u_{Ji} \tag{4.1}$$

Similarly their derivatives are obtained from

$$u_{i,k} = N_{J,k} u_{Ji} \tag{4.2}$$

The lower case index denoted by capital letters denotes the corresponding nodal displacement.

In the N-th iteration step the displacement will be given by

$$u_i^{(N)} = N_J u_{Ji}^{(N)} \tag{4.3}$$

The resulting discretized equations (2.24) can be now written in the form

$$\left(\mathbf{K} + \mathbf{K}^{\mathbf{NL}}\right) \mathbf{u}^{(\mathbf{N})} = \mathbf{p}^{(\mathbf{N}-\mathbf{1})} \tag{4.4}$$

where $\mathbf{K}$ corresponds to the linear part of the equation (3.14) and $\mathbf{K}^{\mathbf{NL}}$ to its non-linear part, which will be linearized for each iteration step and $\mathbf{p}^{(\mathbf{N}-\mathbf{1})}$ denotes the configuration dependent load corresponding to the configuration of the previous iteration step. For this purpose we can write the integrand of the fourth integral (3.14) in the form

$$\frac{1}{2}\Sigma_{ij}\left(u_{k,i}u_{k,j}\right)^{L} = \Sigma_{ij}u_{k,i}^{(N-1)}N_{J,j}u_{Jk}^{(N)} \tag{4.5}$$

or, if the integrand is written in the form

$$\frac{1}{2}u_{k,i}u_{k,j}\Sigma_{ij} = \frac{1}{2}C_{ijlm}u_{n,l}u_{n,m}U_{i,j} \tag{4.6}$$

and for the isotropic material, in which

$$C_{ijlm} = \lambda\delta_{ij}\delta_{lm} + \mu\left(\delta_{il}\delta_{jm} + \delta_{im}\delta_{jl}\right) \tag{4.7}$$

The linearized form of (4.6) with (4.7)

$$\left(\frac{1}{2}C_{ijlm}u_{n,l}u_{n,m}\right)^{L} = \left[\mu\left(u_{n,i}^{(N-1)}N_{J,j} + u_{n,j}^{(N-1)}N_{J,i}\right) + \lambda\delta_{ij}u_{n,m}^{(N-1)}N_{J,m}\right]u_{Jn}^{(N)} \tag{4.8}$$

The linearization of the fifth integral of eq.(3.14) is realized using the 2-nd Piola-Kirchhoff stress in the form

$$S_{ij} = \mu\left(u_{i,j} + u_{j,i}\right) + \lambda\delta_{ij}\left(u_{k,k} + \frac{1}{2}u_{k.l}u_{k,l}\right) + \mu u_{k,i}u_{k,j} \tag{4.9}$$

as follows

$$\begin{aligned} U_{j,i}(S_{ik}u_{j,k})^{L} = & \Big[U_{k,i}S_{ij}^{(N-1)}N_{J,j} + \mu\big(U_{j,k}u_{j,i}^{(N-1)} + U_{j,i}u_{j,k}^{(N-1)}\big)N_{J,i} + \\ & + \mu U_{j,i}u_{j,m}^{(N-1)}\big(u_{k,i}^{(N-1)}N_{J,m} + u_{k,m}^{(N-1)}N_{J,i}\big) + \\ & + \lambda U_{j,i}u_{j,i}^{(N-1)}\big(N_{J,k} + u_{k,m}^{(N-1)}N_{J,m}\big)\Big]u_{Jk}^{(N)} \end{aligned} \tag{4.10}$$

Note that the field variables of the previous, $(N-1)$-th iteration step, are given by use of the nodal displacements computed in that step, whereas the field variables of the current, N-th step, are defined by the shape functions and by the unknown nodal displacements of this step.

In the Newton-Raphson procedure, the increments are computed following eq. (4.4) from

$$\left(\mathbf{K} + \mathbf{K}^{NL}\right) \boldsymbol{\Delta}\mathbf{u}^{(N)} = \boldsymbol{\Delta}\mathbf{p}^{(N)} \tag{4.11}$$

and the displacements in the N-th iteration step are

$$\mathbf{u}^{(N)} = \mathbf{u}^{(N-1)} + \boldsymbol{\Delta}\mathbf{u}^{(N)} \tag{4.12}$$

The iteration is stoped if the quadratic norm of the last displacement increment related to the quadratic norm of the displacements is less then specified value, i.e.

$$e \geq ||\Delta u^{(N)}||/||u^{(N)}|| \tag{4.13}$$

5 Examples

5.1 The first example

In the first example, a simple extension of a square domain of dimensions 1 by 1 by Young modulus E = 1 and plain stress conditions was examined. First, Poisson's ratio equal to zero was assumed. The relation between the tractions, t, and the stretching of the domain, u, is

$$t_1^0 = (u_1 + 1.5u_1^2 + 0.5u_1^3)E \tag{5.1}$$

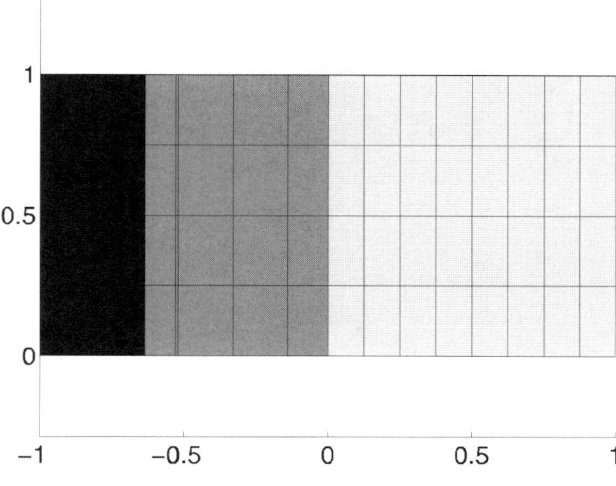

Figure 1. Simple tension with $\nu = 0$

iteration	u_1
1	0.63636
2	0.53039
3	0.52144
4	0.52138

Table 1. Simple extension with $\nu = 0$

Numerically, the converged solution (e = 0.00001) was obtained for t = 1.0 in four iteration steps with following results (Fig. 1 and Tab. 1)

The result in the last step agrees with the analytical result in all digits. Note that initial stiffness can be used for smaller deformations ($t < 0.19$) in this example.

5.2 The second example

For real material with Poisson's ratio not equal to zero, the analytical solution is in the form

$$t_1^0 = \frac{E}{1-\nu^2}(1+u_1)[u_1 + 0.5u_1^2 + \nu(u_2 + 0.5u_2^2)] \tag{5.2}$$

where u_2 is the transverse contraction of the domain. The requirement for the transverse contraction to be zero also leads to t_1 defined by eq. (5.1) in this case.

The numerical FE solution is obtained in four iteration steps (Fig. 2 and Tab. 2)

iteration	u_1	u_2
1	0.66364	-0.33581
2	0.54346	-0.24602
3	0.52292	-0.22460
4	0.52147	-0.22179

Table 2. Simple tension with $\nu = 0.3$

Note that we have to distinguish between true loading and dead loading [Forster, A., Kuhn, G. (1994)]. If pressure is to be prescribed, it must be considered as a nominal traction (related to the deformed surface). The traction is changed in each iteration step, too, and the converged solution (0.80628 0.43218) is reached in the 4-th iteration step.

5.3 The third example

If the domain is fully clamped, the solution corresponding to the second example is obtained in five iteration steps (Fig. 3 and Tab. 3)

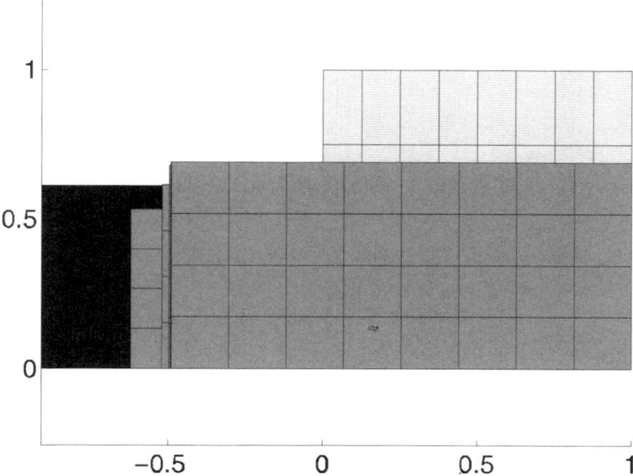

Figure 2. Simple extension with $\nu = 0.3$

iteration	u_1	u_2
1	0.6044	0.2206
2	0.5021	-0.1733
3	0.4796	0.1470
4	0.4758	0.1429
5	0.4754	0.1423

Table 3. Simple tension with one side fully clamped

5.4 The fourth example

In the last example, the pure bending of the beam of dimensions 8 by 1 is demonstrated. The results with Poisson's ratio equal to 0.0 (bending moment of 0.023) and 0.3 (bending moment of 0.0195) are given in (Fig. 4 and Tab. 4) and (Fig. 5 and Tab. 5), respectively. Note that a larger moment is required to get the same bending if the transverse contraction is greater. In this case also, the error measures according to (4.13) are given. u_1 and u_2 denote the displacement components of the upper end point of the beam.

6 Conclusions

This chapter presents the use of the Trefftz functions for the development of MD BEM/FEM based on the reciprocity relations. The stiffness matrix of an element is formulated by non-singular boundary integral equations (BEM). A weak form of the equilibrium is used for the inter-domain connectivity relations.

The formulation is shown for linear elastoplastic problems and for the large strain problems using the total Lagrangian formulation. It leads to the boundary integrals over the original (related) domain computed only once at the beginning of the iterative process and to the domain

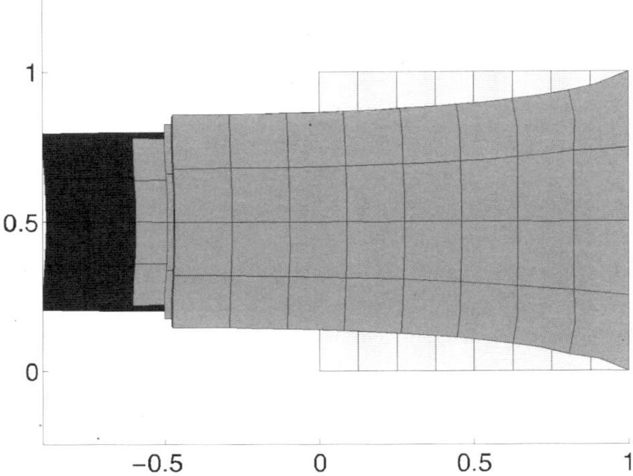

Figure 3. Simple tension with side fully clambed

iteration	u_1	u_2	e
1	0.688	6.877	7.93E-2
2	1.523	5.810	1.84E-2
3	1.746	5.333	4.13E-3
4	1.695	-5.067	2.00E-3
5	1.898	5.221	4.69E-4
6	2.115	5.405	9.91E-4
7	2.274	5.526	4.78E-4
8	2.348	5.580	1.06E-4
9	2.364	5.591	4.91E-6

Table 4. Pure bending $\nu = 0$

integrals which can be treated in different ways: (1) by using the initial stiffness (i.e. modified Newton-Raphson procedure) or, (2) by one load step Newton-Raphson or, (3) by incremental Newton-Raphson procedures, respectively.

Applications for linear problems are shown in the previous papers of the first author et.al. [Kompiš, V., Žmindák, M., Jakubovičová, L. (1999), Kuhn, G., Partheimüller, P., Köhler, O. (1998), Bathe, K.-J. (1996)]. The applications for large strain, large rotation problems given in this paper show better convergence than results obtained by other authors [Maunder, E. A. W., Kompiš, Forster, A., Kuhn, G. (1994)] using BEM or results obtained by commercial FEM. The procedure mostly converged in one increment with few iteration steps even for very large strains and rotations.

Appendix A. Generation of Trefftz (T-)Polynomial Functions in 3D

We will assume an isotropic linear elastic solid without body forces. The displacement field which describes its behaviour under static loading conditions have to satisfy the equilibrium

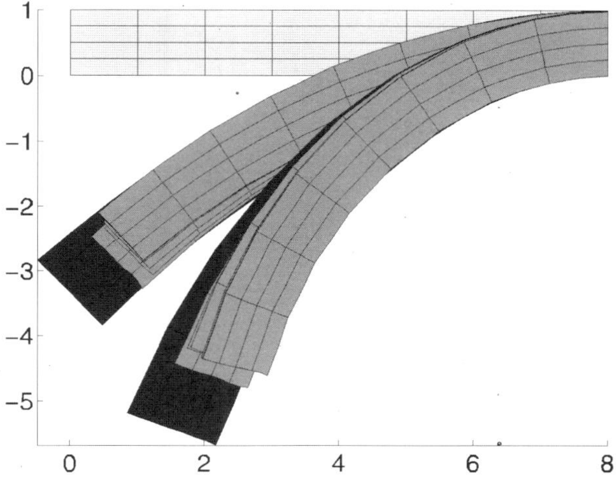

Figure 4. Pure bending $\nu = 0$

iteration	u_1	u_2	e
1	0.907	6.160	6.70E-2
2	1.842	5.790	1.57E-2
3	2.240	5.640	2.34E-3
4	2.341	-5.595	1.63E-4
5	2.375	5.587	1.78E-5
6	2.377	5.581	4.04E-6

Table 5. Pure bending $\nu = 0.3$

equations expressed in displacements by Lame-Navier equations

$$(\lambda + \mu)u_{j,ij} + \mu u_{i,jj} = 0 \tag{A.1}$$

where (i,j=1,2,3) for 3D problems. λ and μ are Lame constants expressed by Young's modulus, E and Poisson's ratio ν

$$\lambda = \frac{E\nu}{(1+\nu)(1-2\nu)} \qquad \mu = \frac{E}{2(1+\nu)} \tag{A.2}$$

We will consider the polynomial approximation of the displacement field

$$\begin{Bmatrix} u_1 \\ u_2 \\ u_3 \end{Bmatrix} = \begin{bmatrix} \mathbf{P(x)} & \mathbf{0} & \mathbf{0} \\ \mathbf{0} & \mathbf{P(x)} & \mathbf{0} \\ \mathbf{0} & \mathbf{0} & \mathbf{P(x)} \end{bmatrix} \begin{Bmatrix} \mathbf{C}^{(1)} \\ \mathbf{C}^{(2)} \\ \mathbf{C}^{(3)} \end{Bmatrix} \tag{A.3}$$

where **P(x)** is full polynomial of the n-th order

$$\mathbf{P(x)} = \left[1\ x_1\ x_2\ x_3\ ...\ x_1^n\ x_1^{n-1}x_2\ ...\ x_2 x_3^{n-1}\ x_3^n\right] \tag{A.4}$$

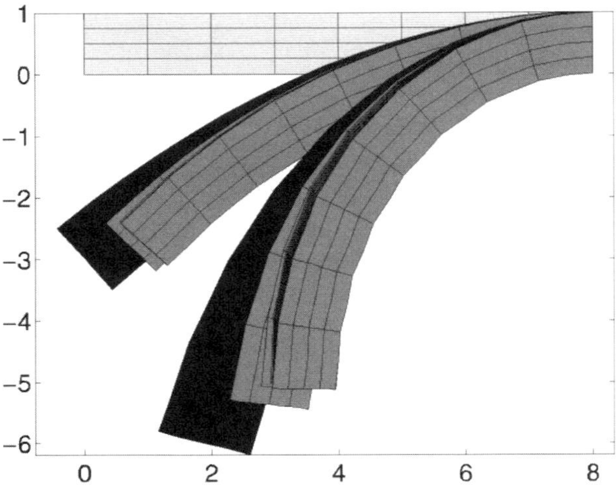

Figure 5. Pure bending $\nu = 0.3$

and $c^{(j)}$ is a vector of unknown coefficients.

The equilibrium equation (A.1) contains the second derivatives of the displacements. Thus, the terms of the zeroth and first order satisfy the homogeneous equation (A.1), i.e. that without the body forces. However, the higher order polynomials cannot be arbitrary in order to satisfy the eq. (A.1). If we conveniently split (A.3) into the form

$$\begin{Bmatrix} u_1 \\ u_2 \\ u_3 \end{Bmatrix} = \begin{bmatrix} \mathbf{A}^{(1)} & \mathbf{0} & \mathbf{0} \\ \mathbf{0} & \mathbf{A}^{(2)} & \mathbf{0} \\ \mathbf{0} & \mathbf{0} & \mathbf{A}^{(3)} \end{bmatrix} \begin{Bmatrix} \mathbf{a}^{(1)} \\ \mathbf{a}^{(2)} \\ \mathbf{a}^{(3)} \end{Bmatrix} + \begin{bmatrix} \mathbf{B}^{(1)} & \mathbf{0} & \mathbf{0} \\ \mathbf{0} & \mathbf{B}^{(2)} & \mathbf{0} \\ \mathbf{0} & \mathbf{0} & \mathbf{B}^{(3)} \end{bmatrix} \begin{Bmatrix} \mathbf{b}^{(1)} \\ \mathbf{b}^{(2)} \\ \mathbf{b}^{(3)} \end{Bmatrix} \tag{A.5}$$

where **B** contains as many terms of each order as there are polynomial terms of two order lower (because the equilibrium equations will be two order lower polynomials obtained by differentiation of approximation displacements field). So for example the third order terms

$$\begin{array}{ccccccc} & & & x_1^3 & & & \\ & & x_1^2 x_2 & & x_1^2 x_3 & & \\ & - & & - & & - & \\ & x_1 x_2^2 & & x_1 x_2 x_3 & & x_1 x_3^2 & \\ x_2^3 & & x_2^2 x_3 & & x_2 x_3^2 & & x_3^3 \end{array} \tag{A.6}$$

will be split so that the upper three terms will be contained in $B^{(1)}$ and the lower part terms in $A^{(1)}$. Cyclically changing the components indices we will obtain the terms for $B^{(j)}$ and $A^{(j)}$ (j=2,3).

To satisfy the equilibrium condition (A.1), we differentiate $B^{(j)}$ and $A^{(j)}$ and set into (1) and obtain the relation

$$\left[\mathbf{M}(x_i)\right]\{\mathbf{b}\} + \left[\mathbf{N}(x_i)\right]\{\mathbf{a}\} = \{\mathbf{0}\} \tag{A.7}$$

Now the vector **b** contains dependent coefficients which have to be expressed through the (independent) coefficients **a** in order to satisfy the equilibrium equations in strong sense.

The problem can be solved for each order separately. It can be done symbolically (e.g. in MAPLE) or numerically. In the last case we choose arbitrary as many or more discrete points as we have the dependent terms and get (A.7) in the form

$$\left[\mathbf{M}({}^{(j)}x_i)\right]\{\mathbf{b}\} = -\left[\mathbf{N}({}^{(j)}x_i)\right]\{\mathbf{a}\} \qquad (A.8)$$

from which the vector of dependent terms, **b**, can be expressed as

$$\{\mathbf{b}\} = -\left[\mathbf{M}^{-1}\right]\left[\mathbf{N}\right]\{\mathbf{a}\} \qquad (A.9)$$

With this we obtain T-polynomial displacements (the third order T-polynomials in the example shown above) in the form

$$\{\mathbf{u}\} = \left(\left[\mathbf{A}(x_i)\right] - \left[\mathbf{B}(x_i)\right]\left[\mathbf{M}\right]^{-1}\left[\mathbf{N}\right]\right)\{\mathbf{a}\} = \left[\mathbf{U}(x_i)\right]\{\mathbf{a}\} \qquad (A.10)$$

Each column of the matrix **U** in (A.10) represents a T-displacement function and the rows are corresponding components. The points for numerical evaluation of equation (A.8) have to be conveniently spaced so that the matrix **M** will be not singular. We can conveniently choose them lying on the sphere of radius equal to 1. Note, that the number of T-functions which can be derived in this form is (2n+1) for 2D problems and $(n+1)^2$ T-functions for 3D problems, where n is the polynomial order.

From the T-displacements corresponding T-stresses are obtained by differentiating (A.10) and setting into (A.11)

$$\sigma_{ij} = \mu(u_{i,j} + u_{j,i}) + \lambda\delta_{ij}u_{k,k} \qquad (A.11)$$

and corresponding T-tractions on the boundaries with the outer normal n_j

$$t_i = \sigma_{ij}n_j \qquad (A.12)$$

where δ_{ij} is the Kronecker delta.

Appendix B. Use of distributed Kelvin tractions as Trefftz functions

The Kelvin traction functions are a basis for BEM and are well known from the literature. The displacements and boundary tractions in the point **x** corresponding to the unit point force in 3D space acting in a point **y** in direction of j-axis are

$$U_{ij}(\mathbf{x},\mathbf{y}) = -\frac{1}{16\pi(1-\nu)\mu r}\{(3-4\nu)\delta_{ij} + r_{,i}r_{,j}\} \qquad (B.1)$$

and

$$T_{ij}(\mathbf{x},\mathbf{y}) = -\frac{1}{8\pi(1-\nu)r^2}\{[(1-2\nu)\delta_{ij} + 3r_{,i}r_{,j}]\frac{\partial r}{\partial n}\}$$
$$(1-2\nu)(r_{,i}n_{,j} - r_{,j}n_{,i}) \qquad (B.2)$$

respectively. Similarly, they are given for plane strain problems as follows

$$U_{ij}(\mathbf{x},\mathbf{y}) = -\frac{1}{8\pi(1-\nu)\mu}\{(3-4\nu)\delta_{ij} + ln(r) - r_{,i}r_{,j}\} \qquad (B.3)$$

and

$$T_{ij}(\mathbf{x},\mathbf{y}) = -\frac{1}{4\pi(1-\nu)r^2}\{[(1-2\nu)\delta_{ij} + 2r_{,i}r_{,j}]\,\frac{\partial r}{\partial n}\}$$

$$(1-2\nu)(r_{,i}n_{,j} - r_{,j}n_{,i}) \tag{B.4}$$

with r being the distance between the points $\mathbf{x}$ and $\mathbf{y}$ and with notation

$$r_{,i} = \frac{\partial r}{\partial x_i(y)} = \frac{r_i}{r} \qquad with \qquad r_i = x_i(y) - x_i(x) \tag{B.5}$$

and

$$\frac{\partial r}{\partial n} = r_{,i}n_i \qquad \frac{r_i n_i}{r} = r_{,n} \tag{B.6}$$

References

Trefftz, E. (1926). Ein Gegenstück zum Ritzschen Verfahren, *Proc. 2nd In nt.Congress of Applied Mechanics*, Zürich.

Kompiš V. (1994). Finite elements satisfying all governing equations inside the element, *Computers & Struct.*, **4**, 273-278.

Kompiš, V., Jakubovičová, L., Konkol F. (1999) Non-singular reciprocity based BEM/FEM formulations, paper presented at *IUTAM/IACEM/IABEM Symposium on Advanced Math. and Comput. Mech. Aspects of the Bound. Elem. Meth.*, Cracow.

Kolodziej, J. A., Uscilowska A. (1997). Trefftz-type procedure for Laplace equation on domains with circular holes, circular inclusions, corners, slits and symmetry, *Computer Assisted Mechanics and Eng. Sciences*, **4**, pp. 501-519.

Jirousek, J., Wróblewski A. (1997). T-elements: State of the art and future trends, *Archives of Comput. Mech.*, **3**, 323-434.

Szybinski, B., Zielinski, A. P. (1995). Alternative T-complete systems of shape functions applied in analytical Trefftz finite elements, *Num. Meth. for Part. Diff. Eqs.*, **11** , 375-388.

Maunder, E. A. W., Ramsay, A. C. A. (1993). Quadratic equilibrium elements, in: J. Robinson (ed.), *FEM Today and Future*, Robinson and Associates,pp. 401-407.

Teixeira de Freitas, J. A., Cismasiu C., Wang, Z. M. (1999). Comparative Analysis of Hybrid-Trefftz Stress and Displacement elements, *Archives of Computational Methods in Engineering*, **6**, 35-59.

Kompiš, V., Fraštia, L. (1997). Polynomial representation of hybrid finite elements, *Computer Assisted Mechanics and Eng. Sciences*, **4**, 521-532.

Cruse, T. A. (1997). An improved boundary integral equation method for three dimensional elastic stress analysis: Polynomial representation of hybrid finite elements, *Computer Assisted Mechanics and Eng. Sciences*, **4**, 521-532.

Cruse, T. A. (1974). An improved boundary integral equation method for three dimensional elastic stress analysis, *Computers & Structures*, **5**, 741-754.

Lachat, J. C., Watson, J. O. (1976). Effective numerical treatment of boundary integral equation: a formulation for three dimensional elasto-statics, *Int. J. Num. Meth. Eng.*, **10**, 991-1005.

Cheung, Y. K., Jin, W. G., Zienkiewicz, O. C. (1989). Direct solution procedure for solution of harmonic problems using complete, non-singular, Trefftz functions, *Commun. In Appl. Numer. Methods*, **5**, 159-169.

Cheung, Y. K., Jin, W. G., Zienkiewicz, O. C. (1991). Solution of Helmholtz equation by Trefftz method, *Int. J. Numer. Meth. Engng.*, **32**, 63-78.

Kompiš, V., Oravec, J., Búry, J. (1999). Reciprocity based FEM, *Strojnícky časopis (Mechanical Engineering)*, **50**, 188-201.

Kompiš, V., Žmindák, M., Jakubovičová, L. (1999). Error estimation in multi-domain BEM (Reciprocity based FEM) In: *ECCM99, European Conference on Computational Mechanics*, Ed. W. Wunderlich, CD-ROM, München.

Kuhn, G., Partheimüller, P., Köhler, O. (1998). Regularization and evaluation of singular domain integrals in boundary element methods, in *Singular Integrals in Boundary Element Methods*, eds V. Sládek, J. Sládek, Comput. Mech. Publ., Southampton, pp. 223-262.

Baláš, J., Sládek, J., Sládek, V. (1989). *Stress Analysis by Boundary Element Method*, Elsevier.

Bausinger, R., Kuhn,G. (1987). *The Boundary Element Method* (in German), Expert Verlag, Germany.

Blacker, T., Belytschko, T. (1994). Superconvergent patch recovery with equilibrium and conjoint interpolant enhancement, *Int. J. Num. Meth. Eng.*, **37**, 517-536.

Niu, Q., Shephard, M. S. (1993). Superconvergent extraction techniques for finite element analysis, *Int. J. Num. Meth. Eng.*, **36**, 811-836.

Zienkiewicz, O. C., Taylor, R. L. (1991). *The Finite Element Method*, Vols. I-II, 4-th Edition, Wiley.

Bathe. K.-J. (1996). *The Finite Element Procedures*, Prentise Hall, Englewood Cliffs.

Kompiš, V., Jakubovičová, L. Errors in modelling high order gradient fields using isoparametric and reciprocity based FEM, to be published.

Maunder, E. A. W., Kompiš, V. Stress recovery techniques based on Trefftz functions, see the CD ROM of this Conference.

Forster, A., Kuhn, G. (1994). A field boundary element formulation for material nonlinear problems at finite strains, *Int. J. Solids Structures*, **31**, 1777-1792.

Prieto, I., Ibán, A. L., Garrido, J. A. (1999). 2D analysis for geometrically non-linear elastic problem using BEM, *Eng. Anal. with Bound. Elements*, **23**, 247-256.

CHAPTER 3

Stress Smoothing in Large Strain, Large Rotation Problems

V. Kompiš and J. Búry

Department of Mechanical Engineering, University of Žilina,
Veľký Diel, 010 26, Žilina, Slovak Republic
e-mail: kompis@fstroj.utc.sk, j@mppserv.utc.sk.

Abstract. Stress smoothing using Trefftz polynomials as interpolation functions in connection with moving least squares (MLS) procedure is shown to be efficient approach for infinitesimal elasticity and plasticity. In the case of large displacements the smoothing procedure based on the classical polynomial interpolation over the domain of influence gives satisfactory smoothing effect in the internal points of the domain, but not in the vicinity of the domain boundaries with prescribed tractions. In such points the static boundary conditions are included into the procedure, which leads to the nonlinear MLS problem.

1 Introduction

It is known that the stress smoothing using interpolation functions which satisfy the governing equations improves the accuracy and the rate of convergence of the stress field. The stress smoothing is usually applied in the post-processing stage of the computation.

The interpolation functions satisfying the governing equations are known as the Trefftz- (or simply T-) functions and can be found for linear problems in the form of polynomials or other types of functions for 2D or 3D problems. They can be introduced also from the fundamental solution, if the source points lie outside the region of interest for the local interpolation.

In nonlinear (finite displacement) problems, the T-functions are nonlinear in the stress-displacement relation and cannot be find exactly. The smoothing property of the interdomain (interelement) field inside the domain is well known and so, the moving least square (MLS) technique using classical interpolation polynomials can be applied to derive the smooth stress field inside the domain. Much larger errors are detected on the domain boundaries, especially those with prescribed tractions and in the regions closed to them. Inclusion of the local static boundary conditions into the smoothing technique enables to improve the accuracy of the stress fields. This technique leads to nonlinear MLS procedure for obtaining the stress field on and near the domain boundaries.

In the second part of this chapter, it is briefly shown, how the T-polynomials can be used in the multi-domain (e.g. FEM) problems in order to obtain a smooth stress field in homogeneous regions for linear problems. In the third part, there is a description of the smoothing technique for obtaining the quantities (displacement gradients, stresses and rotations) derived from the first derivatives of the primary variables (displacements) in large displacement problems.

The technique can be used for computation of the derivatives of primary variables obtained in discrete points by formulations using FEM, BEM and FVM. Similar procedures can be used for many other linear and nonlinear problems of continuum mechanics (thermal fields, magneto-elasto-dynamics, fluid dynamics, etc.) .

2 Stress Smoothing for Infinitesimal Displacement Problems

Stress recovery (stress smoothing) can be considered in the context of an analysis stage in modelling by finite element, multi-domain (MD) BEM, finite volume or meshless methods, when the primary variable obtained in the direct solution are displacements.

We assume here that the displacement formulation were used, i. e. we obtained the displacements as primary variables in the nodal points from the numerical solution. We will suppose that the discrete displacements obtained in this way are accurate enough comparing to the accuracy of their derivatives. It is well known that the stress and strain fields (secondary variables) computed from these displacements in classical way (i.e. elementwise) are of lower accuracy, discontinuous between the elements or subregions for which the displacements were supposed to be continuous and that they converge slower than the primary variables.

It is common practice with discrete data to fit smooth polynomial fields in a least squares process so as to achieve a "best" fit at the data points. Several schemes have been proposed for stress fields by [Zienkiewicz and Zhu (1992)], [Wiberg, Abdulwahab and Zivkas S.(1994)], [Tabbara, Blacker and Belytschko (1994)], [Boroomand and Zienkiewicz (1997)] and [Kvamsdal and Okstad (1998)], which differ as to the selection of data points (e.g. superconvergent points or node points), the properties of the polynomial fields, and the treatment of boundary conditions. The general trend has been to imbue the stress field with stronger equilibrium properties. Moving least squares (MLS) techniques have been introduced for finite element methods in [Blacker and Belytschko (1994)] and further exploited in meshless methods by [Lu, Belytschko and Gu (1994)] and [Fraštia (2000)]. We will describe the use of Trefftz fields in an MLS technique to fit locally the displacement field within a patch of nodal points and to get locally the best approximation for displacement gradient and stress fields.

We call T-functions the functions satisfying the governing (equilibrium) equations inside the domain. We use the T-polynomials (which can be derived using algebraic manipulation or numerically, as described in [Kompiš, Konkol' and Vaško (1999)], or fundamental solution as shown in [Kompiš and Búry (1999)] (very known from the BEM as [Balaš, Sládek and Sládek (1989)] with source points outside of the patch (the distance of the middle point of the patch to the source point equal or greater than the size of the patch is a good choice).

Trefftz fields have also been recently proposed by [Maunder (1999)] to smooth discontinuous stress fields within a patch of elements. Their use in formulation of hybrid in [Kompiš (1994)], [Jirousek and Zielinski (1995)] and [Kompiš and Búry (1999)] and reciprocity based in [Kompiš, Konkol' F. and Vaško (1999)] and [Kompiš, Žmindak and Jakubovičová (1999)] FE leads also to efficient models.

Several procedures of stress smoothing are shown in [Maunder and Kompiš (2000)]. We will suppose following one: We will denote by the point of interest a discrete point where the displacement gradients and stress will be computed by the smoothing procedure. Usually, it will be a nodal point of an element (Fig.1). By the patch domain we will denote the domain containing all nodal points which will be considered to define the Trefftz displacement polynomial in the least square (LS) sense.

Polynomial T-field of displacement is defined within a patch illustrated in Figure 1 in terms of parameters $\boldsymbol{c}$ as in Equation (1):

$$\{\tilde{\mathbf{u}}\} = [\mathbf{U}]\{\mathbf{c}\}. \tag{1}$$

The T-displacements are the columns of $[\boldsymbol{U}]$ and they are chosen to form a basis for the displacement space, and hence they are independent. The differences between discrete values of computed and smoothed displacement at data points are expressed by Equation (2):

$$\{\Delta_1\} = [\hat{U}]\{\boldsymbol{c}\} - \{\boldsymbol{d}\}, \tag{2}$$

where $\{\boldsymbol{d}\}$ contain the data values of displacements, and $\lfloor \boldsymbol{U} \rfloor$ are derived from the matrices in Equation (1) by evaluating rows at the discrete data points.

The MLS method is now illustrated. The task is to minimise

$$\Sigma = \{\Delta_1\}^T\{\Delta_1\} = \{\boldsymbol{c}\}^T[\hat{U}]^T[\hat{U}]\{\boldsymbol{c}\} - 2\{\boldsymbol{c}\}^T[\hat{U}]^T\{\boldsymbol{d}\} + \{\boldsymbol{d}\}^T\{\boldsymbol{d}\}, \tag{3}$$

with respect to variables in $\{\boldsymbol{c}\}$ by solving Equation (4):

$$[\hat{U}]^T[\hat{U}]\{\boldsymbol{c}\} = [\hat{U}]^T\{\boldsymbol{d}\}. \tag{4}$$

Generally the rank $\lfloor \boldsymbol{U} \rfloor\ \lfloor \boldsymbol{U} \rfloor \leq \text{rank}\lfloor \boldsymbol{U} \rfloor$, and $\lfloor \boldsymbol{U} \rfloor\ \lfloor \boldsymbol{U} \rfloor$ is nonsingular if and only if rank $\lfloor \boldsymbol{U} \rfloor$ equals the dimension of $\{\boldsymbol{c}\}$. It is therefore necessary for a unique solution to Equation (4) that dimension $\{\boldsymbol{d}\} \geq$ dimension $\{\boldsymbol{c}\}$.

If the point of interest is the node on the boundary with prescribed tractions, then one approach is to combine the sum of the squares with relative weighting factors and account for the differences in physical dimensions of the terms involved. The tractions in the point are smoothed using T-functions of tractions $[\boldsymbol{T}]$ corresponding to the displacements $[\boldsymbol{U}]$

$$\{\tilde{\boldsymbol{t}}\} = [\boldsymbol{T}]\{\boldsymbol{c}\} \tag{5}$$

and difference between discrete value of computed and smoothed traction at this point is expressed by Equation (6):

$$\{\boldsymbol{\Delta}_2\} = [\hat{\boldsymbol{T}}]\{\boldsymbol{c}\} - \{\boldsymbol{t}\}. \tag{6}$$

Then the MLS method becomes as in Equation (7).

$$\begin{aligned} &\text{minimise } \Sigma = w_1^d \left(\frac{E}{h}\right)^2 \{\boldsymbol{\Delta}_1\}^T \{\boldsymbol{\Delta}_1\} + w_2^d \{\boldsymbol{\Delta}_2\}^T \{\boldsymbol{\Delta}_2\} \Rightarrow \\ &\text{solve } \left[w_1^d \left(\frac{E}{h}\right)^2 \hat{\mathbf{U}}^T \hat{\mathbf{U}} + w_2^d \hat{\mathbf{S}}^T \mathbf{W}_2 \hat{\mathbf{S}} \right] \{\mathbf{c}\} = w_1^d \left(\frac{E}{h}\right)^2 [\hat{\mathbf{U}}^T]\{\mathbf{d}\} + w_2^d [\hat{\mathbf{S}}^T]\{\boldsymbol{\sigma}\} \quad , \end{aligned} \tag{7}$$

where E (Young's modulus) and h (a representative dimension of an element or patch) are introduced to make each term in $\boldsymbol{\Sigma}$ have the same physical dimension, and w_1^d and w_2^d are non-dimensional weighting factors for discrete quantities.

Note that the boundary conditions are taken into account in the single node, the point of interest in the smoothing process. This is because the boundary conditions and nodal displacements do not correspond to the same field in the strong sense. If the point of boundary is a corner point with the traction boundary conditions specified on both sides, the smoothing procedure is not necessary and all components of stress are defined from the boundary conditions.

3 Gradient Smoothing for Large Displacement Problems (the Total Lagrangian Formulation)

In large displacement problems the stress-displacement relation is nonlinear and equilibrium equations are nonlinear in displacements, too. So, it is not possible to find Trefftz functions expressed in displacements, which would satisfy the governing equations in each point, i.e. in the strong sense in some patch domain. Not only stresses, but also the displacement gradients are necessary by the evaluation of nonlinear terms in the stiffness matrix computation and the smoothing procedure is an important stage to ensure the accuracy of computations as shown in Kompiš, Novák and Handrik (2000).

Full polynomials are used instead of the Trefftz polynomials in the smoothing procedures for displacements and derived gradient fields.

We suppose the vector of displacements approximated by

$$\{\boldsymbol{u}(x)\} = [\boldsymbol{P}(x)]\{\boldsymbol{c}\}, \tag{8}$$

where $[\boldsymbol{P}(x)]$ is the full polynomial of co-ordinates $\{\boldsymbol{X}\} = (X_1, X_2, X_3)$ in the original, undeformed configuration of the body and $\{\boldsymbol{c}\}$ is a vector of unknown coefficients.

In this way, each component of the displacement field is approximated by polynomial fields by MLS technique using the discrete nodal displacements obtained from the numerical solution of the problem similarly as in the case of infinitesimal displacements. The T-functions cannot be used in this case, because the governing equations are nonlinear in displacements and so their use would be inefficient. The numerical experiments show that the approach using full polynomials gives satisfactory results inside the domain. The components of the deformation gradient in each point of interest are obtained from the derivatives of displacements (represented by the coefficients of the linear terms of the approximation polynomials (8), if the local coordinate system with the origin located in the point of interest is chosen) as

$$F_{ij} = \delta_{ij} + u_{i,j} = \delta_{ij} + \frac{\partial u_i}{\delta X_j}, \tag{9}$$

where δ_{ij} is Kronecker symbol. The Einstein notation with summation after repeated indices and with index after comma denoting the derivation is used.

Components of the second Piola Kirchhoff stress tensor for isotropic material are given by

$$S_{ij} = \lambda\,\delta_{ij}\left(u_{k,k} + u_{k,l}\,u_{k,l}\right) + \mu\left(u_{i,j} + u_{j,i} + u_{k,i}\,u_{k,j}\right) \tag{10a}$$

or

$$S_{ij} = \lambda\,\delta_{ij}\left(F_{kl}F_{kl} - \delta_{kk}\right) + \mu\left(F_{ki}F_{kj} - \delta_{ij}\right), \tag{10b}$$

where λ and μ are material Lame constants.

If the point of interest is a point of the domain boundary, such approximation does not give satisfactory accuracy and the boundary conditions have to be included into the approximation. Only the traction boundary conditions, which are most important in practical problems are considered here. The tractions related to the undeformed configuration obtained from the displacement gradient are

$$t_j^0 = n_k^0 S_{ki} F_{ji} = n_k^0 \left[\mu \left(u_{k,j} + u_{j,k} + u_{i,k} u_{i,j} + u_{j,i} u_{k,i} + u_{j,i} u_{i,k} + u_{j,i} u_{l,k} u_{l,i} \right) + \right.$$
$$\left. + \lambda \left(\delta_{jk} u_{i,i} + \delta_{jk} u_{i,l} u_{l,i} + u_{j,k} u_{i,i} + u_{j,k} u_{i,l} u_{l,i} \right) \right] \tag{11a}$$

or, those obtained from deformation gradient in shorter form

$$t_j = n_k^0 \left[\frac{\lambda}{2} F_{jk} \left(F_{ml} F_{ml} - F_{ll} - \delta_{ll} \right) + \mu \left(F_{ji} F_{lk} F_{li} - F_{jk} \right) \right] . \tag{11b}$$

The tractions are nonlinear because of large strains and large rotations (Belytschko, Liu and Moran (2000)). The MLS approximation of displacements by polynomial form defined over the discrete nodal displacements in the patch of nodes (domain of influence) and which will simultaneously satisfy the traction boundary conditions in the point of interest has to be solved in an iteration process as follows.

The tractions can be expanded into the Taylor series

$$t_j^{(n)} = t_j^{(n-1)} + \frac{\partial t_j^{(n-1)}}{\partial F_{ml}} \Delta F_{ml} + \left(high\ order\ terms \right). \tag{12}$$

The upper index 0 denoted the relation of tractions to the undeformed configuration was omitted for simplicity and the upper index (n) is now related to iteration step. The expansion (12) will be used in the Newton-Raphson (N-R) iteration process of numerical realisation of nonlinear LS smoothing of displacements when the boundary tractions are taken into account. Only first order terms of the Taylor expansion (12) will be considered.

The necessary derivatives in (12) can be derived from (11b) as follows:

$$\frac{\partial t_j^{(n)}}{\partial F_{ml}} = \left(\lambda C^{(n)} - 2\mu \right) \delta_{jm} n_l^0 + \lambda F_{jk}^{(n)} \left(F_{ml}^{(n)} - \delta_{lm} \right) n_k^0 +$$
$$+ \mu \left(\delta_{jm} F_{ik}^{(n)} F_{il}^{(n)} n_k^0 + F_{ji}^{(n)} F_{mi}^{(n)} n_l^0 + F_{jl}^{(n)} F_{mk}^{(n)} n_k^0 \right) \tag{13}$$

with

$$C^{(n)} = \left(F_{ml}^{(n)} F_{ml}^{(n)} - \delta_{ll} \right) \Big/ 2 \qquad and \qquad \frac{\partial C}{\partial F_{pq}} = F_{pq} . \tag{14}$$

Having choosen the local origin of coordinates in the point of interest, the displacement gradients and realising from Equation (9) that

$$\Delta F_{ij} = \Delta u_{i,j} \tag{15}$$

the following N-R iteration procedure for finding the constants $\{c\}$ in (8) can be defined as follows (2D problem will be described for the sake of abbreviation of the record):

In the n-th iteration step minimise:

$$\left(\begin{bmatrix} \mathbf{P} & \mathbf{0} \\ \mathbf{0} & \mathbf{P} \\ \mathbf{B}_1 & \mathbf{B}_2 \end{bmatrix} \begin{Bmatrix} \Delta\mathbf{c}_1^{(n)} \\ \Delta\mathbf{c}_2^{(n)} \end{Bmatrix} - \begin{Bmatrix} \mathbf{r}_1^{(n)} \\ \mathbf{r}_2^{(n)} \\ \mathbf{r}_t^{(n)} \end{Bmatrix} \right)^2 \tag{16}$$

with

$$c_i^{(n)} = c_i^{(n-1)} + \Delta c_i^{(n)} \tag{17}$$

$$\{\boldsymbol{r}_i^{(n)}\} = \{\boldsymbol{d}_i\} - [\boldsymbol{P}]\{\boldsymbol{c}_i^{(n-1)}\} \tag{18}$$

$$\{\boldsymbol{r}_t^{(n)}\} = \{\bar{\boldsymbol{t}}\} - \{\boldsymbol{t}(\boldsymbol{c}^{(n-1)})\} \tag{19}$$

where i = 1,2 and $\{\boldsymbol{c}_i\}$, $\{\boldsymbol{d}_i\}$ and $\{\boldsymbol{t}\}$ are subvector of the vector of coefficients $\{\boldsymbol{c}\}$ corresponding to the i-th component of displacements, vector of the i-th component of the nodal displacement and the vector of prescribed tractions, respectively. $[\boldsymbol{B}_i]$ are matrices corresponding to the definition (12) and (13).

Solution of the problem (16) is given by:

$$\begin{bmatrix} \boldsymbol{P}^T\boldsymbol{P} + \boldsymbol{B}_1^T\boldsymbol{B}_1 & \boldsymbol{B}_1^T\boldsymbol{B}_2 \\ \boldsymbol{B}_2^T\boldsymbol{B}_1 & \boldsymbol{P}^T\boldsymbol{P} + \boldsymbol{B}_2^T\boldsymbol{B}_2 \end{bmatrix} \begin{Bmatrix} \Delta\boldsymbol{c}_1^{(n)} \\ \Delta\boldsymbol{c}_2^{(n)} \end{Bmatrix} = \begin{Bmatrix} \boldsymbol{P}^T\boldsymbol{r}_1^{(n)} + \boldsymbol{B}_1^T\boldsymbol{r}_t^{(n)} \\ \boldsymbol{P}^T\boldsymbol{r}_2^{(n)} + \boldsymbol{B}_2^T\boldsymbol{r}_t^{(n)} \end{Bmatrix} \tag{20}$$

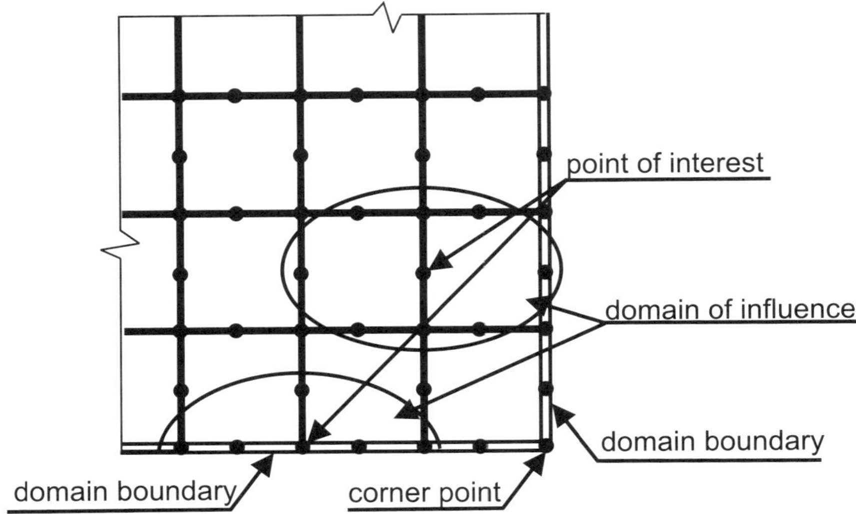

Figure 1: Definitions

Note that in the case of prescribed tractions in the boundary points the bondary conditions define normal and shear components of the Cauchy stress tensor in the surface plane. We did not include weighting factors into the formulation (16) and (20) for simplicity. However, taking the prescribed boundary traction conditions with larger weighting function, this will define the lower importance of approximately computed displacements for evaluation of corresponding stress components and of rotations and displacement gradients in such points. The displacements are important for evaluation of stresses and displacement gradients components in the tangential plane. In the corner points, the only quantities for which the displacements will be decisive are rotations and corresponding displacement gradients.

4 Final Remarks

The nonlinear MLS technique is a useful tool for increasing the accuracy of the smoothed gradient field on and in the vicinity of the domain boundaries. However, the procedure has to be used in the final, i.e. equilibrated state of the computation. Its use during the equilibrium iteration can lead to the element distortion and consequently to the lost of convergence.

References

Balaš J., Sládek V. and Sládek J. (1989), *Stress Analysis by Boundary Element Method*, Elsevier, Amsterdam.

Belytschko T., Liu W. K.and Moran B. (2000), *Nonlinear Finite Elements for Continua and Structures*, J. Wiley.

Blacker T. and Belytschko T. (1994), Superconvergent patch recovery with equilibrium and conjoint interpolant enhancements, *Int. J. Num. Meth. Eng.*, **37**, 517-536.

Boroomand B. and Zienkiewicz O.C. (1997), Recovery by equilibrium in patches (REP). *Int. J. Num. Meth. Eng.*, **40**, 137-164.

Fraštia L. (2000), On certain improvements of element-free Galerkin method, in Proc. of Conference *Numerical Methods in Continuum Mechanics,* CD-ROM ed. Kompiš V., Žmindák M. and Maunder E.A.W., MC Energy, Žilina.

Jirousek J. and Zielinski A.P. (1995), Survey of Trefftz type element formulations, *LSC Int. Report 95/06*, EPFL Lausanne.

Kompiš V. (1994), Finite element satisfying all governing equations inside the element, *Comp. & Struct.* **4**, 273-278.

Kompiš V. and Búry J. (1999), Hybrid-Trefftz finite element formulation based on fundamental solution, in *Discretization Methods in Structural Mechanics*, eds. Mang H.A. and Rammerstorfer F.G., 181-187, Kluwer Acad. Publ.

Kompiš V., Konkoľ F. and Vaško M.(2001), Trefftz-polynomial reciprocity based FE formulation, *Computer Assis. Mech. and Eng. Sci.*, **8**, 385-395.

Kompiš V., Novák P. and Handrik M. (2000) Finite displacement in reciprocity based FE formulation, in Proc. of Conference *Numerical Methods in Continuum Mechanics,* CD-ROM ed. Kompiš V., Žmindák M. and Maunder E.A.W., MC Energy, Žilina.

Kompiš V., Žmindak M., Jakubovičová L.(1999), Error estimation in multi-domain BEM (Reciprocity Based FEM). In ECCM'99, *European Conference on Computational Mechanics*, CD-ROM ed. Wunderlich W., TU Munich.

Kvamsdal T. and Okstad K.M. (1998), Error estimation based on superconvergent patch recovery using statically admissible stress fields. *Int. J. Num. Meth. Eng.*, **42**, 443-472.

Lu Y. Y., Belytschko T. and Gu L. (1994), A new implementation of the element free Galerkin method, *Comput. Meth. Appl. Mech. Engng.*, **113**, 397-414.

Maunder E.A.W. (2001), A Trefftz patch recovery method for smooth stress resultants and applications to Reissner-Mindlin equilibrium plate models, *Computer Assis. Mech. and Eng. Sci.,* **8**, 409-424.

Maunder E.A.W. and Kompiš V. (2000), Stress recovery techniques based on Trefftz functions, in Proc. of Conference *Numerical Methods in Continuum Mechanics,* CD-ROM ed. Kompiš V., Žmindák M. and Maunder E.A.W., MC Energy, Žilina.

Tabbara M., Blacker T. and Belytschko T. (1994), Finite element derivative recovery by moving least square interpolants. *Comput. Meth. Appl. Mech. Engrg.*, **117**, 211-223.

Wiberg N.-E., Abdulwahab F. and Zivkas S. (1994), Enhanced superconvergent patch recovery incorporating equilibrium and boundary conditions. *Int. J. Num. Meth. Eng.*, **37**, 3417-3440.

Zienkiewicz O.C. and Zhu J.Z. (1992), The superconvergent patch recovery and a posteriori error estimates. Part 1: the recovery technique. *Int J. Num. Meth. Eng.*, **33**, 1331-1364.

CHAPTER 4

Special Topics in Multi-Domain BEM Formulations

V. Kompiš

Department of Mechanical Engineering, University of Žilina,
Veľký Diel, 010 26, Žilina, Slovak Republic
e-mail: kompis@fstroj.utc.sk

Abstract.- In this chapter following topics in multi-domain (MD) BEM/reciprocity based FEM formulations in linear elastic problems has been studied: (1) errors in displacement and stress fields are studied in examples of modelling high order gradient and error estimators and (2) modelling of local displacement and stress fields for local loads leading to high field gradients.

1 Introduction

The error estimator is an important tool for obtaining solutions with required accuracy over the whole investigated region by remeshing. The quality of the error estimator can save some intermediate steps of the analysis. In most FE formulations we obtain discontinuous stress fields between the elements. This discontinuity is used as a simple error indicator in many FE programs[ADINA. (1995)]. However, in isoparametric elements, the interelement is discontinuity not the only source of modelling error. The equilibrium equations inside the element are only satisfied in the weak integral sense, and this influences the local error distribution as it is well known. In order to improve the error estimates, the stresses are recovered using polynomial interpolation functions for the nodal displacements over the patches of elements [Hinton, E., Campbell, J. (1974), Zienkiewicz, O. C., Zhu, J. Z. (1987)]. The errors are evaluated as the difference between the stresses obtained from element matrices and those obtained by the stress recovery. Some authors use also equilibrium and boundary conditions to recover more accurate stress fields [Blacker, T., Belytschko, T., (1994) - Wiberg, N.E. (1997)].
In MD formulations using the Boundary Integral Equations (BIE) for the formulation of the sub-domain stiffness matrix it is supposed that the governing equations are satisfied in the strong sense (i.e. in each point) inside the sub-domain and the error of the solution is restricted to the inter-domain and domain boundaries. When the displacements are suggested to be continuous on the inter-domain boundaries, then the boundary traction discontinuities are a reliable error indicator for the global as well as local errors in the solution.
The accuracy of the MD BEM formulations is studied by modelling high degree (in this paper the 6-th degree) polynomial fields defined by the T-polynomials. In this way the accuracy of both stress and displacement fields can be studied. It is shown that an efficient smoothing procedure shown in the Chapter 2 can reduce the errors by one order or even more in the fields with high gradients, when compared to simple averaging of the nodal values.

Another problem dealt in this chapter is the numerical modelling of local effects and especially, problems with concentrated tractions. Such problems arise e.g. in contacting bodies, when the radius of curvature of one contacting surface is much smaller than the radius of curvature of another body in contact. If the difference of radii of curvature is large in one direction, but small in the other, orthogonal direction, the contact is known as a line contact. If the difference of all main radii of curvature in the contact areas is large, the contact is called the point contact. Usually such problems require very fine mesh in order to obtain good approximation of the fields in the vicinity of the contact. Both displacement and stress fields have large gradients in the vicinity of such peculiarities. An enough accurate solution is important for the evaluation of loading capacity of the bodies in contact. Although the concentrated load does not introduce the real situation, it can be good approach of the real conditions in its vicinity according to the Saint Venant principle. In classical numerical models like FEM, the concentrated nodal load does not introduce the concentrated force but a distributed tractions and cannot be used for very local approximation without having very fine mesh model.

2 A Study of Errors in MD BEM Formulation

For the sake of investigation of modelling errors and error estimation in MD formulations we studied the problems on the models of a simple quadrilateral domain with Dirichlet and Neumann boundary conditions. In order to study the local and global errors in the displacement and stress fields, the problem with boundary conditions described by the 6-th order T-polynomials was modelled. The Young modulus equal to 1000 MPa and Poisson ratio equal to 0.3 were chosen in the models. The domain was approximated by eight noded quadrilateral quadratic elements. The displacements in the nodal points of the domain boundary and the tractions on the whole outer boundaries were prescribed according to the exact Trefftz solution for the Dirichlet and the Neumann problem, respectively. In this way the exact errors in both displacement and stress fields could be studied.

The Figure 1 shows the deformed region as defined by the test 6-th degree Trefftz displacement polynomials. The mesh and contours of the displacement (magnitudes of the vector field values), and von Mises stress fields of the test problem are given in Figures 2 and 3, respectively. In all these and following Figures, as well, the maximal values of corresponding field variables in the region are given, so that good information about relative errors can be obtained.

Errors in the calculated fields of the von Misses stress by MD BEM with 10 by 10 elements using quadratic boundary elements for approximation of boundary displacements and boundary tractions are shown in Fig. 4. Corresponding field of displacement errors are given in Figure 5.

Information about the errors like that indicated in Figures 4 and 5 cannot be obtained for general problems. Instead some error estimators are used in the stress recovery phase [Hinton, E., Campbell, J. (1974) - Wiberg, N.E. (1997)]. One of the simplest error estimators used in many commercial programs [ADINA. (1995)] is based on the jumps in the stress fields between neighbour elements/sub-domains. When elements, in which the equilibrium equations are not satisfied inside the elements (in a strong sense), are used in the model, then such error estimators are not reliable, because they do not include the error in the equilibrium equations. We can obtain not

only underestimated errors, but also incorrect distributions of errors [Kompiš, V., Jakubovičová, L. (20C0)].
The errors in displacement field are given in Figure 6 with both prescribed boundary displacements (Fig. 6a) and prescribed boundary tractions (Fig. 6b) in the form of the undeformed mesh distorted by the displacement errors. Such as graph is very useful, because it shows the "numerical inhomogenity" of the approximation functions and about the influence of the boundaries on the distribution of errors.

In this figures the behaviour of errors can be studied locally inside the elements and in the whole domain, too. It is well known that the discontinuities of the first derivative of the displacement fields are basis of the stress jumps between the elements and though, the roughness of displacement errors and errors in stress fields are closely related. From the comparison of the Figures 6a and 6b, it can be seen that the roughness of both fields is comparable, if we exclude the outer rows of elements. This refers to, how important is to include the static boundary conditions into the evaluation of stress fields on the parts of the boundaries with prescribed tractions. Note, that if the boundary tractions are taken into account in the process of stress evaluation (often known as the smoothing process in the post-processing stage), the numerical error noise can be smoothed out.
In order to give the view about the efficiency of the MD BEM formulation some maximal values of the field variables are shown below:

Von Mises stress: 14488.

Error in tractions: 224 (i.e. relative error 0.015).

Error in the averaged stress: 145 (i.e. relative error 0.010).

Error in the smoothed stress: 21 (i.e. relative error 0.0014).

The smoothed stresses were obtained using MLS techniques using quadratic Trefftz interpolation polynomials with the Domain of Influence of the same size as the diagonal of the elements, and using the known boundary tractions in the Point of Interest (POI), when the POI was the node of the domain boundary.

The error in averaged tractions is defined as

$$e_t = \int_{\Gamma_{BE}} \left(\left(t_i^A + t_i^B \right)^T \left(t_i^A + t_i^B \right) \right)^{1/2} \mathbf{d\Gamma} \tag{1}$$

and it is an important parameter for adaptive refinements of the solution. Γ_{BE} is an boundary element on the inter-domain boundaries and the upper indices A and B denote the neighbour quantities of the common inter-domain. If corresponding boundary element lies on the domain boundary, than one of the quantities is the prescribed traction.

3 Modelling of Local Effects

We will consider local boundary traction effects in the solution in the linear problems, i.e., if the principle of superposition can be applied. Similar technique was applied by another technique, the hybrid-displacement FE, using the Trefftz functions for the formulation of stiffness matrix, too[Kompiš, V., Kaukič, M., Žmindák, M. (1995)].

According to the principle of reciprocity the relation between boundary displacements, **u** and boundary tractions, **t** is in the form:

$$\int_{\Gamma} u t^* d\Gamma = \int_{\Gamma} t u^* d\Gamma \tag{2}$$

where the quantities with the star denote corresponding T-functions.

Now we consider the local loading acting in the domain (it can be inside the domain, or on its boundary) and suppose that we have an analytic solution for a similar local problem. The similar local problem can be a solution of the half plane, half space, etc. so, that the real problem can be locally included into the auxiliary problem. This means that we can write

$$\mathbf{u} = \mathbf{u}^0 + \mathbf{u}^l \tag{3}$$

$$\mathbf{t} = \mathbf{t}^0 + \mathbf{t}^l \qquad \textit{and} \qquad \boldsymbol{\sigma} = \boldsymbol{\sigma}^0 + \boldsymbol{\sigma}^l \tag{4}$$

for some part Γ_1 of the surface Γ, with $\Gamma = \Gamma_1 + \Gamma_2$. Here Γ_1 is a part of the boundary Γ, where the local effects are taken into account and Γ_2 is the remaining part of the boundary. The upper indices *0* and *l* denote the smooth and local boundary condition, respectively. Then we can re-write the equation (2) into the form

$$\int_{\Gamma_1} u^0 t^* d\Gamma + \int_{\Gamma_2} u t^* d\Gamma = \int_{\Gamma_1} \bar{t} u^* d\Gamma + \int_{\Gamma_2} \bar{t} u^* d\Gamma - \int_{\Gamma_1} \bar{u}^1 t^* d\Gamma \tag{5}$$

The known quantities are denoted by bar in equation (4). Note that some nodal points can lie on the border between the boundary parts Γ_1 and Γ_2. In such case we can choose, if the unknown part of the displacement vector will be the real or smooth part and add the corresponding integrals to the right (supposed to be known) or left hand side of equation (5).

If a point load will apply in the point $\mathbf{x_P}$, then the equation (5) will be in the form

$$\int_{\Gamma_1} u^0 t^* d\Gamma + \int_{\Gamma_2} u t^* d\Gamma = P(x_P) u^*(x_P) + \int_{\Gamma_2} \bar{t} u^* d\Gamma - \int_{\Gamma_1} \bar{u}^1 t^* d\Gamma \tag{6}$$

Notice, that the boundary Γ_1 is defined so that it contains all the part of sub-domain boundaries where the local gradient is substantial part in the solution. The splitting of the solution is used also in the post-processing according to (4), when the smooth stress field is computed from the discrete displacements in the nodal points and, if the point of interest is on the boundary, or in its vicinity, also corresponding boundary tractions are to be taken into account.

As an example for using the local field technique the 2D ring loaded by concentrated forces acting in its inner boundary is shown in the Figure 7. Only four quadrilateral sub-domains each consisting of four quadratic boundary elements were used for the analysis for modelling a quarter of the double symmetric domain.

4 Conclusions

The exact errors were obtained when the Trefftz polynomials of higher degree were used for test functions. Both geometric and static boundary conditions were examined. This type of test showed the importance of taking into account the boundary conditions in the post-processing (stress recovery) stage. The errors in the recovered stress field by the smoothing procedure are one order lower in the example with higher gradients than those obtained by simple averaging of their nodal values. Smoothing of stress and displacement gradient fields increase the rate of convergence so, that the same rate convergence of these fields is achieved as that in displacement field.

The exact error study can help to evaluate the influence of different modelling procedures on the model accuracy and reliability of error estimates in both displacements and stress distribution.

The second problem dealt in this chapter has to show how the local effects can be effectively modelled using the superposition technique. Although such technique can be used in linear problems only, there are many problems of practical importance, which fall down into this category of problems.The rate of convergence of displacement and stress fields is shown in figure 7. The rate of convergence was studied using the maximal error in corresponding field.The rate of convergence displacements is 4. The rate of convergence of stress in the corner points of elements (i.e. in the points with lowest accuracy, but obtained by the simpliest way) is 1.5 only. The stresses obtained by smoothing procedure has the same rate of convergence as the rate of convergence of primary field, i.e. equal to 4, too. From this study we can see, that by using the smoothing procedure by T-polynomials, the stress field of the some accuracy can be obtained with half fine or even course mesh as the accuracy of the field obtained without smoothing procedure. The smoothing procedure isn't expensive process and so, it is an effective step in each multi-domain solution.

References

ADINA. (1995). Theory and Modeling Guide, ADINA R & D, Inc.

Hinton, E., Campbell, J. (1974). Local and global smoothing of discontinuous finite element functions using a least square method potential problem, *Int. J. Numer. Meth. Eng. & Mach.*, Vol. 8, pp. 461—480.

Zienkiewicz, O. C., Zhu, J. Z. (1987). A simple error estimator and adaptive procedure for practical engineering analysis potential problem, *Int. J. Numer. Meth. Eng. & Mach.*, Vol. 24, pp. 337—357.

Blacker, T., Belytschko, T., (1994). Superconvergent patch recovery with equilibrium and conjoint interpolant enhancement, *Int. J. Num. Meth. Eng.*, Vol. 37, pp. 517—536.

Niu and, Q., Shephard, M.S. (1993). Superconvergent extraction techniques for finite element analysis, *Int. J. Num. Meth. Eng.*, Vol. 36, pp. 811—836.

Ramsay, A.C.A., Maunder, E.A.W. (1996). Effective error estimation from continuous, boundary admissible estimated stress fields, *Comp. & Struct.*, Vol. 61, pp. 331—343.

Wiberg, N.E. (1997). Superconvergent patch recovery: A key to quality assessed FE solutions, *Advan. Eng. Software*, Vol. 28, pp. 85—95.

Kompiš, V., Jakubovičová, L. (2000). Errors in modelling high order gradient field using isoparametric and reciprocity based FEM, *Eng. Modelling*, Vol. 13, No.1-2, pp.27-34.

Kompiš, V., Kaukič, M., Žmindák, M. (1995) Modeling of local effects by hybrid-displacement FE, *J. Comput. Appl. Math.*, Vol. 63, pp. 265-269.

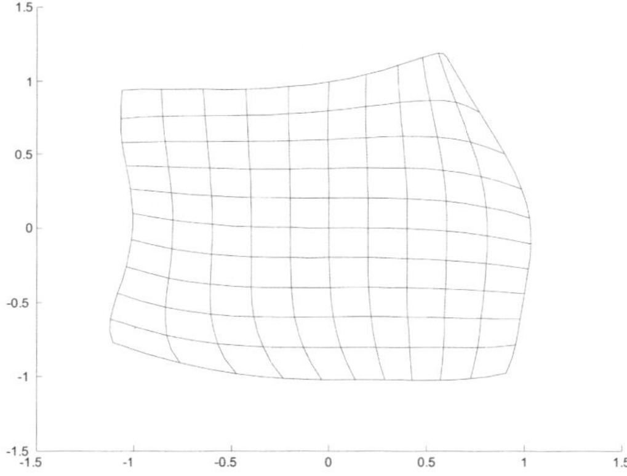

Fig. 1 Deformed region by the test Trefftz polynomials of 6-th order.

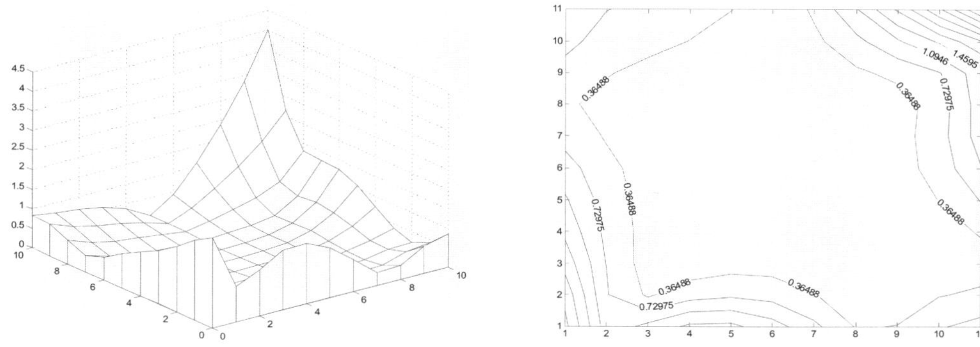

Fig. 2 Countours and mesh of test displacements given by Trefftz displacement of 6-th order.(max. 4.01)

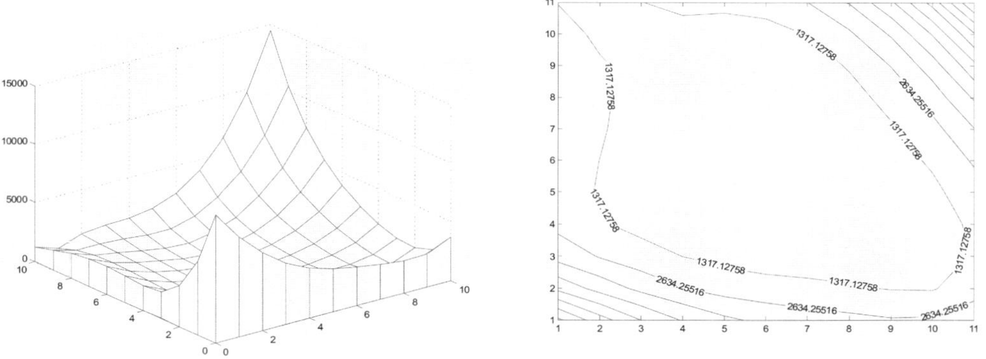

Fig. 3 Countours and mesh of test von Mises streses defined by Trefftz stress polynomials of 6-thorder.

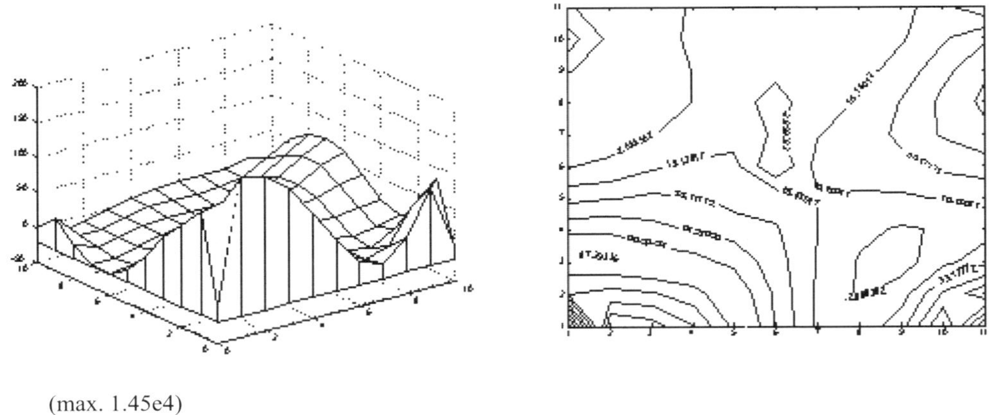

(max. 1.45e4)

Fig. 4 Countours of errors in von Mises field computed by quadratic reciprocity based 10x10 elements. (max. 182.6)

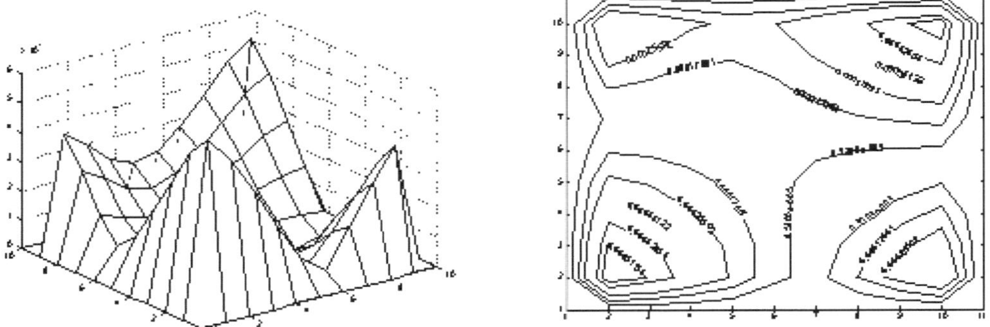

Fig. 5 Countours and mesh of errors in displacement field computed by quadratic reciprocity based elements.(max. 8.76e-4)

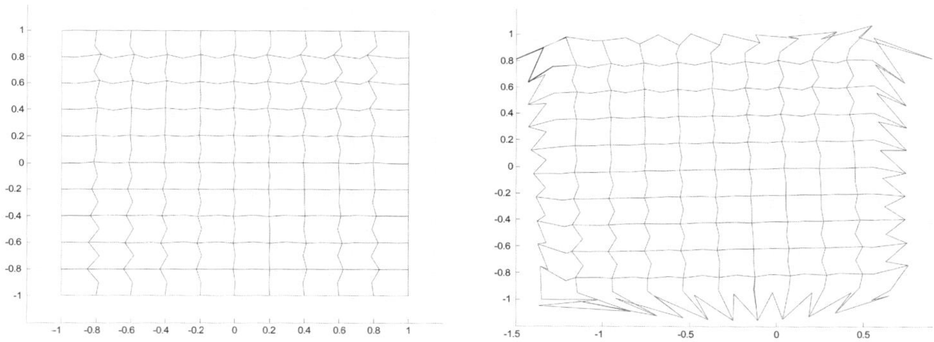

Fig. 6 Errors in displacement field 9a by prescribed boundary displacements and 9b by prescribed boundary tractions using the reciprocity based formulation.

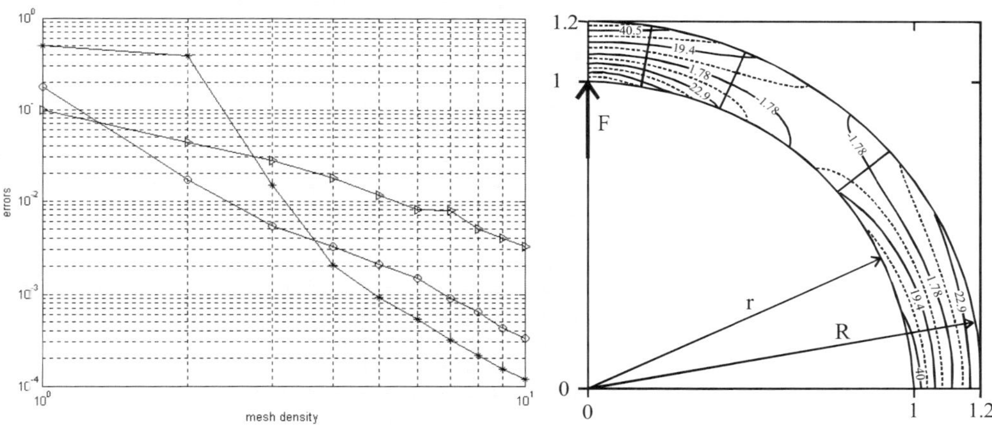

Fig. 7 Convergence of displacement and stress;
>>> - maximum error of the averaged stress
ooo - maximum error of the smoothed stress
*** - error of displacement norm.

Fig. 8 Tangential stress field.

CHAPTER 5

BEM Representation of Diffusion-Convection Equations

V. Kompiš and F. Konkoľ

Department of Mechanical Engineering, University of Žilina,
Veľký Diel, 010 26, Žilina, Slovak Republic
e-mail: kompis@fstroj.utc.sk, konkol@mppserv.utc.sk.

Abstract. In this chapter, the multi-domain boundary element method (MD BEM) will be considered for the problems of computational fluid dynamics (CFD), described by the Navier-Stokes equations. The fluid flow can be convection-dominated in some parts of the domain and diffusion dominated in another parts and it can be steady or unsteady. The multi-domain (MD) BEM representation of different kind of equations used in CFD and their linearization will be introduced. Reciprocity based formulation using non-singular presentation of the Laplace and Helmholz equation is presented for the solution of the non-linear problems.

1 Introduction

The Navier-Stokes equations are usually described in the Eulerian mesh and they introduce a mathematical model of physical conservation laws of mass, momentum and energy. They can be solved by finite difference, finite volume and finite element methods. The BEM has shown some advantages, especially, when the nonlinear terms are negligible, it can transform the solution to the integration over the domain boundaries only and it can better handle with infinite domains. The MD BEM formulation enables to decrease the computational time and also to split simply the domain into the parts where the nonlinear effect are important and where their influence is negligible.

Convection-dominated fluid flows suffer from numerical instabilities. In the BEM the problem can be avoided by the use of Trefftz functions, or Green (fundamental solution) of the appropriate linear differential operators.

In this chapter, the multi-domain boundary element method (MD BEM) will be considered for the problems of computational fluid dynamics (CFD), described by the Navier-Stokes equations.

The fluid flow can be convection-dominated in some parts of the domain and diffusion dominated in another parts. For each of these parts another numerical procedure can be used and so, increase the efficiency of the modelling.

Reciprocity based formulation using non-singular solution of the Laplace and Helmholz equations, respectively, are introduced for the integral equation formulation of the non-linear problems.

2 Integral equation representation of some steady-state CFD equations

In this part we will show how some linear PDE, closely related to CFD, ca be transformed to the boundary integral formulation for their numerical solution.

Let us have the physical problems described by PDE's

$$\phi_{,jj} - b = 0 \tag{1.1}$$

$$\phi_{,jj} + k^2\phi - b = 0 \tag{1.2}$$

respectively, where Φ is a scalar or vector function and b is a scalar or vector constant.

The numerical solution of the equation (1.1) is considered in integral form

$$\int_\Omega (\phi_{,jj} - b)\phi^* d\Omega = 0 \tag{1.3}$$

where Φ^* is a weighting function. Applying the integration per partes and the Green formula, we can write

$$\begin{aligned} \int_\Omega \phi_{,jj}\phi^* d\Omega &= \int_\Omega (\phi_{,j}\phi^*)_{,j} d\Omega - \int_\Omega \phi_{,j}\phi^*_{,j} d\Omega \\ &= \int_\Omega \phi_{,j} n_j \phi^* d\Omega - \int_\Omega \phi_{,j}\phi^*_{,j} d\Omega \\ &= \int_\Gamma \phi_{,n}\phi^* d\Gamma - \int_\Omega \phi_{,j}\phi^*_{,j} d\Omega \end{aligned} \tag{1.4}$$

When the integration per partes and the Green formula is applied to the last integral in (1.4) as

$$\begin{aligned} \int_\Omega \phi_{,j}\phi^*_{,j} d\Omega &= \int_\Omega (\phi\phi^*_{,j})_{,j} d\Omega - \int_\Omega \phi\phi^*_{,jj} d\Omega \\ &= \int_\Gamma \phi\phi^*_{,n} d\Gamma - \int_\Omega \phi\phi^*_{,jj} d\Omega \end{aligned} \tag{1.5}$$

and suppose Φ^* to be a Trefftz function satisfying the homogeneous PDE (Laplace equation)

$$\phi^*_{,jj} = 0 \tag{1.6}$$

then the final form of equation (1.3) can be written in the form:

$$\int_{\Gamma} \phi_{,n} \phi^* d\Gamma - \int_{\Gamma} \phi \phi^*_{,n} d\Gamma - \int_{\Omega} b \phi^* d\Omega = 0 \tag{1.7}$$

which is the boundary integral presentation of the problem (1.1).

Similarly, the integral form of equation (1.2) is

$$\int_{\Omega} (\phi_{,jj} + k^2 \phi - b)\, d\Omega = 0 \tag{1.8}$$

Now following the same way as that used for equation (1.3) we will find

$$\begin{aligned} \int_{\Omega} \left(\phi_{,jj} - k^2 \phi - b\right) \phi^* d\Omega &= \int_{\Gamma} \phi_{,n} \phi^* d\Gamma - \int_{\Gamma} \phi \phi^*_{,n} d\Gamma \\ &+ \int_{\Omega} \phi \left(\phi^*_{,jj} - k^2 \phi^*\right) d\Omega - \int_{\Omega} b \phi^* d\Omega \end{aligned} \tag{1.9}$$

Again, consider the weighting function, Φ^*, to be a Trefftz function, beeing the solution of the homogeneous (Helmholtz) equation

$$\phi^*_{,jj} - \phi^* = 0 \tag{1.10}$$

Then the boundary integral presentation of the problem (1.8) is also the equation (1.7). The only difference between both solutions are the Trefftz weighting function Φ^*, which have to satisfy now the Helmholz equation (1.10), instead of the Laplace equation (1.6) in the former case.

The boundary integral presentation contains both Dirichlet and Neumann boundary conditions and relates their unknown to known parts on the whole domain boundary.

For the MD solution continuity of the primary variable Φ is considered in the strong sense. The inter-domain compatibility and the Neumann boundary condition (usually in the form of flux through the surface or the continuity of tractions) is considered in the weak, Galerkin integral sense

$$\int_{\Gamma_i} \left(\phi^A_{,n} - \phi^B_{,n}\right) w d\Gamma + \int_{\Gamma_t} \left(\phi_{,n} - \bar{\phi}_{,n}\right) w d\Gamma = 0 \tag{1.11}$$

where w is some weighting function defined for each boundary element. Γ_i and Γ_t are inter-domain boundaries and the boundaries with prescribed flux, respectively.

After discretization, the normal derivative, $\Phi_{,n}$ is expressed from (1.7) as a function of the primary variable, Φ and substituted into (1.11). This gives the resulting system of equations for solution of the primary variable Φ.

Stokes equation describing the inviscid incompressible fluid flow in velocity-pressure form

$$v_{i,jj} + p_{,i} + q_i = 0 \tag{1.12}$$

together with incompressibility condition

$$v_{i,i} = 0 \tag{1.13}$$

can be solved by BEM, using the integral representation

$$\int_{\Omega} \left(v_{i,jj} + p_{,i} + q_i\right) v_i^* d\Omega = 0 \text{ and } \int_{\Omega} v_{i,i} p^* d\Omega = 0 \tag{1.14}$$

where v_i^* is a weighting function. In these equations, v_i denotes velocity component and p is modified pressure.

Using transformations according to (1.4) and (1.5) we have

$$\int_{\Omega} v_{i,jj} v_i^* d\Omega = \int_{\Gamma} v_{i,n} v_i^* d\Gamma - \int_{\Gamma} v_i v_{i,n}^* d\Gamma + \int_{\Omega} v_i v_{i,jj}^* d\Omega \tag{1.15}$$

Similarly, the second integral of equation (1.14) can be transformed into

$$\int_{\Omega} p_{,i} v_i^* d\Omega = \int_{\Gamma} p v_i^* n_i d\Gamma - \int_{\Omega} p v_{i,i}^* d\Omega = \int_{\Gamma} p v_n^* d\Gamma - \int_{\Omega} p v_{i,i}^* d\Omega \tag{1.16}$$

with $v_n = v_i n_i$.

Trefftz weighting function using for the boundary integral representation of this problem has to satisfy the following equations

$$v_{i,i}^* = 0 \tag{1.17}$$

and

$$v_{i,jj}^* + p_{,i}^* = 0 \tag{1.18}$$

Note that the Trefftz polynomials satisfying the equations (1.17) and (1.18) will be polynomials order n for velocity components, but the order will be $(n - 1)$ for pressure (see Appendix A in Chapter 2 on the generation of Trefftz polynomials in the chapter on Finite Displacements in Reciprocity Based Multi-Domain BE/FE Formulations in this book). Moreover, after detail examination of the problem, one can find that the T-polynomials will have the same number of independent terms as those in elasticity problems for the same order of the polynomial.

For the evaluation of equation (1.14) the integral

$$\int_{\Omega} v_{i,jj} v_i^* d\Omega = \int_{\Gamma} v_{i,j} v_i n_j d\Gamma - \int_{\Omega} v_{i,j} v_{i,j}^* d\Omega = \int_{\Gamma} v_{i,n} v_i^* d\Gamma - \int_{\Gamma} v_i v_{i,n}^* d\Gamma + \int_{\Omega} v_i v_{i,jj}^* d\Omega \tag{1.19}$$

is obtained after applying the integration per partes and the Gauss' theorem. Index n after comma in these relations denotes the normal derivative $v_{i,n} = v_{i,j} n_j$.

Similarly, the integral

$$\int_{\Omega} p_{,i} v_i^* d\Omega = \int_{\Gamma} p v_i^* v_i d\Gamma - \int_{\Omega} p v_{i,i}^* d\Omega = \int_{\Gamma} p v_n^* d\Gamma - \int_{\Omega} p v_{i,i}^* d\Omega \tag{1.20}$$

Using the basic properties of the Trefftz functions (1.17) and (1.18) and the integral presentations (1.19) – (1.20), the following resulting boundary integral equation is obtained:

$$\int_{\Gamma} v_{i,n} v_i^* d\Gamma - \int_{\Gamma} v_i v_{i,n}^* d\Gamma + \int_{\Gamma} p v_n^* d\Gamma + \int_{\Omega} q_i v_i^* d\Omega \tag{1.22}$$

This single integral equation is the boundary presentation of integrals (1.14).

For the multi-domain formulation, the velocity field is continuous between the elements (subdomains) and the inter-domain equilibrium is satisfied in an integral sense expressed in the terms of the virtual power

$$\int_{\Gamma_t} \delta v_i (t_i - \bar{t}_i) d\Gamma + \int_{\Gamma_i} \delta v_i (t_i^A - t_i^B) d\Gamma = \int_{\Gamma_e} \delta v_i t_i d\Gamma - \int_{\Gamma_t} \delta v_i \bar{t}_i \, d\Gamma = 0 \tag{1.23}$$

where Γ_t , Γ_i and Γ_e denote the part of the domain boundary with prescribed tractions, the inter-domain boundaries and the sub-domain (element) boundaries. A and B denote the neighbor elements with the common boundary and the bar denotes prescribed value.

The tractions $\boldsymbol{t}$ are defined from the total stress vector $\boldsymbol{\sigma}$

$$t_i = \sigma_{ji} n_j \tag{1.24}$$

$$\sigma_{ij} = -p\delta_{ij} + \lambda v_{k,k} \delta_{ij} + \mu (v_{i,j} + v_{j,i}) \tag{1.25}$$

λ and μ are material constants of the thermodynamic state and δ_{ij} is the Kronecker delta.

3 Integral equation representation of parabolic diffusion-convection equations

Let us consider a linear time-dependent diffusion-convection equation describing the transport of an arbitrary conservation field function Φ

$$a\phi_{,jj} - \phi_{,t} + I_u = 0 \tag{2.1}$$

with boundary conditions

$$\begin{aligned} &\phi = \overline{\phi} && on\ \Gamma_1 \quad for \quad t > t_0 \\ &-k\phi_{,j}n_j = \overline{q}_n && on\ \Gamma_2 \quad for \quad t > t_0 \\ &\phi = \phi_0 && in\ \Omega \quad at \quad t = t_0 \end{aligned} \tag{2.2}$$

with notation of derivatives

$$\phi_{,j} = \frac{\partial \phi}{\partial x_j} \qquad and \qquad \phi_{,t} = \frac{\partial \phi}{\partial t} \tag{2.3}$$

a and k are parameters defined according to the considered governing equations with the corresponding constitutive flux law. The last term I_u contains all other terms and can be a function of Φ, too.

For the purpose of numerical solution the governing equation is split into a homogeneous linear and non-homogeneous parts

$$L(\phi) + b = 0 \tag{2.4}$$

where $L(\Phi)$ is a linear differential operator and b stands for pseudo-body force or source term.

The time derivative can be approximated by a finite difference scheme for the time increment $\Delta t = t_N - t_{N-1}$

$$\phi_{,t} \approx (\phi_N - \phi_{N-1}) / \Delta t \tag{2.5}$$

Then the equation (2.1) can be written in a non-homogeneous modified Helmholtz PDE form as

$$\phi_{,jj} - \phi / (a\Delta t) + b = 0 \tag{2.6}$$

The integral representation (see the section above) of this equation is

$$\int_\Gamma \phi \phi^*_{,n} d\Gamma = \int_\Gamma \phi_{,n} \phi^* d\Gamma + \int_\Omega b \phi^* d\Omega \tag{2.7}$$

The variable Φ^* is the modified Helmholtz Trefftz function, i.e. the solution of the homogeneous equation

$$\phi^*_{,jj} - \mu^2 \phi^* = 0 \tag{2.8}$$

In equation (2.8), the derivative

$$(.)_{,n} = \partial(.)/\partial n = (.)_{,j} n_j \tag{2.9}$$

and the parameter μ is defined as

$$\mu^2 = 1/(a\Delta t) \tag{2.10}$$

The solution of the problem (2.8) can be represented by the modified Helmholtz fundamental solution [Škerget L., Hriberšek M. and Kuhn G. (1999)], when the source point will be located outside the sub-domain

$$\phi^* = \frac{1}{a\pi r} \exp(-\mu r) \tag{2.11}$$

$$\phi^*_{,n} = \frac{r_{,n}}{4\pi r^2}(1 + \mu r)\exp(-\mu r) \tag{2.12}$$

for 3D problems, or

$$\phi^* = \frac{1}{2\pi} K_0(\mu r) \tag{2.13}$$

$$\phi^*_{,n} = \frac{r_{,n}}{2\pi} \mu K_1(\mu r) \tag{2.14}$$

for 2D problems. K_0 and K_1 are modified Bessel functions of the second kind and $\mathbf{r}$ is a vector from the source point ξ to the reference point s

$$\mathbf{r} = \mathbf{x}(\xi) - \mathbf{x}(S) \tag{2.15}$$

while r is its magnitude $r = |\mathbf{r}^{\cdot} \,.\, \mathbf{r}|$. The source points can be located arbitrarily outside the sub-domain. Good results can be achieved, if the distance of the source points from the sub-domain is large enough (e.g. equal to the double dimension of the sub-domain) in order to avoid large gradients in the integral kernels and so to justify with lower order numerical integration.

In the vorticity velocity formulation, the pseudo-body force term b contains the convection, source term and initial conditions [Škerget L. and Samec N. (1999)], [Wu J.C. (1982)]

$$b = \frac{1}{a}\left(v_{j,j}\phi + I_u\right) + \mu^2 \phi_{N-1} \tag{2.16}$$

where v_j is a velocity component and Φ introduces the vorticity (scalar in 2D and a component in 3D). The final form of integral equation will be given by

$$\begin{aligned}\int_\Gamma \phi\phi^*_{,n} d\Gamma &= \frac{1}{a}\int_\Gamma \left(a\phi_{,n} - \phi v_n\right)\phi^* d\Gamma + \frac{1}{a}\int_\Omega \phi v_j \phi^*_{,j} d\Omega \\ &+ \frac{1}{a}\int_\Omega I_n \phi^* d\Omega + \mu^2 \int_\Omega \phi_{N-1}\phi^* d\Omega + \int_\Omega v_j \phi_{,j}\phi^* d\Omega\end{aligned} \tag{2.17}$$

where $v_n = v_j n_j$.

The fundamental solutions and also the Trefftz functions can be found only for the linear PDE with constant coefficients. Therefore the velocity v_j has to be decomposed into an average constant vector $\bar{v}_j$, and a perturbed one $\hat{v}_j$

$$v_j = \bar{v}_j + \hat{v}_j \tag{2.18}$$

Using the finite difference approximation, the equation (2.1) can be written in the form

$$a\phi_{,jj} - \bar{v}_{j,j}\phi - \phi / \Delta t + b = 0 \tag{2.19}$$

with the integral representation

$$a\int_\Gamma \phi\phi^*_{,n} d\Gamma = \int_\Gamma \left(a\phi_{,n} - \phi\bar{v}_n\right)\phi^* d\Gamma + \int_\Omega b\phi^* d\Omega \tag{2.20}$$

where

$$\bar{v}_n = \bar{v}_j n_j \tag{2.21}$$

and Φ^* is the T-function satisfying the diffusion-convection equation

$$a\phi^*_{,jj} + \bar{v}_{j,j}\phi^* - \kappa\phi^* = 0 \tag{2.22}$$

which is given for 3D case as

$$\phi^* = \frac{1}{4\pi r a} \exp\left(\frac{\bar{v}_j r_j}{2a} - \mu r \right) \tag{2.23}$$

$$\phi^*_{,n} = \frac{1}{4\pi r^2 a} \left(r_{,n} - \frac{r}{2a} \bar{v}_j + \mu r r_{,n} \right) \exp\left(\frac{\bar{v}_j r_j}{2a} - \mu r \right) \tag{2.24}$$

and for 2D case

$$\phi^* = \frac{1}{2\pi a} K_0(\mu r) \exp\left(\frac{\bar{v}_j r_j}{2a} \right) \tag{2.25}$$

$$\phi^*_{,n} = \frac{1}{2\pi a} \left[\mu r_{,n} K_1(\mu r) - \frac{n_j \bar{v}_j}{2a} K_0(\mu r) \right] \exp\left(\frac{\bar{v}_j r_j}{2a} \right) \tag{2.26}$$

where

$$\mu^2 = \left(\frac{\bar{v}}{2a} \right)^2 + \frac{\kappa}{a} = \left(\frac{\bar{v}}{2a} \right)^2 + \frac{1}{a \Delta t} \qquad \textit{with} \qquad \bar{v}^2 = \bar{v}_j \bar{v}_j \tag{2.27}$$

The pseudo-body force, *b,* includes convective flux for the perturbed velocity field $\hat{v}_j$ only, source term and initial conditions, e.g.

$$b = -\hat{v}_{j,j} \phi + I_u + \phi_{N-1} / \Delta t \tag{2.28}$$

The resulting integral presentation is identical to equation (2.17) with only $\hat{v}_j$ is to be inserted instead of v_j into the first volume integral.

The MD formulation for this type of problems was shown in the previous part (see eq. (1.11)).

References

Škerget L., Hriberšek M. and Kuhn G. (1999).: Computational fluid dynamics by boundary-domain integral method, *Internat. J. Num. Meth. Engng.*, **46,** 1291-1311.

Škerget L. and Samec N. (1999). BEM for non-Newtonian fluid flow, *Eng. Analysis with Bound. Elem.*, **23**, 435-442 .

Wu J.C. (1982).: Problem of General Viscous Flow. *Developments in BEM*, vol.2, Chapter 2. Elsevier Applied Science Publication, London.

CHAPTER 6

A Symmetric Hybrid Boundary Element Method for Acoustical Problems

Lothar Gaul, Marcus Wagner and Wolfgang Wenzel

Institute A of Mechanics, University of Stuttgart, Germany

Abstract. The boundary element method (BEM) can be effectively used to solve mixed boundary value problems, as well as fluid-structure interaction problems. Among the advantages of BEM are the reduction of the problem dimension by one, high accuracy and the efficient treatment of problems with infinite domains. The direct BEM leads to a system of equations with dense and unsymmetric matrices. Thus, coupling with symmetric domain discretization techniques, such as finite element methods, is numerically inefficient. For this reason, alternative approaches are use herein derived from variational principles leading to symmetry by construction. By relaxing continuity between the fields on the boundary and those in the domain two multi-field variational principles have been formulated. Emerging from these variational principles, two Hybrid Boundary Element Methods (HBEM) were developed, the Hybrid Displacement Method and the Hybrid Stress Method, respectively. The choice of weighted fundamental solutions as domain approximations selected independently from polynomial approximations of boundary variables allow to map domain integrals in boundary integrals. Symmetric sytems of equations for frequency and time domain are obtained.

1 Introduction

For the investigation of acoustic scattering and radiation in time and frequency domain, as well as in bounded and unbounded domains, several numerical approaches are in use.

An alternative approach to the direct Boundary Element Method (BEM) (Gaul and Fiedler, 1997) emerges from variational principles of solid mechanics and leads to symmetry by construction. Hamilton's principle and the dynamical Hellinger-Reissner potential (Washizu, 1982) are used in this paper to obtain two multi-field variational principles for time and frequency domain, respectively. Emerging from these variational principles, two Hybrid Boundary Element Methods (HBEM) were developed for elastodynamics and acoustics, the HSBEM (Dumont, 1987, 1989) and the HDBEM (Gaul and Wagner, 1996). Hamilton's principle in a velocity potential formulation for acoustics is a one-field principle. By relaxing continuity between the field on the boundary and in the domain a multi-field variational principle has been formulated, as a starting point for a discretization procedure. The first time domain formulation of the HBEM was presented by Dumont and de Oliveira (1993) for two-dimensional elasticity.

In the following an exact frequency-dependent formulation, employing the HSBEM is used to solve exterior acoustical problems. A frequency-dependent Hermitian stiffness equation for the fluid is obtained. In the second part a time domain formulation of the HDBEM is introduced, resulting in an equation of motion with a symmetric mass and stiffness matrix.

with the four independent fields, the potential gradient $\hat{\phi}_{,i}$, the two independent potential fields $\hat{\phi}$ and $\hat{\phi}^{\nabla}$, and a Lagrangian multiplier $\hat{\lambda}$ introducing the static part of the Helmholtz equation. Additionally, essential boundary conditions have to be considered $\hat{\phi} - \hat{\bar{\phi}} = 0 \quad \text{on} \quad \Gamma_{\phi}$. Note that the potential field $\hat{\phi}$ is assumed only on the boundary, whereas $\hat{\phi}_{,i}$ exists on the boundary and in the domain.

Applying the fundamental lemma to the vanishing first variation yields an interpretation of the Lagrangian multiplier as $\hat{\lambda} \equiv \hat{\phi}^{\nabla}$. Moreover, the governing field equations are obtained, the natural boundary conditions, as well as compatibility conditions of the two potential fields. Considering that the variations on the boundaries vanish yields

$$\begin{aligned} \delta\hat{\mathcal{H}}^{\mathrm{R}}[\hat{\phi}_{,i},\hat{\phi}^{\nabla},\hat{\phi}] &= \int\limits_{\Omega} \rho\,[\hat{\phi}^{\nabla}_{,ii} + \kappa^2\hat{\phi}^{\nabla}]\delta\hat{\phi}^{\nabla^*}\,\mathrm{d}\Omega - \int\limits_{\Gamma} \rho\,(\hat{\phi} - \hat{\phi}^{\nabla})\delta\hat{\psi}^*\,\mathrm{d}\Gamma \\ &\quad - \int\limits_{\Gamma} \rho\,(\hat{\psi} - \hat{\bar{\psi}})\delta\hat{\phi}^*\,\mathrm{d}\Gamma = 0\,, \end{aligned} \tag{6}$$

as a starting point for the derivation of the HSBEM. If the Dirichlet boundary condition is enforced as subsidiary condition, Eq. (6) completely describes the acoustical field problem.

The numerical solution requires approximation functions for the field variables. The approximation of the boundary variable $\hat{\phi}$ is carried out with a shape function vector $\boldsymbol{N}$ and a nodal vector $\check{\boldsymbol{\phi}}$

$$\hat{\phi}(x,\omega) = \boldsymbol{N}^{\mathrm{T}}(x)\check{\boldsymbol{\phi}}(x,\omega)\,. \tag{7}$$

The approximations of the potential $\hat{\phi}^{\nabla}$ and gradient $\hat{\phi}_{,i}$ in the domain Ω, as well as the flux $\hat{\psi}$ on the boundary, are given as a superposition of n fundamental solutions $\Phi(r^{(k)},\omega)$, evaluated at the loadpoint $\xi^{(k)}$ that is collocated with the boundary node x ($r^{(k)} = |x - \xi^{(k)}|$). The fundamental solutions are weighted by source strengths γ_i, $i = 1..n$. Hence, the approximation of the potential field is obtained as

$$\hat{\phi}^{\nabla}(x,\omega) = \boldsymbol{\Phi}^{\mathrm{T}}(r^{(i)},\omega)\,\boldsymbol{\gamma}(\omega) \quad \text{and} \quad \hat{\psi}(x,\omega) = \boldsymbol{\Psi}^{\mathrm{T}}(r^{(i)},\omega)\,\boldsymbol{\gamma}(\omega) \tag{8}$$

The gradient of $\boldsymbol{\Phi}$ is multiplied by the outward normal to yield the flux approximation $\boldsymbol{\Psi}$ on the boundary. By modifying the domain Ω such that small spheres with radii ϵ, centered at the load points $\boldsymbol{\xi}$ where the fundamental solutions are singular, are subtracted, the domain Ω' with boundary Γ' is introduced. The properties of the Dirac distribution acting at the points now outside of the considered domain let the domain integral in Eq. (6) vanish in the limit $\Omega' \to \Omega$, if the domain approximations are introduced, yielding a boundary integral formulation in the absence of sources. Inserting the approximations given by Eqs. (7) and (8) yields a discretized formulation of the vanishing first variation

$$\lim_{\epsilon \to 0} \Bigg\{ \delta\boldsymbol{\gamma}^{\mathrm{H}} \Bigg(\underbrace{\int\limits_{\Gamma'} \boldsymbol{\Psi}^*\boldsymbol{\Phi}^{\mathrm{T}}\,\mathrm{d}\Gamma}_{\boldsymbol{F}}\,\boldsymbol{\gamma} - \underbrace{\int\limits_{\Gamma'} \boldsymbol{\Psi}^*\boldsymbol{N}^{\mathrm{T}}\,\mathrm{d}\Gamma}_{\boldsymbol{H}}\,\check{\boldsymbol{u}} \Bigg)$$

with the four independent fields, the potential gradient $\hat{\phi}_{,i}$, the two independent potential fields $\hat{\phi}$ and $\hat{\phi}^{\nabla}$, and a Lagrangian multiplier $\hat{\lambda}$ introducing the static part of the Helmholtz equation. Additionally, essential boundary conditions have to be considered $\hat{\phi} - \hat{\bar{\phi}} = 0 \quad \text{on} \quad \Gamma_{\phi}$. Note that the potential field $\hat{\phi}$ is assumed only on the boundary, whereas $\hat{\phi}_{,i}$ exists on the boundary and in the domain.

Applying the fundamental lemma to the vanishing first variation yields an interpretation of the Lagrangian multiplier as $\hat{\lambda} \equiv \hat{\phi}^{\nabla}$. Moreover, the governing field equations are obtained, the natural boundary conditions, as well as compatibility conditions of the two potential fields. Considering that the variations on the boundaries vanish yields

$$\delta\hat{\mathcal{H}}^{\mathrm{R}}[\hat{\phi}_{,i}, \hat{\phi}^{\nabla}, \hat{\phi}] = \int_{\Omega} \rho\,[\hat{\phi}^{\nabla}_{,ii} + \kappa^2 \hat{\phi}^{\nabla}]\delta\hat{\phi}^{\nabla^*}\,\mathrm{d}\Omega - \int_{\Gamma} \rho\,(\hat{\phi} - \hat{\phi}^{\nabla})\delta\hat{\psi}^*\,\mathrm{d}\Gamma$$
$$- \int_{\Gamma} \rho\,(\hat{\psi} - \hat{\bar{\psi}})\delta\hat{\phi}^*\,\mathrm{d}\Gamma = 0\,, \tag{6}$$

as a starting point for the derivation of the HSBEM. If the Dirichlet boundary condition is enforced as subsidiary condition, Eq. (6) completely describes the acoustical field problem.

The numerical solution requires approximation functions for the field variables. The approximation of the boundary variable $\hat{\phi}$ is carried out with a shape function vector $\boldsymbol{N}$ and a nodal vector $\check{\boldsymbol{\phi}}$

$$\hat{\phi}(x,\omega) = \boldsymbol{N}^{\mathrm{T}}(x)\check{\boldsymbol{\phi}}(x,\omega)\,. \tag{7}$$

The approximations of the potential $\hat{\phi}^{\nabla}$ and gradient $\hat{\phi}_{,i}$ in the domain Ω, as well as the flux $\hat{\psi}$ on the boundary, are given as a superposition of n fundamental solutions $\Phi(r^{(k)},\omega)$, evaluated at the loadpoint $\xi^{(k)}$ that is collocated with the boundary node x ($r^{(k)} = |x - \xi^{(k)}|$). The fundamental solutions are weighted by source strengths γ_i, $i = 1..n$. Hence, the approximation of the potential field is obtained as

$$\hat{\phi}^{\nabla}(x,\omega) = \boldsymbol{\Phi}^{\mathrm{T}}(r^{(i)},\omega)\,\boldsymbol{\gamma}(\omega) \quad \text{and} \quad \hat{\psi}(x,\omega) = \boldsymbol{\Psi}^{\mathrm{T}}(r^{(i)},\omega)\,\boldsymbol{\gamma}(\omega) \tag{8}$$

The gradient of $\boldsymbol{\Phi}$ is multiplied by the outward normal to yield the flux approximation $\boldsymbol{\Psi}$ on the boundary. By modifying the domain Ω such that small spheres with radii ϵ, centered at the load points $\boldsymbol{\xi}$ where the fundamental solutions are singular, are subtracted, the domain Ω' with boundary Γ' is introduced. The properties of the Dirac distribution acting at the points now outside of the considered domain let the domain integral in Eq. (6) vanish in the limit $\Omega' \to \Omega$, if the domain approximations are introduced, yielding a boundary integral formulation in the absence of sources. Inserting the approximations given by Eqs. (7) and (8) yields a discretized formulation of the vanishing first variation

$$\lim_{\epsilon \to 0} \Bigg\{ \delta\boldsymbol{\gamma}^{\mathrm{H}} \Bigg(\underbrace{\int_{\Gamma'} \boldsymbol{\Psi}^* \boldsymbol{\Phi}^{\mathrm{T}}\,\mathrm{d}\Gamma}_{\boldsymbol{F}}\,\boldsymbol{\gamma} - \underbrace{\int_{\Gamma'} \boldsymbol{\Psi}^* \boldsymbol{N}^{\mathrm{T}}\,\mathrm{d}\Gamma}_{\boldsymbol{H}}\,\check{\boldsymbol{u}} \Bigg)$$

$$+\,\delta\check{\phi}^{\mathrm{H}}\Bigg(\underbrace{\int\limits_{\Gamma_t'}\boldsymbol{N}^{*}\bar{\psi}\,\mathrm{d}\Gamma}_{\boldsymbol{f}}-\underbrace{\int\limits_{\Gamma'}\boldsymbol{N}^{*}\boldsymbol{\Psi}^{\mathrm{T}}\,\mathrm{d}\Gamma}_{\boldsymbol{H}^{\mathrm{H}}}\,\boldsymbol{\gamma}\Bigg)\Bigg\}=0\,, \tag{9}$$

where $()^{\mathrm{H}}$ denotes the complex conjugate transpose. The vector $\boldsymbol{f}$ of equivalent nodal fluxes contains the transformations of the given Neumann data to the nodes of the discretization. The matrix $\boldsymbol{H}$ is known as the double-layer potential. The frequency-dependent matrix can be expanded in a static part $\boldsymbol{H}_0$ and a frequency-dependent part $\boldsymbol{H}_\omega$, $\boldsymbol{H}=\boldsymbol{H}_0+\boldsymbol{H}_\omega$. For the following, the feature of a frequency-dependent fundamental solution, that it tends to the behavior of the adjoint static fundamental solution in the limit, when the angular frequency *or* the distance between field and load point vanish is recalled (Gaul and Wagner, 1999). Due to this only the main diagonal of the static part $\boldsymbol{H}_0$ contains a strong singularity, which can be evaluated as a finite-part integral. The Hermitian matrix $\boldsymbol{F}$ contains the product of the weakly and strongly singular fundamental solutions at the load points i and j. This matrix can also be expanded in the same way $\boldsymbol{F}=\boldsymbol{F}_0+\boldsymbol{F}_\omega$. An entry $F_{0_{ii}}$ of the static part is made up of a hypersingular integral over the boundary extension Γ^{ϵ}, similar to the integral-free term in the conventional BEM, and the remaining boundary Γ^{c}. The second integral can be evalutated as finite-part integral, whereas the entries over Γ^{ϵ} cannot be evaluated directly, since the transformation to spherical coordinates does not regularize a hypersingularity. The computation of these terms will be discussed later.

Applying the fundamental lemma yields a matrix equation for the unknowns γ and $\check{\phi}$. Solving for $\check{\phi}$ one obtains

$$\boldsymbol{K}\,\check{\phi}=\boldsymbol{f}\,, \tag{10}$$

with the Hermitian stiffness matrix $\boldsymbol{K}=\boldsymbol{H}^{\mathrm{H}}\boldsymbol{F}^{-1}\boldsymbol{H}$.

For the computation of the matrix $\boldsymbol{F}$ the hypersingular main diagonal of the static part $\boldsymbol{F}_0$ must be evaluated, where the load points of the Dirichlet and Neumann fundamental solutions coincide. For this reason orthogonality properties of this matrix (Dumont, 1989) are used to obtain the main diagonal entries without any further integrations. A physical interpretation starts from a virtual work expression for the static matrix $\boldsymbol{F}_0$ and results in

$$\check{\phi}^{\nabla}=\boldsymbol{F}_0\gamma\,, \tag{11}$$

with the new potential field $\check{\phi}^{\nabla}$, corresponding to Eq. (4). From Eq. (9) the following equations are obtained, considering Eq. (11)

$$\boldsymbol{H}_0\phi=\check{\phi}^{\nabla}\quad\text{and}\quad\boldsymbol{H}_0{}^{\mathrm{T}}\gamma=\boldsymbol{f}\,. \tag{12}$$

It is well-known, that $\boldsymbol{H}_0$ is a singular matrix, with the right and left nullspace $\boldsymbol{w}$ and $\boldsymbol{v}$, respectively

$$\boldsymbol{H}_0\boldsymbol{w}=\boldsymbol{0}\quad\text{and}\quad\boldsymbol{v}^{\mathrm{T}}\boldsymbol{H}_0=\boldsymbol{0}\,. \tag{13}$$

Taking the concept of constant potentials into account, these properties can be used to deduce that if γ belongs to the left nullspace of $\boldsymbol{H}_0$, $\gamma\in\mathcal{V}$, it transforms $\boldsymbol{f}$ into the null element of the range of $\boldsymbol{H}_0$, $\boldsymbol{f}\in\emptyset$. In this case it follows from Eq. (13) that the potential $\check{\phi}$ belongs to the nullspace of $\boldsymbol{H}_0$, $\check{\phi}\in\mathcal{W}$ and therefore $\check{\phi}^{\nabla}$ transforms into the null element, $\check{\phi}^{\nabla}\in\emptyset$. Finally, it can be deduced from Eq. (11) that the left and right nullspace of $\boldsymbol{F}_0$ are identical to the left nullspace of $\boldsymbol{H}_0$

$$\boldsymbol{F}_0 \boldsymbol{v} = \boldsymbol{v}^{\mathrm{T}} \boldsymbol{F}_0 = \boldsymbol{0}\,. \tag{14}$$

This equation is uniquely solvable in a potential problem for the unkown main diagonal entries of $\boldsymbol{F}_0$.

3 HDBEM in Time Domain

The HBEM formulation is derived from Hamilton's principle stating that the solution of a boundary value problem requires stationarity of the time integral over the Lagrangian density $\Pi(\phi)$ given for a compressible fluid in the domain Ω.

Characteristic for the HBEM is the separation of the potential field into a domain field ϕ and boundary field $\tilde{\phi}$. To maintain the compatibility on the boundary the difference $\phi - \tilde{\phi}$ is integrated into Hamiltons principle by the Lagrange multiplier method and leads to a three field functional. The generalized functional consists of terms for the kinetic and potential energy, the Neumann boundary conditions and the compatibility term of the domain and boundary variables

$$\Pi(\psi, \tilde{\phi}, \lambda) = \int_{t_0}^{t_1} \left[\frac{1}{2} \int_{\Omega} \rho\, (\phi_{,i})^2 \, \mathrm{d}\Omega - \frac{1}{2} \int_{\Omega} \rho \, \frac{1}{c^2} (\dot{\phi})^2 \, \mathrm{d}\Omega + \int_{\Gamma_v} \rho\, \dot{\tilde{\phi}}\, \bar{u}_n \, \mathrm{d}\Gamma + \int_{\Gamma} \lambda (\tilde{\phi} - \phi) \, \mathrm{d}\Gamma \right] \mathrm{d}t\,. \tag{15}$$

with Dirichlet boundary conditions $\phi = \bar{\phi}$ on Γ_ϕ. The first Variation of the functional reads

$$\delta\Pi(\phi, \tilde{\phi}, \lambda) = \int_{t_0}^{t_1} \left[- \int_{\Omega} \rho \left(\phi_{,ii} - \frac{1}{c^2} \ddot{\phi} \right) \delta\phi \, \mathrm{d}\Omega + \int_{\Gamma} (\rho\, v_n - \lambda)\, \delta\phi \, \mathrm{d}\Gamma + \int_{\Gamma_v} (-\rho\, \bar{\psi} + \lambda)\, \delta\tilde{\phi} \, \mathrm{d}\Gamma + \int_{\Gamma} \rho\, (\tilde{\phi} - \phi)\, \delta\lambda \, \mathrm{d}\Gamma \right] \mathrm{d}t = 0\,. \tag{16}$$

Applying the fundamental lemma the Lagrangian multiplier can be identified as $\lambda = \rho\, \tilde{\psi}$. In a time domain formulation the variables are separated in time and space dependency. The domain variables are approximated by a superposition of spatial static fundamental solutions weighted by time depend parameters. Davi and Milazzo (1994) adopted this approach in 2D-elasticity. Therefore the approximation of the variables in the domain reads

$$\phi(x,t) = \boldsymbol{\Phi}^T(r^{(i)})\, \boldsymbol{\gamma}(t)\,, \quad \psi(x,t) = \boldsymbol{\Psi}^T(r^{(i)})\, \boldsymbol{\gamma}(t) \tag{17}$$

where $\boldsymbol{\Phi}$ is a vector containing the static fundamental solutions with loadpoints located at nodes of a boundary discretization. The time dependent parameters $\gamma_i(t)$ are unknowns. The potential and velocity fields on the boundary, $\tilde{\phi}$ and $\tilde{\psi}$, respectively, are approximated by shape functions and corresponding time-dependent nodal values analogous to (7).

Inserting the approximations into the generalized functional (16) yields

$$\delta\Pi(\phi,\tilde{\phi},\tilde{\psi}) = \int_{t_0}^{t_1} \Bigg[-\gamma^{\mathrm{T}} \int_{\Omega} \rho \nabla^2 \boldsymbol{\Phi}\boldsymbol{\Phi}^{\mathrm{T}}\,\mathrm{d}\Omega\,\delta\gamma + \ddot{\gamma}^{\mathrm{T}} \int_{\Omega} \frac{\rho}{c^2} \boldsymbol{\Phi}\boldsymbol{\Phi}^{\mathrm{T}}\,\mathrm{d}\Omega\,\delta\gamma - \delta\gamma^{\mathrm{T}} \int_{\Gamma} \rho \boldsymbol{\Phi}\boldsymbol{\Psi}^{\mathrm{T}}\,\mathrm{d}\Gamma\,\gamma$$
$$+\delta\gamma^{\mathrm{T}} \int_{\Gamma} \rho \boldsymbol{\Phi}\boldsymbol{N}^{\mathrm{T}}\,\mathrm{d}\Gamma\,\hat{\psi} - \int_{\Gamma_\psi} \rho \bar{\psi}\,\boldsymbol{N}^{\mathrm{T}}\,\mathrm{d}\Gamma\,\delta\hat{\phi} - \hat{\psi}^{\mathrm{T}} \int_{\Gamma_\psi} \rho \boldsymbol{N}\boldsymbol{N}^{\mathrm{T}}\,\mathrm{d}\Gamma\,\delta\hat{\phi}$$
$$+\delta\hat{\psi}^{\mathrm{T}} \int_{\Gamma} \rho \boldsymbol{N}\boldsymbol{N}^{\mathrm{T}}\,\mathrm{d}\Gamma\,\hat{\phi} - \delta\hat{\psi}^{\mathrm{T}} \int_{\Gamma} \rho \boldsymbol{N}\boldsymbol{\Phi}^{\mathrm{T}}\,\mathrm{d}\Gamma\,\gamma \Bigg]\,\mathrm{d}t = 0\,. \tag{18}$$

The first domain integral in (18) vanishes when the fundamental solutions are inserted in $\nabla^2\boldsymbol{\Phi}$ and the loadpoints ξ^i are excluded by modifing the boundary as presented in (de Figueiredo and Brebbia, 1991). The modified domain converges to the original domain by reducing the radius of the excluded spheres in a limiting process.

A consequence of the approximation (17) is that the second domain integral $\int_\Omega \boldsymbol{\Phi}\boldsymbol{\Phi}^{\mathrm{T}}\,\mathrm{d}\Omega$ in (18) remains after modifying the domain. In the case of a harmonic form of the differential operator for the remaining part it is easy to transform the domain integral into a boundary integral. When a function W is defined by $\nabla^2 W(r^{(i)}) + \Phi(r^{(i)}) = 0$ and the weak form of this term with the weighting function $\Phi(r^{(i)})$ is created, Greens formula yields

$$\int_{\Omega} (\nabla^2 W + \Phi)\Phi\,\mathrm{d}\Omega = \int_{\Gamma} \frac{\partial W}{\partial \boldsymbol{n}} \Phi\,\mathrm{d}\Gamma - \int_{\Gamma} W\Psi\,\mathrm{d}\Gamma + \int_{\Omega} W\nabla^2\Phi\,\mathrm{d}\Omega + \int_{\Omega} \Phi\Phi\,\mathrm{d}\Omega\,. \tag{19}$$

Forcing the left hand side of (19) to zero leads to

$$\int_{\Omega} \Phi\,\Phi\,\mathrm{d}\Omega = \int_{\Gamma} W\,\Psi\,\mathrm{d}\Gamma - \int_{\Gamma} \frac{\partial W}{\partial \boldsymbol{n}}\,\Phi\,\mathrm{d}\Gamma - \int_{\Omega} W\,\nabla^2\Phi\,\mathrm{d}\Omega\,. \tag{20}$$

Considering the modified domain with excluded loadpoints the expression $\nabla^2\boldsymbol{\Phi}$ in the last integral of (20) can be replaced by the property of the fundamental solution and therefore the domain integral vanishes. The remaining relation replaces the domain integral by boundary integrals. The integral kernels have the same singular behaviour as the single- und double layer potential for the function $W(r^{(i)})$ and their directional derivatives are non-singular functions. If the matrices

$$\begin{aligned} \boldsymbol{V} &= \lim_{\epsilon\to 0} \int_{\Gamma'} \frac{1}{c^2}\left(\boldsymbol{W}\boldsymbol{\Psi}^{\mathrm{T}} - \frac{\partial \boldsymbol{W}}{\partial \boldsymbol{n}}\boldsymbol{\Phi}^{\mathrm{T}}\right)\mathrm{d}\Gamma\,, \quad \boldsymbol{F} = \lim_{\epsilon\to 0} \int_{\Gamma'} \boldsymbol{\Phi}\boldsymbol{\Psi}^{\mathrm{T}}\,\mathrm{d}\Gamma\,, \\ \boldsymbol{G} &= \lim_{\epsilon\to 0} \int_{\Gamma'} \boldsymbol{\Phi}\boldsymbol{N}^{T}\,\mathrm{d}\Gamma\,, \quad \boldsymbol{L} = \int_{\Gamma} \boldsymbol{N}\boldsymbol{N}^{T}\,\mathrm{d}\Gamma\,, \quad \boldsymbol{f} = \int_{\Gamma_\psi} \boldsymbol{N}\bar{\psi}\,\mathrm{d}\Gamma \end{aligned} \tag{21}$$

are inserted in (16), the principle reads

$$\delta\Pi(\phi,\tilde{\phi},\tilde{\psi}) = \int_{t_0}^{t_1} \rho\Big\{\delta\gamma^{\mathrm{T}}\Big[-\boldsymbol{V}\ddot{\gamma} - \boldsymbol{F}\gamma + \boldsymbol{G}\hat{\psi}\Big] - \delta\hat{\phi}^{\mathrm{T}}\Big[\boldsymbol{f} - \boldsymbol{L}^{\mathrm{T}}\hat{\psi}\Big]$$

$$+ \delta\hat{\psi}^{\mathrm{T}}\left[\boldsymbol{L}\hat{\phi} - \boldsymbol{G}^{\mathrm{T}}\gamma\right]\Bigg\}\mathrm{d}t = 0\,. \tag{22}$$

The vector $\boldsymbol{f}$ contains the equivalent values of the normal velocity on the boundary.

By postulating that the first variation vanishes with independent parameters $\delta\gamma$, $\delta\hat{\phi}$ und $\delta\hat{\psi}$, the matrix equations

$$\boldsymbol{V}\ddot{\gamma} + \boldsymbol{F}\gamma - \boldsymbol{G}\hat{\psi} = 0\,, \tag{23}$$

$$\boldsymbol{f} - \boldsymbol{L}^{\mathrm{T}}\hat{\psi} = 0 \quad \text{auf} \quad \Gamma_N\,, \tag{24}$$

$$\boldsymbol{L}\hat{\phi} - \boldsymbol{G}^{\mathrm{T}}\gamma = 0 \tag{25}$$

can be obtained. The relation between the second time derivative of the superposition parameters and the nodal values can be derived from (25). Solving equations (23) to (25) for $\ddot{\hat{\phi}}$ respectivly $\hat{\phi}$ leads to the equation of motion

$$\boldsymbol{M}\ddot{\hat{\phi}} + \boldsymbol{K}\hat{\phi} = \boldsymbol{f} \tag{26}$$

with the symmetric positiv definite mass matrix $\boldsymbol{M} = \boldsymbol{L}^{\mathrm{T}}\boldsymbol{G}^{-1}\boldsymbol{V}(\boldsymbol{G}^{-1})^{\mathrm{T}}\boldsymbol{L}$ and the symmetric stiffnes matrix $\boldsymbol{K} = \boldsymbol{L}^{\mathrm{T}}\boldsymbol{G}^{-1}\boldsymbol{F}(\boldsymbol{G}^{-1})^{\mathrm{T}}\boldsymbol{L}$.

4 Numerical Examples

As a numerical example, the sound field in a car compartment is considered. The compartment is excited in the foot region by a vibrating panel, vibrating with constant normal velocity v_n = 100m/s at a frequency of f = 200Hz. Fig. 1 shows the resulting sound pressure field.

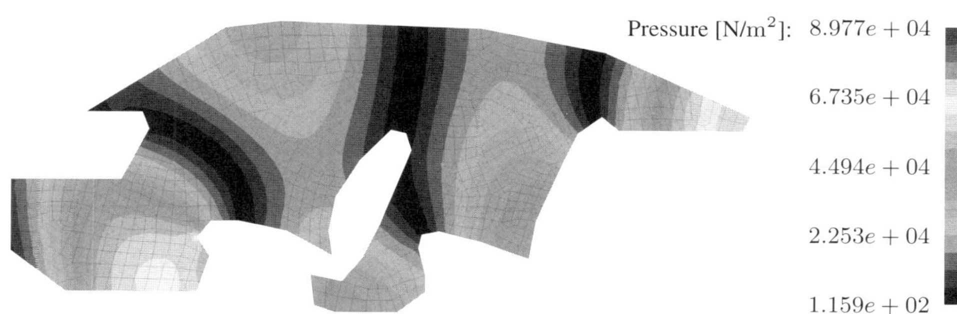

Figure 1. Magnitude of the sound pressure in car compartment at f = 200Hz.

As a second example a time answer of a 2d waveguide filled with water is calculated. The waveguide with reflecting neumann boundary conditions is exited with a step load on one side. The result is obtained by mode superposition and shown in Fig. 2.

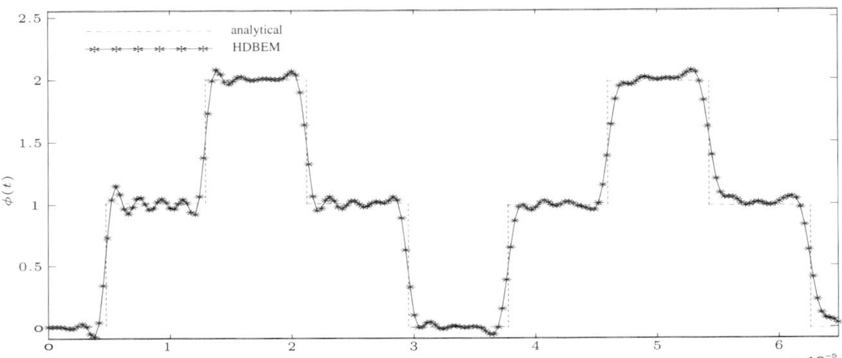

Figure 2. Velocity versus time in the middle of the wave guide

5 Conclusions

The fundamentals of hybrid boundary integral formulations in dynamics for the frequency and time domain are presented. Since a pure boundary formulation is obtained, the *dimension of the field problem is reduced by one*. The resulting system matrices are dense but *symmetric*. The HSBEM is developed from the dynamical Hellinger-Reissner potential using fundamental solutions for the approximation of the domain variable, giving rise to matrix $\boldsymbol{F}$ with a hypersingular main diagonal. The main diagonal cannot be evaluated mathematically, thus a criterion for the evaluation of the main diagonal has been applied, starting from physical interpretations.

On the other hand, a new BE time domain formulation has been introduced which is derived from Hamilton's principle. Fundamental solutions can be used for the approximation of domain variables. Because fundamental solutions identically fulfill the field equation of the fluid, domain integrals can be replaced by boundary integrals. Thus only discretization of the boundary is required for the numerical solution, yielding symmetric matrices. The computation of the matrix entries is expensive because weakly, strongly, as well as hypersingular integrals have to be treated. However, differently from the known symmetric time domain formulations, the matrices have to be calculated only once.

References

Davi, G., and Milazzo, A. (1994). A symmetric and positive definite variational BEM for 2–d free vibration analysis. *Engineering Analysis with Boundary Elements* 14:343–348.

de Figueiredo, T., and Brebbia, C. (1991). A new variational boundary element method for potential problems. *Engineering Analysis with Boundary Elements* 8:45–50.

Dumont, N., and de Oliveira, R. (1993). The hybrid boundary element method applied to time-dependent problems. In Brebbia, C., and Rencis, J., eds., *Boundary Elements XV: Fluid Flow and Computational Aspects*, volume 1, 363–376. London: Computational Mechanics Publications.

Dumont, N., and de Oliveira, R. (1997). The exact dynamic formulation of the hybrid boundary element method. In *Proceedings of XVIII CILAMCE*, volume 1, 357–364.

Dumont, N. (1987). The hybrid boundary element method. In *Boundary elements IX*, 117–130. Berlin: Springer.

Dumont, N. (1989). The hybrid boundary element method: An alliance between mechanical consistency and simplicity. *Applied Mechanics Review* 42(11):54–63.

Gaul, L., and Fiedler, C. (1997). *Methode der Randelemente in Statik und Dynamik.* Wiesbaden/Braunschweig: Friedrich Vieweg & Sohn.

Gaul, L., and Wagner, M. (1996). A new hybrid symmetric boundary element method in elastodynamics. In Saranen, J., and Constanda, C., eds., *4th International Conference on Integral Methods in Science and Engineering*.

Gaul, L., and Wagner, M. (1999). Zur Dynamik der Fluid-Struktur Interaktion. *Zeitschrift für angewandte Mathematik und Mechanik* 79(S2):289–290.

Washizu, K. (1982). *Variational Methods in Elasticity and Plasticity.* Oxford: Pergamon Press.

CHAPTER 7

A FEM/HBEM Approach for the Simulation of Noise Radiation from a Rolling Tire

Lothar Gaul, Marcus Wagner, Matthias Fischer,[1] Udo Nackenhorst[2] and Bodo Nolte[3]*

[1] Institute A of Mechanics, University of Stuttgart, Germany
[2] Institute for Structural and Computational Mechanics, University of Hanover, Germany
[3] Institute of Mathematics, University of the Federal Armed Forces Hamburg, Germany

Abstract. The noise radiation from rolling tires is studied numerically by using a sequential finite-element (FEM) - hybrid boundary-element (HBEM) approach of the field equations. The equations of motion for the rolling wheel are developed in the frame of an Arbitrary Eulerian-Lagrangian description with a time independent formulation for steady state rolling and a spatial description of vibrations. The calculation of steady state rolling is strongly nonlinear due to the nonlinear material behavior of the tire composites, large displacements with loads depending on the shape and the contact problem itself. This is treated by an incremental-iterative approach including the computation of the contact normal and shear tractions. After the configuration of steady state rolling is known, a modal analysis is performed in this deformed state. For the eigen-analysis small linear vibratory displacements are superimposed on the large elastic deformations of the fixed control state.
The noise radiation caused by the vibration modes is computed by the symmetric hybrid boundary element method (Gaul and Wagner, 1996; Wagner, 2000; Gaul et al., 2000) . The relative normal velocities at the wheel surface are the Neumann data of the acoustic domain. After selecting a road impedance the sound pressure distribution on the wheel and the road surface are calculated. The sound field in the domain surrounding the wheel and road is determined hereafter by an efficient field-point algorithm inherent in the HBEM.

1 Introduction

With increasing traffic density the reduction of traffic noise becomes more and more important. Success in reducing sound sources in cars has been achieved. This holds for example for the main engine noise. Thus, nowadays the sound of the rolling tires is the main source of noise radiation. This effect is dominating when cars exceed a speed of 40 km/h and trucks a speed of 65 km/h. This rolling noise is mainly caused by the vibrations of the tires (Bschorr, 1986). These vibrations are excited by the physical contact of the rolling tires with the road surface. As a result airborne sound is generated.

The special excitation mechanisms are deterministic on the one hand but as well can be stochastic. E.g. an excitation can be caused by the periodical contact of the tire treads and the

* Support by Deutsche Forschungsgemeinschaft DFG under SFB 404 and Friedrich-und-Elisabeth-BOYSEN-Stiftung is gratefully acknowledged.

roughness of the road surface. Thus, one step to overcome the noise problem is to optimize the tire tread design in an acoustical sense. Modern treads are designed with uneven spacing generating a broader and lower excitation spectrum. This technique may be used to decrease the sound pressure level up to 5 dB (Tonhauser, 1996). However, this technique has been tested to the extent of its limits. Further reduction techniques may include the following:

a) The structural dynamic behavior of the wheel or the tires needs to be improved acoustically. In this context it has to be ensured that other important tire characteristics, including security or economical aspects will not be negatively influenced.
b) The whole tire–road system needs to be optimized because the characteristic quality of the road surface implies a great potential for noise reduction (Stenschke and Jäcker, 1996) .

Due to the complicated structure of pneumatic tires with strongly non–linear mechanical response the exact mechanisms of all noise sources are not well understood yet. Of course, the sound pressure level of a passing vehicle is measured easily, but these measurements at a few fixed positions are not significant for the entire noise emission. More complete measurements of the noise radiation of rolling tires are possible only in the laboratory (Saemann, 1997). Another way to investigate the sound noise origin is to use numerical simulation methods. Besides, the numerical simulation enables the observation of single design–parameter influences on the noise radiation as well as of safety and economy aspects.

2 The Simulation Procedure

To explain the technique for the numerical simulation of rolling noise in this paper an illustrative way has been chosen rather than a detailed mathematical description for which is referred to Nackenhorst and Nolte (1998) . The technique splits into three main steps:

1. The computation of the strongly nonlinear problem of a steady state rolling wheel in contact with the road by a finite element approach,
2. A modal analysis for the free vibratory response of the rolling wheel in its stationary rolling contact state,
3. The computation of the noise radiation from the wheel by HBEM.

In the first step the equilibrium of a steady state rolling wheel is computed by use of a special finite element approach which is described in more details in the next section. This problem is strongly nonlinear due to the nonlinear material behavior of tire composites, large displacements, loads depending on the shape and the contact problem itself. Therefore, an incremental – iterative procedure is necessary for this analysis, expressed by the equations

$$ {}^{t}\mathbf{K}_{\text{eff}}\left({}^{t}\phi\right)\,\Delta\phi = {}^{t+\Delta t}\mathbf{f}^{(e)} - {}^{t}\mathbf{f}^{(i)} + {}^{t}\mathbf{f}^{(c)}\,, \tag{1} $$

$$ {}^{t+\Delta t}\phi = {}^{t}\phi + \Delta\phi\,. \tag{2} $$

Herein ${}^{t}\mathbf{K}_{\text{eff}}\left({}^{t}\phi\right)$ is the effective stiffness matrix which depends on the actual state of deformation ${}^{t}\phi$ and the contact state. The target of the Newton–Raphson like procedure is to find the equilibrium of the applied external forces ${}^{t+\Delta t}\mathbf{f}^{(e)}$, the internal forces ${}^{t}\mathbf{f}^{(i)}$ due to the divergence of the stress tensor and the contact forces ${}^t\mathbf{f}^{(c)}$. For a realistic modeling of the tire behavior more than 50.000 unknowns have to be solved. This is done on modern unix–workstations in a reasonable amount of time. However, this is only possible by using a special relative kinematics formulation described in the next section. Additionally, stable and efficient contact algorithms are needed.

After the configuration $\bar{\phi}$ of steady state rolling is investigated a modal–analysis is performed in this deformed configuration. By doing this, small linear vibrations φ are superimposed on the large elastic deformation $\bar{\phi}$, going along with the relative velocities $\dot{\varphi}$. In addition, the contact conditions are kept fixed for this eigen–analysis. For this an eigenvalue problem

$$\left\{\lambda_i \begin{bmatrix} \mathbf{M} & \mathbf{0} \\ \mathbf{0} & \mathbf{I} \end{bmatrix} + \begin{bmatrix} \mathbf{G} & \mathbf{K}_{\text{eff}}(\bar{\phi}) \\ -\mathbf{I} & \mathbf{0} \end{bmatrix}\right\} \begin{bmatrix} \dot{\varphi} \\ \varphi \end{bmatrix}_i = \mathbf{0} \tag{3}$$

has to be solved. It is meaningful that the eigenvalues λ_i and eigenvectors φ_i are complex valued quantities due to the gyroscopic effects expressed by the skew–symmetric matrix $\mathbf{G}$. Furthermore, $\mathbf{M}$ is the mass matrix and $\mathbf{I}$ is the identity matrix in eq. (3).

From the computed relative velocities $\dot{\varphi}$ on the wheel surface, e.g. from the calculated eigenmodes or a modal superposition of these, the relative velocity components perpendicular to the wheel $\mathbf{v}_n$ are extracted. These velocity components are the Neumann input data for the third step, the computation of the noise radiation by using the hybrid boundary element method (Wagner, 2000; Gaul et al., 2000) which is outlined shortly in the following.

The derivation starts from the Hellinger-Reissner potential for acoustics in frequency domain ($(\,)^*$ denotes a conjugate complex variable)

$$\begin{aligned} \hat{\mathcal{H}}^{\mathrm{R}}[\hat{\phi}_{,i}, \hat{\phi}^{\nabla}, \hat{\phi}] = & \int_{\Omega} \frac{1}{2}\rho \left(\hat{\phi}_{,i}\hat{\phi}^*_{,i} - \kappa^2 \hat{\phi}^{\nabla}\hat{\phi}^{\nabla^*} + 2((\hat{\phi}_{,i})_{,i})(\hat{\phi}^{\nabla})^*\right) \mathrm{d}\Omega \\ & - \int_{\Gamma_\psi} \rho(\hat{\psi} - \hat{\bar{\psi}})\hat{\phi}^* \mathrm{d}\Gamma - \int_{\Gamma_\phi} \rho\hat{\bar{\phi}}\hat{\psi}^* \mathrm{d}\Gamma, \end{aligned} \tag{4}$$

with three independent fields, the potential gradient $\hat{\phi}_{,i}$ and the two independent potential fields $\hat{\phi}$ and $\hat{\phi}^{\nabla}$. The flux on the boundary is denoted by $\hat{\psi}$.

The numerical solution requires approximation functions for the field variables. The approximation of the boundary variable $\hat{\phi}$ is carried out with a shape function vector $\boldsymbol{N}$ and a nodal vector $\check{\boldsymbol{\phi}}$ equivalent to the direct BEM. The approximations of the potential $\hat{\phi}^{\nabla}$ and gradient $\hat{\phi}_{,i}$ in the domain Ω, as well as the flux, are given as a weighted superposition of n fundamental solutions $\Phi(r^{(i)}, \omega)$, where $r^{(i)} = |\mathbf{x} - \boldsymbol{\xi}^{(\mathbf{i})}|$ denotes the Euclidian distance between the load point $\boldsymbol{\xi}^{(i)}$ and the field point $\mathbf{x}$. The load points are collocated with the nodes of the boundary discretization. The fundamental solutions are weighed by unknown parameters γ_i, $i = 1..n$. Hence, the approximation of the potential field, the gradient field in the domain and the flux on the boundary are given as

$$\hat{\phi}^{\nabla} = \boldsymbol{\Phi}^{\mathrm{T}}\boldsymbol{\gamma}\,, \quad \hat{\phi}_{,j} = \boldsymbol{\Phi}^{\mathrm{T}}_{,j}\boldsymbol{\gamma} \quad \text{and} \quad \hat{\psi} = (\boldsymbol{\Phi}^{\mathrm{T}}_{,j} n_j)\boldsymbol{\gamma}\,. \tag{5}$$

The discretization of the vanishing first variation of eq. (4) yields

$$\begin{bmatrix} -\boldsymbol{F} & \boldsymbol{H} \\ \boldsymbol{H}^{\mathrm{H}} & \mathbf{0} \end{bmatrix} \begin{bmatrix} \boldsymbol{\gamma} \\ \check{\boldsymbol{\phi}} \end{bmatrix} = \begin{bmatrix} \mathbf{0} \\ \boldsymbol{f} \end{bmatrix}. \tag{6}$$

because the fundamental solutions identically fulfill the Helmholtz equation. This is why the domain integrals vanish. The load point singularity in the fundamental solutions is taken care of by excavating the domain around the load point by a sphere and letting the radius vanish in a limiting process. Particularly, the second equation of eq. (6) yields

$$\gamma = \boldsymbol{H}^{\mathrm{H}} \boldsymbol{f} . \tag{7}$$

In a Neumann-problem $\boldsymbol{f} = -\int_{\Gamma} \boldsymbol{N} \mathbf{v}_n \, \mathrm{d}\Gamma$ is given completely. Thus, by eq. (7) the weighting parameters γ can be obtained and then inserted in the domain approximation eq. (5) yielding the complete field solution. For this, only the complex conjugate transpose $(\,)^H$ of the matrix $\boldsymbol{H} = \int_{\Gamma} (\boldsymbol{\Phi}^*_{,j} n_j) \boldsymbol{N}^T \, \mathrm{d}\Gamma$ needs to be calculated and no further boundary integrations are necessary.

In a first approach the road surface influence can be bounded as hard–walled, which is efficiently implemented by a mirroring technique of the fundamental solution as discussed in the following. A more realistic description of the reflection of sound waves on the road surface can be modeled by prescribed impedances, but then the road needs to be discretized as well. The discretization has to be truncated at a certain distance from the wheel. This approach is currently under investigation. The HBEM offers also for this case a promising approach since it turns out that in such a half-space problem the involved system matrices are analytic (Dumont and Wagner, 2000).

3 Numerical Simulation of Stationary Rolling Wheels

For the description of rolling in relative kinematics an Arbitrary Lagrangian Eulerian (ALE) formulation has been chosen, which is depicted schematically in fig. 1. The motion Ψ is decomposed into the rigid body motion χ which is described in Eulerian coordinates, i.e. the mesh is fixed in space, and into the deformation ϕ described in Lagrangian coordinates. The material time derivative of the motion,

$$\frac{\mathrm{d}\boldsymbol{\Psi}}{\mathrm{d}t} = \frac{\partial \phi}{\partial t} + \mathrm{Grad}\phi \cdot \dot{\chi} \, , \tag{8}$$

then splits into the relative velocity and into the convective velocity. In the steady state case the relative velocity vanishes. Thus, the stationary rolling state is modeled independent of the time and no expensive time discretization is necessary. Furthermore, the fine discretization in space necessary for the contact analysis can be concentrated in the contact region only and the mesh can be coarse over the remaining circumference. And moreover, the relative dynamics is described in space immediately, as it is needed for the computation of the sound radiation.

For illustration, the ALE–description can be interpreted as the introduction of an observer riding on the axis of the rolling wheel. The ALE–observer monitors the motion of an arbitrary particle which at the moment is located at the fixed point $\mathbf{x}$ but it does not collect the history of the material particles. Therefore, special formulations and algorithms for the treatment of the contact conditions, especially for the computation of the contact shear tractions, have to be worked out. For details the reader is referred to e.g. Nackenhorst (1993) .

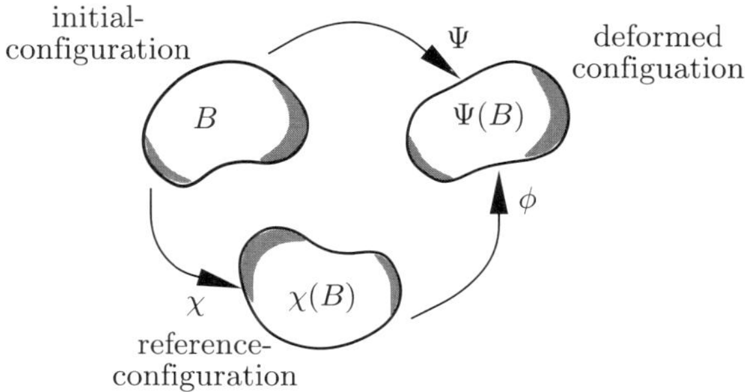

Figure 1. The ALE mapping of motion.

4 Numerical Analysis of Sound Radiation

For the analysis of sound radiation from vibrating structures the HBEM has the following decisive advantages: a) The Sommerfeld radiation condition is fulfilled inherently, b) Only the acoustically relevant boundaries, i.e. sound generating or reflecting surfaces, have to be discretized. This reduces the dimension of the problem by one and c) no discretization of the infinite road surface is necessary as long as it can be modeled hard–walled.

The basic field equations of the acoustical problem are the wave equations for the sound pressure and for the particle velocity. A transformation into the frequency domain leads to the Helmholtz equation which turns out to be the Euler equation of the Hellinger-Reissner potential 4. A boundary integral formulation is obtained from the Hellinger Reissner potential by approximating the domain variables by weighted superposition of test functions Φ which identically fulfill the Holmholtz equation. The test function

$$\Phi = \frac{1}{4\pi r} e^{-ikr} \tag{9}$$

is called the fundamental solution for the full space (Gaul and Fiedler, 1997). Herein r is the Euclidean distance between the field point $\mathbf{x}$ and the so called load point $\boldsymbol{\xi}$ as shown in fig. 2. Because the Sommerfeld radiation condition is fulfilled a priori by use of the test function eq. (9), no waves are reflected from the boundary Γ_∞ at infinity. Additionally, by the use of the modified test function

$$\Phi_h(r, r') = \frac{1}{4\pi} \left(\frac{1}{r} e^{-ikr} + \frac{1}{r'} e^{-ikr'} \right) \tag{10}$$

a hard–walled surface of the road is implemented without discretization of the road surface. This can be visualized as a superposition of two sound fields, the actual sound field generated by the radiator and a reflected wave from the road surface as illustrated in fig. 2. This is called the mirror technique. In conclusion only the surface of the tire has to be discretized.

5 Example: A Simple Wheel

For demonstrating the computation procedure, the noise radiation caused by fundamental vibration modes of a simple wheel model in rolling contact has been analyzed. In the first calculation

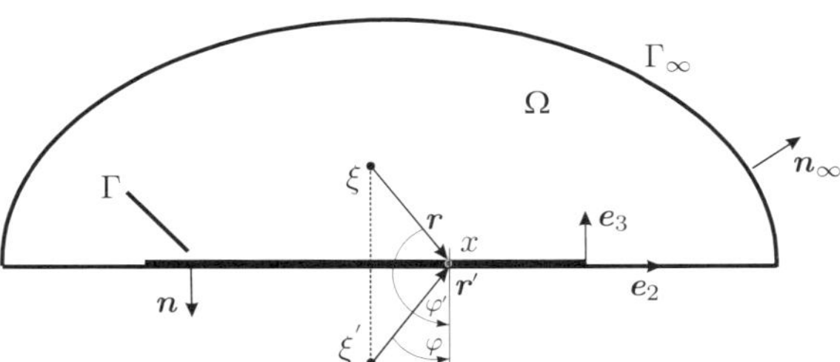

Figure 2. The mirror technique.

step, the wheel rotation is initiated up to a speed of 40 km/h and pressed against the rigid road surface. In this particular deformed state of stationary rolling, the computation of the eigenfrequencies and the eigenvectors is carried out by eq. (3). Gyroscopic effects have been neglected here, thus only real eigensolutions have been computed. From the eigenvectors the relative velocities have been obtained and their components normal to the surface have been extracted. These normal components of the relative velocity are the Neumann boundary data of a subsequent BEM–analysis for computing the sound pressure on the surface of the tire by the linear system in eq. (6). In case that only a field-point solution is of interest, this equation has not to be solved since the given problem is of Neumann-type. To obtain the solution at arbitrary field points in the domain the sound pressure has been computed by solving eq. (7). This provides a significant numerical advantage of the HBEM compared to a direct BEM (Gaul and Fiedler, 1997; Nackenhorst and Nolte, 1998) approach. In figs. 3 and 4 a) the results for a typical eigen-

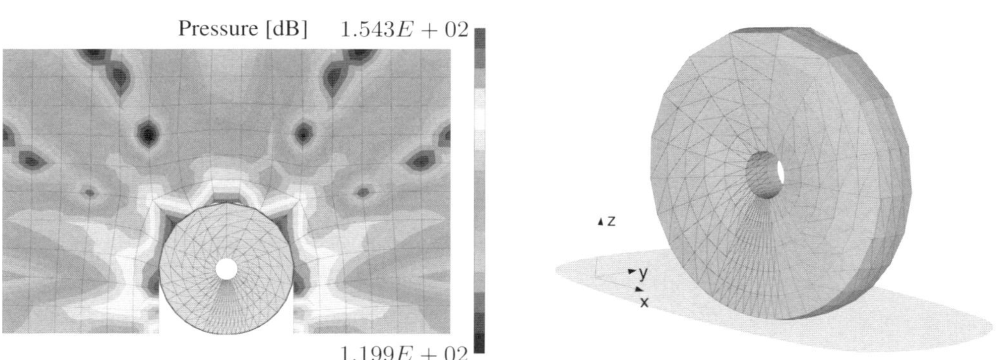

Figure 3. Noise radiation at 1279.3 Hz, full-space solution with SYSNOISE.

mode at 1279.3 Hz are depicted. In this example the road surface has been neglected. The first picture shows the computation with the commercial code SYSNOISE (1996) and the second figure shows the result with the HBEM. The results coincide very good. The HBEM method is more effective since only the $\boldsymbol{H}$-matrix has to be set up in comparison to the direct BEM approach in SYSNOISE where also the single-layer potential must be computed. Note that because the sound

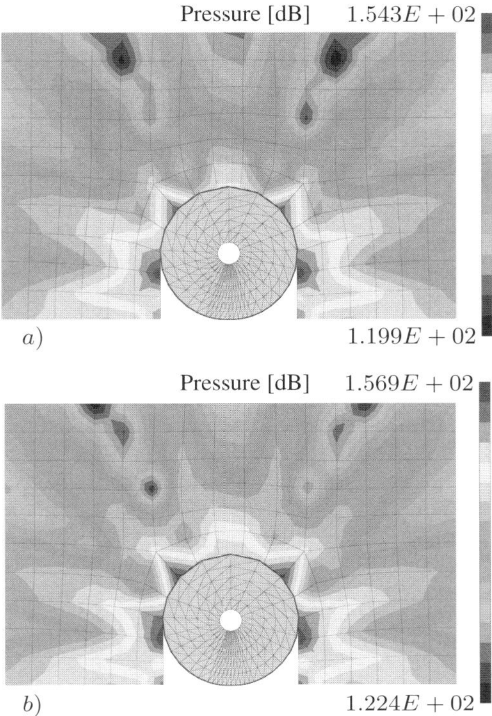

Figure 4. Noise radiation at 1279.3 Hz, a) full-space solution, b) hard–walled road surface.

pressure calculation in this example is based on normalized mode-shapes the given dB-values of the sound pressure only lead to a qualitative measure.

To demonstrate the influence of the road, an analysis with a hard-walled boundary condition is shown in fig. 4 b) taking into account the half-space fundamental solution of eq. (10). In comparison to the full-space solution (fig. 4 a)) an evidently different sound radiation characteristic is obtained and the maximum sound pressure level is approximately 2.6 dB higher than in the full-space solution. The amplification is smaller than a doubling of the sound pressure as expected from the mirroring technique in eq. (10) because the radiated pressure wave does not hit the road perpendicular. Generally it can be seen that the influence of the road can not be neglected in a realistic simulation.

6 Conclusions

The aim of the presented research activities is the development of numerical simulation strategies for noise generation analysis. In this paper a first step for a better understanding of rolling noise has been described. Based on eigenmodes, which are determined in the deformed state of the stationary rolling wheel by finite element techniques, sound radiation analysis is carried out by the hybrid boundary element method. This allows the simulation of the radiation characteristics of single mode shapes. The computation strategy has been demonstrated by the analysis of a simple model wheel.

The next step will be the examination of excitation spectra from the road surface roughness and the simulation of the real forced vibration behavior of rolling wheels by a modal superposition approach. Simultaneously, the reflection behavior of the rough road surface has to be measured and taken into account by impedance boundary conditions in the simulation.

References

Bschorr, O. (1986). Reduktion von Reifenlärm. *Automobil–Industrie* 6/86:721–729.

Dumont, N. A., and Wagner, M. (2000). The hybrid stress boundary element method applied to half-space problems in acoustic fluid-structure interaction. In *Congresso Ibero Latino Americano sobre Métodos Computacionais para Engenharia.*

Gaul, L., and Fiedler, C. (1997). *Methode der Randelemente in Statik und Dynamik.* Wiesbaden/Braunschweig: Friedrich Vieweg & Sohn.

Gaul, L., and Wagner, M. (1996). A new hybrid symmetric boundary element method in elastodynamics. In Saranen, J., and Constanda, C., eds., *4th International Conference on Integral Methods in Science and Engineering.*

Gaul, L., Wagner, M., and Wenzel, M. (2000). Hybrid boundary element methods in frequency and time domain. In von Estorff, O., ed., *Boundary Elements in Acoustics, Advances and Applications*, 121–164. Southampton: WIT Press.

Nackenhorst, U., and Nolte, B. (1998). Numerische simulation von rollgeräuschen. In Nixdorff, K., ed., *Proceedings Anwendungen der Akustik in der Wehrtechnik.*

Nackenhorst, U. (1993). On the finite element analysis of steady state rolling contact. In Aliabadi, M., and Brebbia, C., eds., *Contact Mechanics, Computational Techniques.* Southampton: Computational Mechanics Publications.

Saemann, E.-U. (1997). Schallquellenortung am Reifen mittels akustischer Holographie im Zeitbereich. In *Reifen–Fahrbahn–Fahrwerk, VDI–Berichte 1350.* Düsseldorf, Germany: VDI–Verlag.

Stenschke, R., and Jäcker, M. (1996). Einfluß von Reifen und Fahrbahnen auf das Reifen/Fahrbahn–Geräusch von Kraftfahrzeugen und administrative Maßnahmen zur Verminderung. In *Proceedings Innovative Ideen zur Minderung der Kfz-Rollgeräusche.*

SYSNOISE. (1996). *User's Manual, Version 5.3A.* Leuven, Belgium: LMS Numerical Technologies N.V.

Tonhauser, J. (1996). Einfluß des Reifen-/Fahrbahngeräusches auf das Außengeräusch von PKW; Stand der Technik. In *Proceedings Innovative Ideen zur Minderung der Kfz-Rollgeräusche.*

Wagner, M. (2000). *Die hybride Randelementmethode in der Akustik und zur Struktur-Fluid-Interaktion.* Ph.D. Dissertation, Institute A of Mechanics, University of Stuttgart.

CHAPTER 8

About the Numerical Solution of the Equations of Piezoelectricity

Martin Kögl[1] and Lothar Gaul[2]

[1] Departamento de Engenharia de Estruturas e Fundações, Escola Politécnica, Universidade de São Paulo, Brazil

[2] Institute A of Mechanics, University of Stuttgart, Germany

Abstract. In this paper, the numerical solution of piezoelectric problems by means of two discretization methods – the Finite Element Method (FEM) and the Boundary Element Method (BEM) – is described. At first, using the equations of elastostatics, the similarities and differences of the methods are explained and some of their advantages and disadvantages are pointed out. After this, the piezoelectric formulations of both methods are introduced. A numerical example serves to demonstrate the excellent agreement of the FEM and BEM results as well as to show the superiority of the BEM in the calculation of elastic stresses and the electric field.

1 Introduction

The increasing importance and use of piezoelectric materials in engineering sciences calls for reliable numerical methods that are capable of accurately modelling and solving problems involving piezoelectrics. Besides the FEM, Boundary Element Methods are a very powerful tool for the numerical solution of field problems of mathematical physics, since they offer some inherent advantages over FEM like the discretization of the boundary only, as well as an improved accuracy in stress calculations. In this article, the formulations of both methods are compared, and some of their advantages and disadvantages are pointed out.

2 Finite Element and Boundary Element Methods

This section gives a brief comparison of the Finite Element and Boundary Element Methods. For a more detailed introduction into FEM, the interested reader is referred to the books of Bathe (1996) and Zienkiewicz and Taylor (1991; 1994). The BEM is described in detail in Brebbia et al. (1984) and Gaul and Fiedler (1996).

2.1 General Features

The most striking difference between FEM and BEM, and one of the important advantages of the latter, concerns the discretization. While in FEM the complete domain has to be discretized, in BEM only the discretization of the boundary is required, as seen in Figure 1. Depending on the complexity of the actual structure and load case under investigation, this simplified discretization can lead to important time savings in the mesh creation and modification process. Other areas in which the BEM possesses certain advantages are problems involving infinite or semi-infinite domains (e.g. acoustics, soil-structure interaction, etc.), as well as stress concentration problems. This is partly due to the use of fundamental solutions, which are *analytical* free space solutions of the governing differential field equations, and can therefore accurately represent far-fields and stresses.

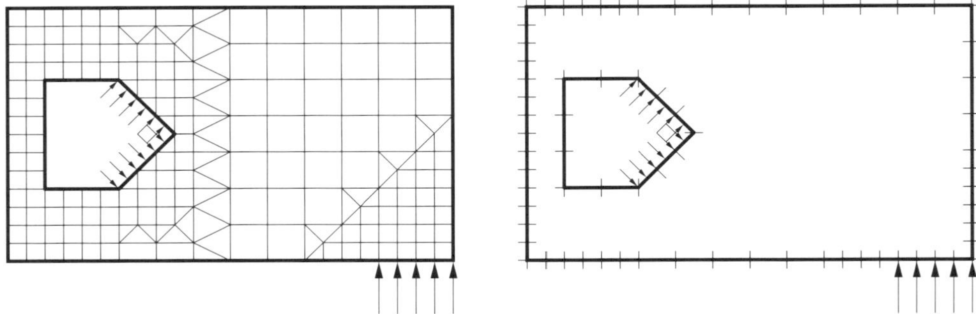

Figure 1. Discretization with finite elements (left) and boundary elements (right)

On the other hand, the fact that a fundamental solution is required poses some problems, e.g., for inhomogeneous and non-linear analyses, where special methods such as the Dual Reciprocity Method (DRM) (Partridge et al., 1992) have to be applied. Since the FEM does not rely on fundamental solutions, it has less difficulties in handling these kinds of problems. Also, in structural mechanics, Finite Element Methods have the advantage of leading to symmetric, positive definite, and sparsely populated matrices, so that special time-saving solvers can be used, whereas the BEM matrices are usually non-symmetric and fully populated.

2.2 Comparison of FE and BE Formulations

This section aims at demonstrating the similarities and differences between the methods by introducing them in a comparative manner, as shown in Figure 2. This is done for the case of linear elastostatics without body forces, in order to concentrate on the essentials. The starting point for both formulations is a weighted residual statement, in which the governing differential equation and boundary conditions are weighted with test functions w. Integration by parts leads to the weak statement, which is the basis of the FEM; further integration by parts results in the inverse statement, which forms the basis of the BEM.

The next steps show some significant differences between the methods. The standard FEM as derived here is a pure displacement formulation, in which only the displacements appear as unknowns. These displacements are approximated by functions which have to fulfill the forced boundary conditions. The stresses and tractions are calculated only after the solution of the displacement field. This involves derivation of the shape functions, which results in a loss of accu-

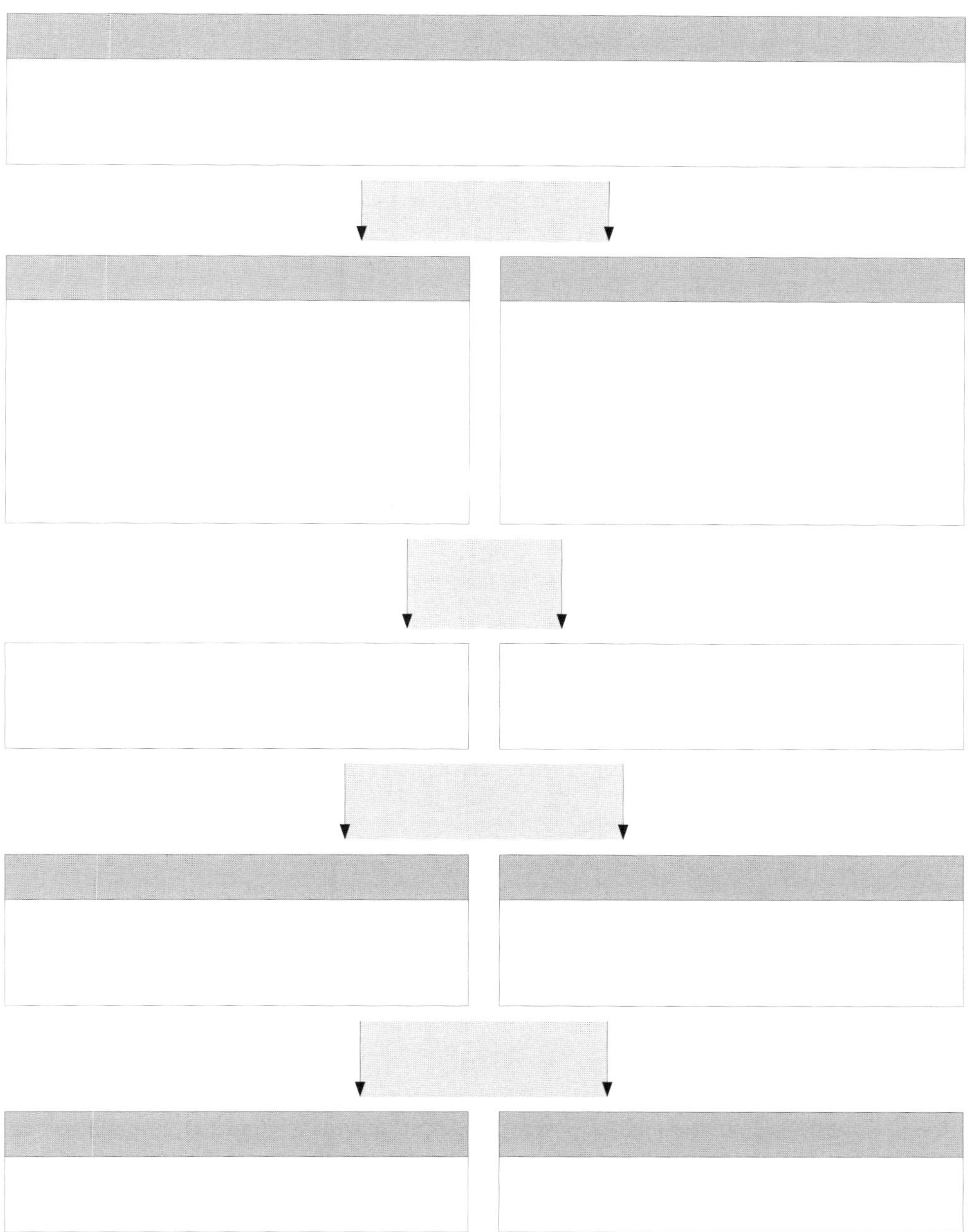

Figure 2. Comparison of Finite Element Method and Boundary Element Method

racy for stress and traction results. On the other hand, the Boundary Element Method is a mixed formulation, containing both displacements and tractions as unknowns. Hence, apart from the displacements, the traction field, too, has to be approximated. The approximation functions now have to fulfill the natural boundary conditions. This is why the results for the tractions obtained with BEM are usually more accurate than those obtained with FEM.

Another difference between the methods lies in the choice of the weighting function. The classical FEM is a Bubnov-Galerkin method, using the same set of functions for the approximation of the displacement field as for the weighting functions. On the other hand, the BEM is a Petrov-Galerkin method, where the weighting functions are chosen as the fundamental solutions of the governing differential equations. This leads to an elimination of the domain integral by virtue of the sifting property of the Dirac impulse, which results in a pure boundary formulation for the unknown displacement and traction fields.

3 Basic Equations of Piezoelectricity

The elastic and electric field of a piezoelectric body are described in terms of elastic displacements u_k and electric potential φ by the differential equations (for more details, see e.g. Kögl (2000) and Tiersten (1969))

$$C_{ijkl}\, u_{k,li} + e_{lij}\, \varphi_{,li} = \rho \ddot{u}_j \,, \tag{1}$$

$$e_{ikl}\, u_{k,li} - \epsilon_{il}\, \varphi_{,li} = 0 \,, \tag{2}$$

with mass density ρ, elasticity tensor C_{ijkl}, piezoelectric tensor e_{lij}, and permittivity tensor ϵ_{il}. Here and in the following, Einstein's summation convention is applied, the comma denotes partial differentiation with respect to the spatial coordinates, and the superimposed dot denotes partial differentiation with respect to time.

The stress tensor σ_{ij} and electric displacement vector D_i are obtained from the constitutive equations

$$\sigma_{ij} = C_{ijkl}\, \varepsilon_{kl} - e_{lij}\, E_l \,, \tag{3}$$

$$D_i = e_{ikl}\, \varepsilon_{kl} + \epsilon_{il}\, E_l \,, \tag{4}$$

where $\varepsilon_{kl} = 1/2(u_{k,l} + u_{l,k})$ denotes the strain tensor and $E_l = -\varphi_{,l}$ denotes the electric field vector. The fluxes are given by

$$t_j = \sigma_{ij}\, n_i \,, \qquad q = D_i\, n_i \,, \tag{5}$$

with traction vector t_j, charge flux density q, and direction vector n_i.

Using a condensed notation, where lower case subscripts range from 1 to 3, and upper case subscripts from 1 to 4, it is possible to combine corresponding elastic and electric quantities into generalized piezoelectric vectors and tensors, which simplifies the FEM and BEM formulations. These generalized tensors are defined as follows:

$$\Sigma_{iJ} := \begin{cases} \sigma_{ij} \,, & j = J = 1,2,3 \\ D_i \,, & J = 4 \end{cases} \,, \qquad Z_{Kl} := \begin{cases} \varepsilon_{kl} \,, & k = K = 1,2,3 \\ -E_l \,, & K = 4 \end{cases} \,, \tag{6}$$

$$U_K := \begin{cases} u_k\,, & k = K = 1,2,3 \\ \varphi\,, & K = 4 \end{cases}\,, \qquad T_J := \begin{cases} t_j\,, & j = J = 1,2,3 \\ q\,, & J = 4 \end{cases}\,, \tag{7}$$

$$C_{iJKl} := \begin{cases} C_{ijkl}\,, & j = J = 1,2,3;\; k = K = 1,2,3 \\ e_{lij}\,, & j = J = 1,2,3;\; K = 4 \\ e_{ikl}\,, & J = 4; \qquad\qquad k = K = 1,2,3 \\ -\epsilon_{il}\,, & J = 4; \qquad\qquad K = 4 \end{cases}\,. \tag{8}$$

In this condensed notation, the motion and electric field of a piezoelectric body can now be described by

$$\mathcal{L}_{JK}\, U_K(x_i) = \mathcal{B}_{JK} U_K(x_i)\,, \qquad x_i \in \Omega \subset I\!R^3 \tag{9}$$

with boundary conditions

$$U_J = \bar{U}_J \qquad \text{on} \qquad \Gamma_U\,, \tag{10}$$

$$T_J = \bar{T}_J \qquad \text{on} \qquad \Gamma_T\,, \tag{11}$$

and initial conditions

$$U_J(t=0) = U_J^0\,, \tag{12}$$

$$\dot{U}_J(t=0) = \dot{U}_J^0\,. \tag{13}$$

In Eqn (9), $\mathcal{L}_{JK} := C_{iJKl}\partial_i\partial_l$ is the elliptic operator of static piezoelectricity, and $\mathcal{B}_{JK} := \rho\,\tilde{\delta}_{JK}\,\partial_t^2$ is a differential operator, which describes the effects of inertia. A modified Kronecker delta has been introduced, accounting for the fact that the electric field is assumed to be quasi-static:

$$\tilde{\delta}_{JK} := \begin{cases} \delta_{jk}\,, & j = J = 1,2,3\,, \quad k = K = 1,2,3 \\ 0\,, & \text{otherwise} \end{cases}\,. \tag{14}$$

4 FEM and BEM for Piezoelectric Continua

4.1 Finite Element Method

Finite Element Methods for dynamic piezoelectricity are already well established (see e.g. Allik and Hughes (1970) and Lerch (1990)). Therefore, only a brief review is given here, using the condensed notation. The starting point of the FE formulation is the following variational principle (Allik and Hughes, 1970):

$$\int_\Omega \left(\Sigma_{iJ}\,\delta Z_{Ji} + \rho\,\ddot{u}_i\,\delta u_i\right)\,\mathrm{d}\Omega = \int_{\Gamma_T} \bar{T}_I\,\delta U_I\,\mathrm{d}\Gamma\,. \tag{15}$$

In order to apply the FEM, the piezoelectric body is subdivided into a finite number N_E of domain elements, the so-called *finite elements*, as shown in Figure 1. These elements are mapped onto reference elements by means of a series of shape functions Φ^q and nodal coordinates $\check{x}_i^q$

$$x_i(\xi,\eta,\zeta) \approx \sum_{q=1}^{N} \Phi^q(\xi,\eta,\zeta)\,\check{x}_i^q\,, \tag{16}$$

where N is the number of nodes of the respective element, $\check{(\cdot)}$ denotes nodal values, and ξ, η, and ζ are the local coordinates in the reference element. The integrals in Eqn (15) can now be expressed by a sum of integrals over the elements.

In most Finite Element analyses, the isoparametric concept is used, which means that the field variables are approximated by the same shape functions Φ^q as used for the mapping of the elements:

$$U_I(\xi,\eta,\zeta) \approx \sum_{q=1}^{N} \Phi^q(\xi,\eta,\zeta)\, \check{U}_I^q \,. \tag{17}$$

The field variables are now replaced by these approximations, and by assembling the contributions of all finite elements one obtains a system of ordinary differential equations in time

$$\boldsymbol{M}\ddot{\check{\boldsymbol{U}}} + \boldsymbol{K}\check{\boldsymbol{U}} = \check{\boldsymbol{Q}} \,. \tag{18}$$

The mass matrix in Eqn (18) has the form

$$\boldsymbol{M} = \begin{bmatrix} \boldsymbol{M}_{uu} & 0 \\ 0 & 0 \end{bmatrix} , \tag{19}$$

reflecting the absence of electric inertia effects due to the quasi-static approximation of the electric field.

4.2 Boundary Element Methods

While the FEM for dynamic piezoelectricity is already well established, a BEM formulation has only recently been developed by the authors (Kögl and Gaul, 1999; Kögl and Gaul, 2000). This Boundary Element formulation starts with the piezoelectric reciprocity relation

$$\int_{\Omega} (\mathcal{L}_{JK}U_K\, U^*_{MJ} - \mathcal{L}_{JK}U^*_{MK}\, U_J)\ \mathrm{d}\Omega = \int_{\Gamma} (U^*_{MJ}\, T_J - T^*_{MJ}\, U_J)\ \mathrm{d}\Gamma \,, \tag{20}$$

where U^*_{MJ} and T^*_{MJ} are the generalized displacement and traction fundamental solutions of the static piezoelectric operator, defined by

$$\mathcal{L}_{JK}U^*_{MK} = -\delta_{JM}\, \delta(x_i,\xi_i) \qquad \text{and} \qquad T^*_{MJ} := C_{iJKl}\, U^*_{MK,l}\, n_i \,. \tag{21}$$

By replacing the expressions in the domain integral with Eqns (9) and (21), one obtains a piezoelectric representation formula

$$U_M(\xi) = \int_{\Gamma} (U^*_{MJ}\, T_J - T^*_{MJ}\, U_J)\ \mathrm{d}\Gamma - \int_{\Omega} U^*_{MJ}\mathcal{B}_{JK}U_K\ \mathrm{d}\Omega \,. \tag{22}$$

The remaining domain integral, which results from the use of the static fundamental solution and contains the effects of inertia, can be transformed to the boundary using the Dual Reciprocity Method. To this end, the domain term is approximated by a series of tensor functions f^q_{JN} and unknown coefficients α^q_N

$$\mathcal{B}_{JK}U_K(x) \approx \sum_{q=1}^{N} f^q_{JN}(x)\, \alpha^q_N \,, \tag{23}$$

so that, by substituting the approximation (23) into Eqn (22), a new representation formula for the generalized displacement field can be obtained

$$U_K(\xi) = \int\limits_\Gamma (U^*_{KJ}\,T_J - T^*_{KJ}\,U_J)\,\mathrm{d}\Gamma + \sum_{q=1}^{N} \left(U^q_{KN}(\xi) + \int\limits_\Gamma (T^*_{KJ}\,U^q_{JN} - U^*_{KJ}\,T^q_{JN})\,\mathrm{d}\Gamma \right) \alpha^q_N \,, \tag{24}$$

representing $U_K(\xi),\ \xi \in \Omega$ in terms of field quantities on the boundary only. The particular solutions U^q_{JN} and T^q_{JN}, which appear in Eqn (24), are defined by

$$\mathcal{L}_{JK}\,U^q_{KN} = f^q_{JN} \qquad \text{and} \qquad T^q_{JN} := C_{iJKl}\,U^q_{KN,l}\,n_i \,. \tag{25}$$

From Eqn (24) the internal displacements $U_K(\xi)$ can be calculated at any point $\xi \in \Omega$ if the boundary variables and the coefficients α^q_N are known.

The next step is the discretization of the model into finite elements on the boundary, the so-called *boundary elements*, as shown in Figure 1. Similar to Finite Element analyses, the boundary elements are mapped onto reference elements, and the field variables U_I and T_I are approximated by a series of shape functions and nodal values

$$x_i = \sum_{q=1}^{N} \Phi^q\,\check{x}^q_i \,, \qquad U_I = \sum_{q=1}^{N} \Phi^q\,\check{U}^q_I \,, \qquad T_I = \sum_{q=1}^{N} \Phi^q\,\check{T}^q_I \,, \tag{26}$$

where N is the number of local nodes of the respective boundary element. Now, the load point ξ in the representation formula (24) is transferred to the boundary, which leads to the boundary integral equation (BIE). Finally, the approximations (26) can be introduced, so that a system of equations

$$\boldsymbol{M}\ddot{\check{\boldsymbol{U}}} + \boldsymbol{H}\check{\boldsymbol{U}} = \boldsymbol{G}\check{\boldsymbol{T}} \tag{27}$$

can be assembled by summation over the boundary elements. This system is similar to Eqn (18) obtained in Finite Element analysis, but as usual in BEM analysis the system matrices are not symmetric. In addition, the piezoelectric mass matrix obtained with the present BE formulation contains a non-diagonal entry

$$\boldsymbol{M} = \begin{bmatrix} \boldsymbol{M}_{uu} & \boldsymbol{0} \\ \boldsymbol{M}_{\varphi u} & \boldsymbol{0} \end{bmatrix} , \tag{28}$$

resulting from the fact that the particular solution fields are coupled. However, this term does not present any problems in the solution process, as shown in the following section.

5 Solution of the Systems of Equations

Comparing the systems (18) and (27) obtained with the piezoelectric FE and BE formulations, one sees that the BEM equations are of a mixed type, containing unknown generalized displacements and fluxes, while the FEM system contains only generalized displacements. By subdividing the vectors and matrices into known and unknown parts, denoted by the superscripts ‘k’ and 'u', it is possible to eliminate the unknown fluxes $\boldsymbol{T}^{\mathrm{u}}$ from the equations (27) (the symbol $\check{(\cdot)}$ denoting nodal values is omitted in the following). This leads to

$$\boldsymbol{M}^{\mathrm{u}}\ddot{\boldsymbol{U}}^{\mathrm{u}}(t) + \boldsymbol{K}^{\mathrm{u}}\boldsymbol{U}^{\mathrm{u}}(t) = \boldsymbol{Q}^{\mathrm{k}}(t) \,, \tag{29}$$

where the matrices $\boldsymbol{M}^{\mathrm{u}}$, $\boldsymbol{K}^{\mathrm{u}}$ and vector $\boldsymbol{Q}^{\mathrm{k}}$ result from the condensation (for a more detailed description of the solution process, see Kögl (2000)). Eqn (29) has now the same form as the FEM system in Eqn (18).

Next, the unknown displacement vector is split into an elastic part, denominated by the subscript 'u', and an electric part, denominated by the subscript 'φ':

$$\begin{bmatrix} \boldsymbol{M}^{\mathrm{u}}_{uu} & \boldsymbol{0} \\ \boldsymbol{M}^{\mathrm{u}}_{\varphi u} & \boldsymbol{0} \end{bmatrix} \begin{bmatrix} \ddot{\boldsymbol{u}}^{\mathrm{u}}(t) \\ \ddot{\boldsymbol{\varphi}}^{\mathrm{u}}(t) \end{bmatrix} + \begin{bmatrix} \boldsymbol{K}^{\mathrm{u}}_{uu} & \boldsymbol{K}^{\mathrm{u}}_{u\varphi} \\ \boldsymbol{K}^{\mathrm{u}}_{\varphi u} & \boldsymbol{K}^{\mathrm{u}}_{\varphi\varphi} \end{bmatrix} \begin{bmatrix} \boldsymbol{u}^{\mathrm{u}}(t) \\ \boldsymbol{\varphi}^{\mathrm{u}}(t) \end{bmatrix} = \begin{bmatrix} \boldsymbol{Q}^{\mathrm{k}}_{u}(t) \\ \boldsymbol{Q}^{\mathrm{k}}_{\varphi}(t) \end{bmatrix} . \tag{30}$$

When $\boldsymbol{M}^{\mathrm{u}}_{\varphi u} \equiv 0$, Eqn (30) applies to FEM, too.

It is now possible to eliminate the electric degrees of freedom, reducing the piezoelectric system (30) to a purely elastic one

$$\bar{\boldsymbol{M}}^{\mathrm{u}}\ddot{\boldsymbol{u}}^{\mathrm{u}}(t) + \bar{\boldsymbol{K}}^{\mathrm{u}}\boldsymbol{u}^{\mathrm{u}}(t) = \bar{\boldsymbol{Q}}^{\mathrm{k}}(t) \,, \tag{31}$$

where the condensed matrices $\bar{\boldsymbol{M}}^{\mathrm{u}}$ and $\bar{\boldsymbol{K}}^{\mathrm{u}}$ and vector $\bar{\boldsymbol{Q}}^{\mathrm{k}}$ contain the influence of the electric field on the elastic deformation. Eqn (31) can be solved with standard time-stepping algorithms such as Houbolt, Newmark, Wilson-θ, etc., yielding the unknown displacements at every time step. When the displacement field is known, the calculation of the unknown electric variables and fluxes at the current time step is straightforward.

6 Example

The piezoelectric body shown in Fig. 3 is subjected to a Heaviside type loading at the specified face. The material used in the computations is a piezoelectric ceramic, poled in x_3-direction, with transversely isotropic material properties. Its elastic and piezoelectric moduli, and relative permittivities, are given by

$$\begin{array}{lll} C_{11} = 107600 \text{ MPa} \,, & C_{33} = 100400 \text{ MPa} \,, & C_{12} = 63120 \text{ MPa} \,, \\ C_{13} = 63850 \text{ MPa} \,, & C_{44} = 19620 \text{ MPa} \,, & \\ e_{31} = -9.6 \text{ N/Vm} \,, & e_{33} = 15.1 \text{ N/Vm} \,, & e_{15} = 12.0 \text{ N/Vm} \,, \\ \epsilon^{\mathrm{rel}}_{11} = 1936 \,, & \epsilon^{\mathrm{rel}}_{33} = 2109 \,, & \end{array} \tag{32}$$

the mass density is $\rho = 7800$ kg/m^3. Also, the following relations hold between the moduli: $C_{22} = C_{11}$, $C_{23} = C_{13}$, $C_{66} = 0.5(C_{11} - C_{12})$, $e_{32} = e_{31}$, $e_{24} = e_{15}$, $\epsilon^{\mathrm{rel}}_{22} = \epsilon^{\mathrm{rel}}_{11}$.

The transient response has been calculated using both the FEM and the BEM formulation presented in Section 4. The discretization of the body is shown in the figure. It consists of 375 finite elements and 350 boundary elements. Two FEM computations have been performed, the first using 8-node brick elements (*linear FEM*, 2304 dof), the second with 20-node brick elements (*quadratic FEM*, 8304 dof). For the BEM calculation, 4-node boundary elements have been used,

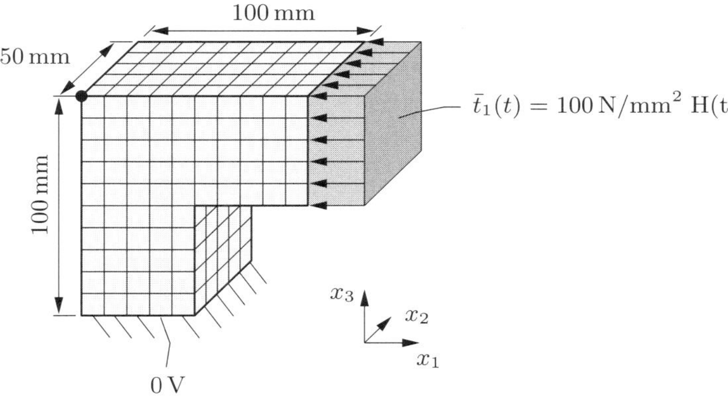

Figure 3. Piezoelectric body with discretization and boundary conditions

along with 99 internal nodes, resulting in a total of 1804 degrees of freedom. The use of internal nodes is not strictly necessary but enhances the accuracy of the results, since it improves the interpolation procedure introduced in the Dual Reciprocity formulation (see e.g. Kögl (2000)).

For the time integration, a damped Newmark algorithm is chosen, using a time step size of $\Delta t = 5\ \mu\mathrm{s}$ and parameters $\delta = 0.7$ and $\alpha = 0.5$ as proposed by Kögl and Gaul (1999). The time histories of the electric potential φ and velocity $\dot{u}_1$ at point $(0, 0, 100)$ are shown in Figure 4. Excellent agreement between the linear FEM and DR-BEM computations can be observed in both cases, with only a slight deviation from the more accurate results of the FEM computation using quadratic element shape functions.

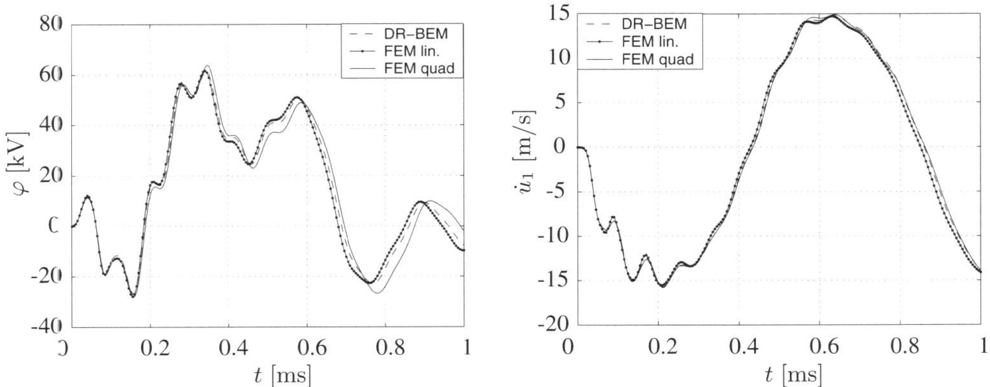

Figure 4. Electric potential $\varphi(t)$ and velocity $\dot{u}_1(t)$ at point (0,0,100)

Figure 5 shows the results obtained for the electric field E_1, and elastic stress σ_{13} at a point with the coordinates $(37.5, 12.5, 62.5)$. Taking the quadratic Finite Element calculations as reference, it can be observed that the Boundary Element computations with linear element shape functions are much more accurate than the corresponding linear Finite Element computations, for reasons which have already been explained in Section 2. This shows that the improved accuracy of the Boundary Element Methods in the calculation of flux quantities is maintained in the present piezoelectric DR-BEM.

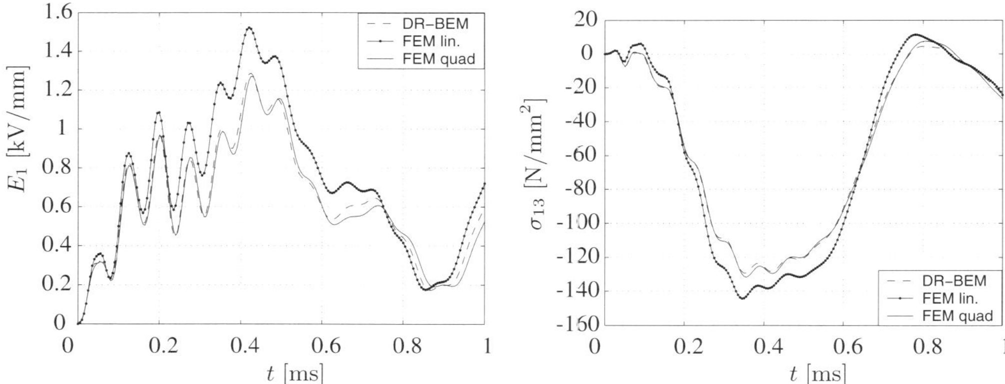

Figure 5. Electric field $E_1(t)$ and elastic stress $\sigma_{13}(t)$ at point (37.5,12.5,62.5)

7 Conclusions

A comparative study and introduction to the numerical calculation of piezoelectric continua with Finite Elements and Boundary Elements has been given. In a numerical example, excellent agreement could be demonstrated between the Finite Element results and those obtained using a Boundary Element formulation recently introduced by the authors. It could be observed that for the electric field and stress calculations, the results obtained by the new BEM formulation are more accurate than the results from Finite Element computations when using a comparable grid. This BEM formulation therefore presents a powerful numerical tool and an alternative to FEM for the computation of transient piezoelectric phenomena, especially for electric field and stress analysis.

References

Allik, H., and Hughes, T. J. R. (1970). Finite element method for piezoelectric vibration. *Int. J. Num. Meth. Eng.* 2:151–157.

Bathe, K.-J. (1996). *Finite element procedures.* New Jersey: Prentice-Hall.

Brebbia, C., Telles, J., and Wrobel, L. (1984). *Boundary element techniques.* Berlin: Springer-Verlag.

Gaul, L., and Fiedler, C. (1996). *Boundary Element Methods in Statics and Dynamics (in German).* Braunschweig: Verlag Vieweg.

Kögl, M., and Gaul, L. (1999). Dual reciprocity boundary element method for three-dimensional problems of dynamic piezoelectricity. In *Boundary Elements XXI*, 537–548. Southampton: Computational Mechanics Publications.

Kögl, M., and Gaul, L. (2000). A Boundary Element Method for transient piezoelectric analysis. *Engineering Analysis with Boundary Elements* 24(7-8):591–598.

Kögl, M. (2000). *A Boundary Element Method for Dynamic Analysis of Anisotropic Elastic, Piezoelectric, and Thermoelastic Solids.* PhD thesis, Universität Stuttgart.

Lerch, R. (1990). Simulation of piezoelectric devices by two- and three-dimensional finite elements. *IEEE Transactions on ultrasonics, ferroelectrics, and frequency control* 37(2):233–247.

Partridge, P. W., Brebbia, C. A., and Wrobel, L. C. (1992). *The dual reciprocity boundary element method.* Southampton: Computational Mechanics Publications.

Tiersten, H. F. (1969). *Linear piezoelectric plate vibrations.* New York: Plenum Press.

Zienkiewicz, O. C., and Taylor, R. L. (1991). *The finite element method*, volume 2. London: McGraw-Hill, fourth edition.

Zienkiewicz, O. C., and Taylor, R. L. (1994). *The finite element method*, volume 1. London: McGraw-Hill, fourth edition.

CHAPTER 9

Transient Viscoelastodynamic Boundary Element Formulations

Lothar Gaul[1] and Martin Schanz[2]

[1] Institute A of Mechanics, University of Stuttgart, Germany
[2] Institute of Applied Mechanics, Technical University of Braunschweig, Braunschweig, Germany

Abstract. As an alternative to domain discretization methods, the boundary element method (BEM) provides a powerful tool for the calculation of dynamic structural response in frequency and time domain. Field equations of motion and boundary conditions are cast into boundary integral equations (BIE), which are discretized only on the boundary. Fundamental solutions are used as weighting functions in the BIE which fulfil the Sommerfeld radiation condition, i.e., the energy radiation into a surrounding medium is modelled correctly. Therefore, infinite and semi-infinite domains can be effectively treated by the method. The soil represents such a semi-infinite domain in soil-structure-interaction problems. The response to vibratory loads superimposed to static pre-loads can often be calculated by linear viscoelastic constitutive equations. Conventional viscoelastic constitutive equations can be generalized by taking fractional order time derivatives into account.
Here, two time domain BEM approaches including generalized viscoelastic behaviour are compared with the Laplace domain BEM approach and subsequent numerical inverse transformation. One of the presented time domain approaches uses an analytical integration of the elastodynamic BIE in a time step. Viscoelastic constitutive properties are introduced after Laplace transformation by means of an elastic-viscoelastic correspondence principle. The transient response is obtained by inverse transformation in each time step. The other time domain approach is based on the so-called 'convolution quadrature method'. In this formulation, the convolution integral in the BIE is numerically approximated by a quadrature formula whose weights are determined by the same Laplace transformed fundamental solutions used in the first method and a linear multistep method.
A numerical study of wave propagation problems in 3-d viscoelastic continuum is performed for comparing the three BEM formulations.

1 Introduction

The Boundary Element Method (BEM) has become a widely used numerical tool in statics and dynamics, e.g., Gaul and Fiedler (1997). A review about the efforts in dynamics is published by Beskos (1997, 1987). Main advantages of the method are the reduction of the problem dimension by one and the implicit fulfilment of the radiation condition for unbounded domains. Due to the last advantage the BEM seems to be preferable for the calculation of infinite or semi-infinite domains, e.g., soil.

Modelling soil as an elastic solid can be only a crude approximation of the real material. A better description of the material properties can be achieved using viscoelastic constitutive equations. In these equations, normally, the stress strain relation is a differential or integral equation of hereditary type (Christensen, 1971). Improved curve fitting of measured material properties by constitutive equations with fewer parameters is achieved with the concept of fractional differ-integration (Bagley and Torvik, 1986). A review about fractional calculus applied to dynamic problems has recently been given by Rossikhin and Shitikova (1997).

Viscoelastic boundary element formulations are mostly published for the quasi-static case (see, e.g, Tanaka, 1987 and Sim and Kwak, 1988 and Carini et al., 1991), or in dynamics using a frequency or Laplace domain representation of the governing integral equation. These formulations are developed by applying the elastic-viscoelastic correspondence principle to the elastodynamic boundary element formulation, e.g., Kobayashi and Kawakami (1985) for a frequency domain or Manolis and Beskos (1981) for a Laplace domain formulation. In these formulations, the complex moduli concept is used which can be extended to fractional operator viscoelasticity allowing not only integer powers of the frequency, e.g., Gaul et al. (1988).

Calculation of transient response, however, requires the inverse transformation. Since all numerical inversion formulas depend on a proper choice of their parameters (Narayanan and Beskos, 1982), a direct evaluation in time domain seems to be preferable. But, formulations directly in time domain require the knowledge of viscoelastic fundamental solutions which are not yet known for the general visco-elastodynamic case. Only for a simple Maxwell model, a solution has been obtained analytically by Gaul and Schanz (1992) and has been implemented in a boundary element formulation (Schanz and Gaul, 1993). Based on the frequency domain fundamental solution with subsequent inverse transformation, a 1-d solution has been proposed by Wolf and Dabre (1986). In the 3-d case, Gaul and Schanz (1994b) developed a formulation for a generalized (with fractional derivatives) 3-parameter model using the Laplace transformed fundamental solution which is inverted within each time step. Recently, Schanz and Antes (1997b) published a viscoelastic formulation based on the so-called 'convolution quadrature method' proposed by Lubich (1988a). In this formulation, the convolution integral is numerically approximated by a quadrature formula whose weights are determined by the Laplace transform of the fundamental solution and a linear multistep method. The extension of the above mentioned formulations to the case of different viscoelastic behaviour of the deviatoric and the hydrostatic part of the stress-strain relation, leading to a time dependent Poisson's ratio, is done by Schanz (1999).

Here, all three approaches, calculation in Laplace domain and the two time domain formulations, are presented. Results of forced wave propagation in a 3-d rod are compared. Finally, the wave propagation in an elastic concrete foundation slab bonded to viscoelastic semi-infinite soil is studied.

2 Constitutive Equation

Decomposition of the stress tensor σ_{ij} or the strain tensor ε_{ij} into the the hydrostatic part $\delta_{ij}\sigma_{kk}/3$ or $\delta_{ij}\varepsilon_{kk}/3$ and the deviatoric part s_{ij} or e_{ij} yields

$$\sigma_{ij} = \frac{1}{3}\sigma_{kk}\delta_{ij} + s_{ij} \quad \text{and} \quad \varepsilon_{ij} = \frac{1}{3}\varepsilon_{kk}\delta_{ij} + e_{ij} \text{ with } s_{ii} = e_{ii} = 0\,, \tag{1}$$

respectively. Two independent sets of constitutive equations for viscoelastic materials exist after this decomposition

$$\sum_{k=0}^{N} p'_k \frac{\mathrm{d}^k}{\mathrm{d}t^k} s_{ij} = \sum_{k=0}^{M} q'_k \frac{\mathrm{d}^k}{\mathrm{d}t^k} e_{ij} \qquad \sum_{k=0}^{N} p''_k \frac{\mathrm{d}^k}{\mathrm{d}t^k} \sigma_{ii} = \sum_{k=0}^{M} q''_k \frac{\mathrm{d}^k}{\mathrm{d}t^k} \varepsilon_{ii}. \tag{2}$$

More flexibility in fitting measured data in a large frequency range is obtained by replacing the integer order time derivatives by fractional order time derivatives (Gaul et al., 1991).

The derivative of fractional order α is defined by

$$\frac{\mathrm{d}^{\alpha}x(t)}{\mathrm{d}t^{\alpha}} = \frac{1}{\Gamma(1-\alpha)}\frac{\mathrm{d}}{\mathrm{d}t}\int\limits_{0}^{t}\frac{x(t-\tau)}{\tau^{\alpha}}\mathrm{d}\tau \quad 0 \le \alpha < 1 \tag{3}$$

with the Gamma function $\Gamma(1-\alpha) = \int_0^{\infty} e^{-x}x^{-\alpha}\,\mathrm{d}x$, as the inverse operation of fractional integration attributed to Riemann and Liouville (Oldham and Spanier, 1974). A different definition based on generalized finite differences is given by Grünwald (1867)

$$\frac{\mathrm{d}^{\alpha}x(t)}{\mathrm{d}t^{\alpha}} = \lim_{N\to\infty}\left\{\left(\frac{t}{N}\right)^{-\alpha}\sum_{j=0}^{N-1}\frac{\Gamma(j-\alpha)}{\Gamma(-\alpha)\Gamma(j+1)}x\left(t\left[1-\frac{j}{N}\right]\right)\right\}. \tag{4}$$

This discrete definition is more convenient in constitutive equations treated by time stepping algorithms and the equivalence of definition (3) and (4) can be shown.

The fractional derivatives in equations (3) appear complicated in time domain. However, Laplace transform reveals the useful result

$$\mathscr{L}\left\{\frac{\mathrm{d}^{\alpha}x(t)}{\mathrm{d}t^{\alpha}}\right\} = s^{\alpha}\mathscr{L}\{x(t)\} - \sum_{k=0}^{n-1}s^{k}\frac{\mathrm{d}^{\alpha-1-k}}{\mathrm{d}t^{\alpha-1-k}}x(0), \quad n-1<\alpha\le n \tag{5}$$

where s is the Laplace variable.

With the definition (3) or (4) the generalized viscoelastic constitutive equations are given by

$$\sum_{k=0}^{N}p'_k\frac{\mathrm{d}^{\alpha_k}}{\mathrm{d}t^{\alpha_k}}s_{ij} = \sum_{k=0}^{M}q'_k\frac{\mathrm{d}^{\beta_k}}{\mathrm{d}t^{\beta_k}}e_{ij} \qquad \sum_{k=0}^{N}p''_k\frac{\mathrm{d}^{\alpha_k}}{\mathrm{d}t^{\alpha_k}}\sigma_{ii} = \sum_{k=0}^{M}q''_k\frac{\mathrm{d}^{\beta_k}}{\mathrm{d}t^{\beta_k}}\varepsilon_{ii}\,. \tag{6}$$

For $N = M = 1$ and fractional time derivatives of order α and β a possible representation of a uniaxial stress strain equation with 5 parameters (E Young's modulus, p, q viscoelastic parameters and the orders of the fractional derivatives α and β)

$$p\frac{\mathrm{d}^{\alpha}}{\mathrm{d}t^{\alpha}}\sigma + \sigma = E\left(\varepsilon + q\frac{\mathrm{d}^{\beta}}{\mathrm{d}t^{\beta}}\varepsilon\right) \tag{7}$$

is given. For wave propagation problems, e.g., in unbounded domains, the wave velocity is important. In viscoelasticity it is defined with the initial modulus $E(0)$ of the relaxation function (Christensen, 1971)

$$c = \sqrt{\frac{E(0)}{\rho}}\,. \tag{8}$$

For the determination of the initial modulus, the initial value theorem of Laplace transform $\lim_{t\to 0} f(t) = \lim_{s\to\infty} s\mathscr{L}\{f(s)\}$ is used. This theorem and the definition of the initial modulus lead to

$$E(0) = \lim_{s\to\infty} E\frac{1+qs^{\beta}}{1+ps^{\alpha}} = E\lim_{s\to\infty}\frac{\frac{1}{s^{\alpha}}+qs^{\beta-\alpha}}{\frac{1}{s^{\alpha}}+p}\,. \tag{9}$$

The limiting process gives an finite value only for the restriction $\alpha = \beta$. For solid viscoelastic materials this restriction is in accordance with experimental data.

Finally, according to Bagley and Torvik (1986) a credible model of the viscoelastic phenomenon should predict nonnegative internal work and a nonnegative rate of energy dissipation. To satisfy these restrictions, constraints on the remaining parameters of the model are developed. This produces the constraints (see, Schanz, 1994)

$$E \geq 0 \qquad q > p > 0 \qquad 0 < \alpha = \beta < 2\,. \tag{10}$$

A powerful tool for calculating viscoelastic behaviour from a known elastic response is the elastic-viscoelastic correspondence principle. According to this principle (see, e.g., Flügge, 1975) the viscoelastic solution is calculated from the analytical elastic solution by replacing the elastic moduli in the Fourier or Laplace transformed domain by the transformed impact response functions of the viscoelastic material model. The viscoelastic solution is then obtained by inverse transformation.

As an example for calculating viscoelastic behaviour with the elastic-viscoelastic correspondence principle a free-fixed 1-d rod (Young's modulus E, density ρ, cross section A) is used (Fig. 1). The solution for the displacement $u(x,t)$ can be determined with inverse Laplace transformation form the elastic solution in Laplace domain (see, Graffi, 1954)

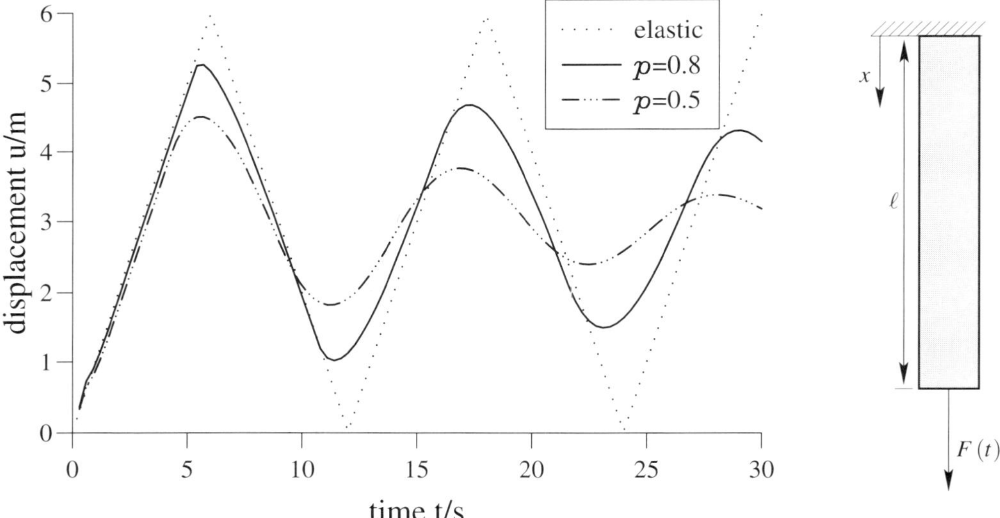

Figure 1. Displacement of a free-fixed 1-d rod versus time for different viscoelastic parameter p

$$\hat{u}(x,s) = \frac{\hat{F}(s)}{A\rho c}\,\frac{\sinh\left(\frac{s}{c}x\right)}{s\cosh\left(\frac{s}{c}\ell\right)}\,, \tag{11}$$

after inserting the elastic-viscoelastic correspondence

$$E \rightarrow E\frac{1+qs^{\alpha}}{1+ps^{\alpha}} \qquad \Rightarrow \qquad c \rightarrow c\sqrt{\frac{1+qs^{\alpha}}{1+ps^{\alpha}}} \, . \tag{12}$$

In the equations above $\hat{()}$ denotes the Laplace transformed functions. Whereas the elastic solution can be obtained by an analytical inversion, the viscoelastic one is performed by an numerical inverse transformation. Here, the method of Talbot (1979) is used. For $E = 1\frac{N}{m^2}, A = 1m^2, q = 1sec^{-0.8}, \alpha = 0.8$ and $\ell = 3m$ the displacement $u(\ell,t)$ is plotted versus time for different viscoelastic parameters p in Fig. 1.

For 3-d continua the elastic-viscoelastic correspondence principle is as simple as in the 1-d case. If the same damping mechanisms are assumed in hydrostatic and deviatoric stress strain state, the corresponding 3-d constitutive equations are gained by replacing the uniaxial stress and strain by the hydrostatic and the deviatoric states. For the above mentioned generalized three parameter model (7) the elastic-viscoelastic correspondence is the same as in the 1-d case (12) with unchanged real Poisson's ratio ν. The last condition is only valid if the same damping mechanisms are present (Schanz, 1999).

As for the above mentioned 1-d rod example, construction of fundamental solutions by the elastic-viscoelastic correspondence principle leads to viscoelastic solutions in Laplace or Fourier domain, firstly. Afterwards, a numerical inverse transformation is necessary.

3 Viscoelastic BE–Formulations

Assuming homogeneity and linear viscoelastic material the dynamic behaviour of a waveguide as well as the wave propagation in an unbounded domain Ω will be determined. In the following, Latin indices take the values 1,2, and 1,2,3 in 2-d and 3-d, respectively, where summation convention is implied over repeated indices, and commas and over-dots denote spatial and time differentiation, respectively. On the boundary $\Gamma = \Gamma_p \cup \Gamma_u$ of the domain Ω, tractions $p_i(\mathbf{x},t)$ and displacements $u_j(\mathbf{x},t)$ are prescribed, respectively, i.e.

$$\sigma_{ij}n_j = p_i(\mathbf{x},t) \quad \text{for } t>0,\ \mathbf{x}\in\Gamma_p, \qquad u_i(\mathbf{x},t) = q_i(\mathbf{x},t) \quad \text{for } t>0,\ \mathbf{x}\in\Gamma_u\,, \tag{13}$$

while the initial conditions $u_i(\mathbf{x},0) = 0$ and $\dot{u}_i(\mathbf{x},0) = 0, \mathbf{x} \in \Omega\cup\Gamma$ are assumed to be zero. In eqn. (13), σ_{ij} is the stress tensor and n_j means the outward normal vector on the boundary Γ. Neglecting the body force effects, the dynamic extension of Betti's reciprocal work theorem combining two states of displacements and tractions (U_{ij}, P_{ij}) and (u_j, p_j) leads to the integral equation

$$c_{ij}(\mathbf{y})u_j(\mathbf{y},t) = \int_{\Gamma} U_{ij}(\mathbf{x},\mathbf{y},t) * p_j(\mathbf{x},t)\,\mathrm{d}\Gamma - \oint_{\Gamma_x} P_{ij}(\mathbf{x},\mathbf{y},t) * u_j(\mathbf{x},t)\,\mathrm{d}\Gamma\,, \tag{14}$$

where $*$ denotes the convolution with respect to time. U_{ij} and P_{ij} are the displacements and tractions, respectively, due to a unit impulse in the direction x_i, i.e., the time-dependent fundamental solution of the full space. For the linear viscoelastic material described in section 2, e.g., the generalized 3-parameter model (7), the integral free terms $c_{ij}(\mathbf{y})$ are identical to those of elastostatics and dependent only on the local geometry at $\mathbf{y}$ and on Poisson's ratio ν, i.e., for $\mathbf{y}$ on a smooth boundary $c_{ij}(\mathbf{y}) = \frac{\delta_{ij}}{2}$. However, this is valid only for the assumption of the same viscoelastic behaviour of the deviatoric and hydrostatic stress strain state.

When $\mathbf{x}$ approaches $\mathbf{y}$, the kernel $U_{ij}(\mathbf{x},\mathbf{y},t)$ is weakly singular and $P_{ij}(\mathbf{x},\mathbf{y},t)$ is strongly singular, i.e., the second integral in eqn. (14) exists only in the Cauchy principal value sense.

According to the boundary element method the boundary surface Γ is discretized by E isoparametric elements Γ_e where F polynomial spatial shape functions $N_e^f(\mathbf{x})$ are defined. Hence, with the time dependent nodal values $u_j^{ef}(t)$ and $p_j^{ef}(t)$ the following representation is adapted

$$u_j(\mathbf{x},t) = \sum_{e=1}^{E}\sum_{f=1}^{F} N_e^f(\mathbf{x})\, u_j^{ef}(t), \quad p_j(\mathbf{x},t) = \sum_{e=1}^{E}\sum_{f=1}^{F} N_e^f(\mathbf{x})\, p_j^{ef}(t). \tag{15}$$

Inserting these 'ansatz' functions in eqn. (14) gives

$$c_{ij}(\mathbf{y})\, u_j(\mathbf{y},t) = \sum_{e=1}^{E}\sum_{f=1}^{F}\left\{ \int_{\Gamma_e} U_{ij}(\mathbf{x},\mathbf{y},t)\, N_e^f(\mathbf{x})\, \mathrm{d}\Gamma * p_j^{ef}(t) - \oint_{\Gamma_e} P_{ij}(\mathbf{x},\mathbf{y},t)\, N_e^f(\mathbf{x})\, \mathrm{d}\Gamma * u_j^{ef}(t)\right\}. \tag{16}$$

In what follows, usually, the fundamental solutions must be known. Only for the simplest viscoelastic model, the Maxwell material, the fundamental solutions are available analytically (Gaul and Schanz, 1992). For more realistic models the fundamental solutions are only known in Laplace domain. Until now, three different boundary element formulations have been proposed. These are summarized here and afterwards described in more detail:

- Formulation 1:
  Uses the boundary element method in Laplace domain and a subsequent numerical inverse transformation, proposed, e.g., by Ahmad and Manolis (1987).
- Formulation 2:
  Time shape functions for the unknown displacements and tractions are chosen and the convolution of these functions with the elastodynamic fundamental solution is integrated analytically. The result is transformed in Laplace domain, where the elastic-viscoelastic correspondence principle is applied. An numerical inverse transformation leads to a time domain boundary element formulation (Gaul and Schanz, 1994a).
- Formulation 3:
  Uses the so called 'convolution quadrature method' developed by Lubich (1988a) to evaluate the convolution in eqn. (16). This method only requires the fundamental solutions in Laplace domain (Schanz, 1999).

In all three formulations the spatial integration of the regular integrals over the boundary elements is performed by Gaussian quadrature. The weakly singular integrals are regularized with polar coordinate transformation and the strongly singular integrals with the method proposed by Guiggiani and Gigante (1990). Finally, the solution of the resulting algebraic system of equations is solved by a direct solver.

Formulation 1: The Laplace transform of the integral equation (16) is

$$c_{ij}(\mathbf{y})\,\hat{u}_j(\mathbf{y},s) = \sum_{e=1}^{E}\sum_{f=1}^{F}\left\{\int_{\Gamma_e}\hat{U}_{ij}(\mathbf{x},\mathbf{y},s)\,N_e^f(\mathbf{x})\,\hat{p}_j^{ef}(s)\,\mathrm{d}\Gamma - \oint_{\Gamma_e}\hat{P}_{ij}(\mathbf{x},\mathbf{y},s)\,N_e^f(\mathbf{x})\,\hat{u}_j^{ef}(s)\,\mathrm{d}\Gamma\right\}. \tag{17}$$

The viscoelastic fundamental solutions $\hat{U}_{ij}(\mathbf{x},\mathbf{y},s)$ and $\hat{P}_{ij}(\mathbf{x},\mathbf{y},s)$ are obtained by applying the elastic-viscoelastic correspondence to the elastodynamic solutions (Cruse and Rizzo, 1968)

$$\begin{aligned}\hat{U}_{ij}(\mathbf{x},\mathbf{y},s) = \frac{1}{4\pi\rho}\Bigg\{&\frac{1}{r^2}\left(\frac{3r_ir_j}{r^3}-\frac{\delta_{ij}}{r}\right)\left[\frac{s\frac{r}{c_1}+1}{s^2}e^{-\frac{r}{c_1}s}-\frac{s\frac{r}{c_2}+1}{s^2}e^{-\frac{r}{c_2}s}\right]\\ &+\frac{r_ir_j}{r^3}\left[\frac{1}{c_1^2}e^{-\frac{r}{c_1}s}-\frac{1}{c_2^2}e^{-\frac{r}{c_2}s}\right]+\frac{\delta_{ij}}{rc_2^2}e^{-\frac{r}{c_2}s}\Bigg\},\end{aligned} \tag{18}$$

where $r=\sqrt{r_ir_i}$ with $r_i=x_i-y_i$. The corresponding fundamental traction components needed in eqn. (17) are obtained via

$$\hat{P}_{ij}=\rho\left(c_1^2-2c_2^2\right)\hat{U}_{jm,m}\delta_{ik}n_k+\rho c_2^2\left(\hat{U}_{ji,k}n_k+\hat{U}_{jk,i}n_k\right) \tag{19}$$

with the Kronecker symbol δ_{ik}. The correspondence principle simply replaces the elastic wave velocities by the viscoelastic ones

$$c_1=\sqrt{\frac{E(1-\nu)}{\rho(1-2\nu)(1+\nu)}}\rightarrow c_1\sqrt{\frac{1+qs^\alpha}{1+ps^\alpha}}\qquad c_2=\sqrt{\frac{E}{\rho 2(1+\nu)}}\rightarrow c_2\sqrt{\frac{1+qs^\alpha}{1+ps^\alpha}} \tag{20}$$

in eqn. (18) and (19). As next step, the solution is transformed back to time domain. Preferable methods for this application are those of Durbin (1974) or Crump (1976), according to the experience reported in Manolis and Beskos (1981).

Formulation 2: As has been shortly summarized above, this formulation uses the elastodynamic fundamental solutions in time domain, e.g., for the displacements

$$\begin{aligned}U_{ij}(\mathbf{x},\mathbf{y},t,\tau)=\frac{1}{4\pi\rho}\Bigg\{&(t-\tau)f_0(r)\left[H\left(t-\tau-\frac{r}{c_1}\right)-H\left(t-\tau-\frac{r}{c_2}\right)\right]\\ &+\frac{f_1(r)}{c_1^2}\delta\left(t-\tau-\frac{r}{c_1}\right)+\frac{f_2(r)}{c_2^2}\delta\left(t-\tau-\frac{r}{c_2}\right)\Bigg\}\end{aligned} \tag{21}$$

with abbreviated functions depending on spatial coordinates only

$$f_0(r)=\frac{3r_{,i}r_{,j}-\delta_{ij}}{r^3}\qquad f_1(r)=\frac{r_{,i}r_{,j}}{r}\qquad f_2(r)=\frac{\delta_{ij}-r_{,i}r_{,j}}{r}. \tag{22}$$

The solution for the traction is calculated by the time domain expression corresponding to (19). The elapsed time t is discretized by N equal time-increments Δt. The simplest nontrivial choice for the time shape functions, ensuring that no terms drop out in the boundary integral equation after one differentiation, are linear shape functions for the displacements $u_j^{ef}(t)$ and constant shape functions for the tractions $p_j^{ef}(t)$

$$u_j^{ef}(t) = \left(u_{j(m-1)}^{ef} \frac{t_m - \tau}{\Delta t} + u_{jm}^{ef} \frac{\tau - t_{m-1}}{\Delta t} \right) \qquad p_j^{ef}(t) = 1 \cdot p_{jm}^{ef} . \tag{23}$$

The actual time step is m. Inserting these 'ansatz' function in eqn. (16) analytical time integration can be carried out within each time step because of the properties of the Dirac and Heaviside functions. This integration leads to piecewise defined functions. For the sake of brevity the integration is indicated only for the first term on the right side of eqn. (16)

$$\int_{t_{m-1}}^{t_m} U_{ij}(\mathbf{x},\mathbf{y},t-\tau)\,\mathrm{d}\tau = \frac{1}{4\pi\rho} \begin{cases} 0 & t < t_{m-1} + \frac{r}{c_1} \\ \frac{f_0(r)}{2}\left[(t-t_{m-1})^2 - \left(\frac{r}{c_1}\right)^2 \right] + \frac{f_1(r)}{c_1^2} & t_{m-1} + \frac{r}{c_1} < t < t_m + \frac{r}{c_1} \\ f_0(r)\left[t t_m - t t_{m-1} - \frac{t_m^2}{2} + \frac{t_{m-1}^2}{2} \right] & t_m + \frac{r}{c_1} < t < t_{m-1} + \frac{r}{c_2} \\ \frac{f_0(r)}{2}\left[\left(\frac{r}{c_2}\right)^2 - (t_m - t)^2 \right] + \frac{f_2(r)}{c_2^2} & t_{m-1} + \frac{r}{c_2} < t < t_m + \frac{r}{c_2} \\ 0 & t_m + \frac{r}{c_2} < t \ . \end{cases} \tag{24}$$

Now, after performing the convolution, function (24) and the corresponding function for the tractions are transformed in the Laplace domain

$$\begin{aligned} &\int_0^\infty \sum_{m=1}^{n} \int_{t_{m-1}}^{t_m} U_{ij}(\mathbf{x},\mathbf{y},t-\tau)\,\mathrm{d}\tau \; e^{-st}\,\mathrm{d}t \\ &\qquad = \frac{1}{4\pi\rho} \sum_{m=1}^{n} \Bigg\{ \int_{t_{m-1}+\frac{r}{c_1}}^{t_m+\frac{r}{c_1}} \left[\frac{f_0(r)}{2} \left((t-t_{m-1})^2 - \left(\frac{r}{c_1}\right)^2 \right) + f_1(r) \right] e^{-st}\,\mathrm{d}t \\ &\qquad\quad + \int_{t_m+\frac{r}{c_1}}^{t_{m-1}+\frac{r}{c_2}} f_0(r) \left(t\,t_m - t\,t_{m-1} - \frac{t_m^2}{2} + \frac{t_{m-1}^2}{2} \right) e^{-st}\,\mathrm{d}t \\ &\qquad\quad + \int_{t_{m-1}+\frac{r}{c_2}}^{t_m+\frac{r}{c_2}} \left[\frac{f_0(r)}{2} \left(\left(\frac{r}{c_2}\right)^2 - (t-t_m)^2 \right) + f_2(r) \right] e^{-st}\,\mathrm{d}t \Bigg\} \\ &\qquad = \frac{1}{4\pi\rho} \sum_{m=1}^{n} \left(e^{-st_{m-1}} - e^{-st_m} \right) \Bigg\{ f_0(r) \frac{1}{s^2} \left[\left(\frac{r}{c_1}\right) e^{-\frac{r}{c_1}s} - \left(\frac{r}{c_2}\right) e^{-\frac{r}{c_2}s} \right] \\ &\qquad\quad + f_1(r) \frac{1}{s} e^{-\frac{r}{c_1}s} + f_2(r) \frac{1}{s} e^{-\frac{r}{c_2}s} \Bigg\}. \end{aligned} \tag{25}$$

In Laplace domain the elastic-viscoelastic correspondence (20) is applied. Causality of the solution implies that no response is present prior to the arrival of the waves. This physical requirement is assumed for the numerical inversion of eqn. (25). The correspondence relation

$$\mathscr{L}\{f\}e^{-st_{m-1}} - \mathscr{L}\{f\}e^{-st_m} \bullet\!\!-\!\!\circ\, f(t-t_{m-1})H(t-t_{m-1}) - f(t-t_m)H(t-t_m) \tag{26}$$

is taken into account as well. The inverse transformation therefore leads to a function defined piecewise in the time ranges of eqn. (24) with the viscoelastic wave velocities

$$c_1^{viscoelastic} = c_1\sqrt{\frac{q}{p}} \qquad c_2^{viscoelastic} = c_2\sqrt{\frac{q}{p}}\,, \tag{27}$$

defined in accordance to eqn. (8). The inversion is carried out numerically by the method of Talbot (1979), because the method of Crump (1976) produces convergence problems and the method of Durbin (1974) is very time consuming (for details see, Gaul and Schanz, 1999).

Formulation 3: In this formulation, the convolution between the fundamental solutions and the corresponding nodal values in eqn. (16) is performed numerically with the so-called 'convolution quadrature method' proposed by Lubich (1988b, 1988a). The quadrature formula approximates a convolution integral

$$y(t) = f(t) * g(t) = \int_0^t f(t-\tau)\,g(\tau)\,\mathrm{d}\tau \tag{28}$$

by the finite sum

$$y(n\Delta t) = \sum_{j=0}^{n} \omega_{n-j}(\Delta t)\,g(j\Delta t)\,, \qquad n = 0,1,\ldots,N\,. \tag{29}$$

The integration weights $\omega_n(\Delta t)$ are the coefficients of the power series for the function $\hat{f}\left(\frac{\gamma(z)}{\Delta t}\right)$ at the point $\frac{\gamma(z)}{\Delta t}$. Herein, $\gamma(z)$ is the quotient of the characteristic polynomials of a linear multistep method, e.g., the backward differentiation formula of second order $\gamma(z) = \frac{3}{2} - 2z + \frac{1}{2}z^2$. The coefficients are calculated by the integral

$$\omega_n(\Delta t) = \frac{1}{2\pi i}\int_{|z|=\mathscr{R}} \hat{f}\left(\frac{\gamma(z)}{\Delta t}\right) z^{-n-1}\,\mathrm{d}z \approx \frac{\mathscr{R}^{-n}}{L}\sum_{\ell=0}^{L-1}\hat{f}\left(\frac{\gamma\left(\mathscr{R}e^{i\ell\frac{2\pi}{L}}\right)}{\Delta t}\right)e^{-in\ell\frac{2\pi}{L}}\,, \tag{30}$$

with $\mathscr{R}$ being the radius of a circle in the domain of analyticity of $\hat{f}(z)$. The integral in eqn. (30) is approximated by a trapezoidal rule with L equal steps $\frac{2\pi}{L}$, after transformation to polar coordinates. Details of the convolution quadrature method can be found in Lubich (1988b, 1988a). An example of this quadrature formula related to the boundary element method is presented in Schanz and Antes (1997a).

The quadrature formula (29) is applied to eqn. (16). The result is the following boundary element time-stepping formulation for $n = 0,1,\ldots,N$

$$c_{ij}(\mathbf{y})\,u_j(\mathbf{y},n\Delta t) = \sum_{e=1}^{E}\sum_{f=1}^{F}\sum_{k=0}^{n}\left\{\omega_{n-k}^{ef}\left(\hat{U}_{ij},\mathbf{y},\Delta t\right)p_j^{ef}(k\Delta t) - \omega_{n-k}^{ef}\left(\hat{P}_{ij},\mathbf{y},\Delta t\right)u_j^{ef}(k\Delta t)\right\} \tag{31}$$

with the weights corresponding to eqn. (30)

$$\omega_n^{ef}\left(\hat{U}_{ij},\mathbf{y},\Delta t\right) = \frac{\mathscr{R}^{-n}}{L}\sum_{\ell=0}^{L-1}\int_{\Gamma_e}\hat{U}_{ij}\left(\mathbf{x},\mathbf{y},\frac{\gamma\left(\mathscr{R}e^{i\ell\frac{2\pi}{L}}\right)}{\Delta t}\right)N_e^f(\mathbf{x})\,\mathrm{d}\Gamma\; e^{-in\ell\frac{2\pi}{L}}\,, \tag{32}$$

$$\omega_n^{ef}\left(\hat{P}_{ij},\mathbf{y},\Delta t\right) = \frac{\mathscr{R}^{-n}}{L}\sum_{\ell=0}^{L-1}\oint_{\Gamma_e}\hat{P}_{ij}\left(\mathbf{x},\mathbf{y},\frac{\gamma\left(\mathscr{R}e^{i\ell\frac{2\pi}{L}}\right)}{\Delta t}\right)N_e^f(\mathbf{x})\,\mathrm{d}\Gamma\; e^{-in\ell\frac{2\pi}{L}}\,. \tag{33}$$

Note that the calculation of the quadrature weights (32) and (33) is only based on the Laplace transformed fundamental solutions. Therefore, applying the elastic-viscoelastic correspondence principle to the fundamental solutions (18) and (19) leads to a visco-elastodynamic boundary element formulation in time domain. The calculation of the integration weights (32) and (33) is performed very fast with a technique similar to the Fast Fourier Transform (FFT).

4 Numerical Examples

The propagation of waves in a 3–d continuum has been calculated by the presented three viscoelastic boundary element formulations. The first part of the examples compare numerical results of the three methods. As well, the wave propagation in a 3-d rod is compared with the 1-d solution. In the second part, the wave propagation in an elastic foundation slab bonded to a viscoelastic half-space is calculated.

4.1 Comparison of the Methods

The problem geometry, material data and the associated boundary discretization of the 3-d rod are shown in Fig. 2. The rod is taken to be fixed on one end, and is excited by a pressure jump according to a unit step function $H(t)$ on the other free end. The remaining surfaces are traction free. Linear spatial shape functions on 72 triangles are used. Fig. 3 shows the longitudinal dis-

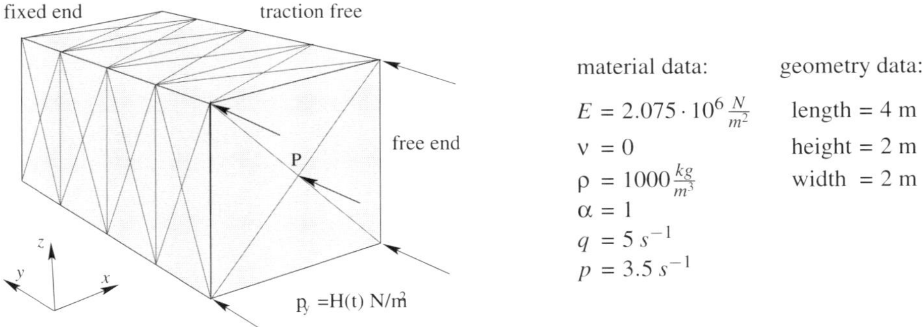

Figure 2. Step function excitation of a free–fixed rod

placement in the centre of the free end cross section (point P) versus time for all three presented formulations and the 1-d solution. The difference between the results, also compared to the 1-d

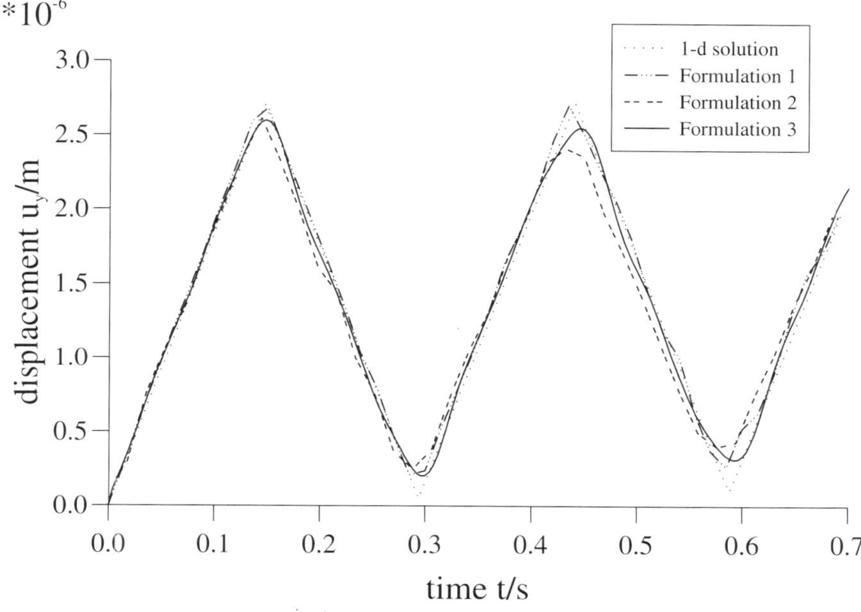

Figure 3. Longitudinal displacement of point P

solution, is quite small. The time step size Δt is chosen optimal for formulation 2 and 3, respectively. This means for formulation 2 the value $\beta = \frac{c_1 \Delta t}{r_e}$, with the characteristic element-length r_e, is in the range $0.6 < \beta < 1$ (see, Schanz, 1994) and for formulation 3 $\beta \approx 0.15$ (see, Schanz, 2001). However, formulation 3 is not as sensitive to an optimal choice of β as formulation 2. This means, if e.g. $\beta = 0.6$ is used in formulation 3 the results differ not much from the results with $\beta = 0.15$. Below critical values, $\beta = 0.6$ in formulation 2 and $\beta = 0.1$ in formulation 3, both formulations become unstable.

A proper parameter choice is also necessary for formulation 1. In this formulation the real part of the selected Laplace variable s must be tuned such that the solution behaves well. A bad choice of this parameter can lead to wrong results. Furthermore, this parameter is problem dependent, so that no general rule can be given how to choose this parameter.

From the viewpoint of CPU-time formulation 3 must be preferred, because it is nearly ten times faster than formulation 2. An comparison of CPU-time with the Laplace domain method is not general valid, because the amount of CPU-time depends linearly on the used amount of frequencies. However, this amount is closely related to the problem.

Summarizing the problems of all three formulations, the conclusion of a comparison is that formulation 3 is a very effective formulation and simple as well to treat viscoelastic problems.

Nevertheless, both other formulations can be used as well, but their parameters must be chosen very carefully.

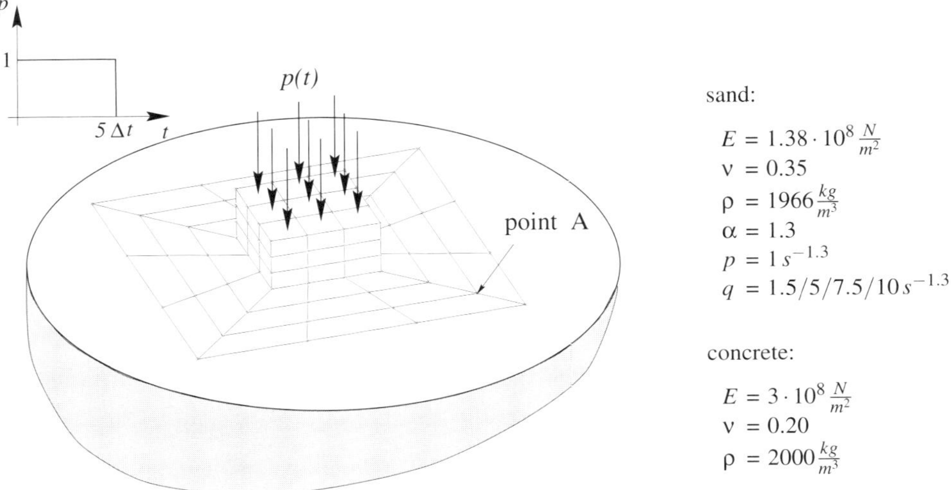

Figure 4. Elastic concrete slab on viscoelastic half-space: boundary element discretization, material data and loading function

4.2 Wave Propagation in an Unbounded Half-Space

The propagation of waves in an elastic concrete foundation slab ($2m \times 2m \times 1m$) bonded on a viscoelastic soil half-space has been calculated by formulation 2. Both domains are coupled by a substructure technique based on displacement- and traction-continuity at the interface. The assumption of welded contact does not allow the nonlinear effect of partial uplifting.

The problem geometry and the associated boundary discretization are shown in Fig. 4. The soil discretization is truncated after a distance of the foundation length. The surface of the foundation slab is excited by a positive and negative pressure jump according to Fig. 4. Linear spatial shape functions have been used. Similar to the Courant criterion the time step size Δt has been chosen close to the time needed by the viscoelastic compression wave to travel across the largest element.

In Fig. 5, vertical surface displacement at point A on the half-space surface is plotted versus time for different values of the constitutive parameter q. Obviously, the wave speeds increase for higher values of q, because of a stiffening of the material with growing influence of viscosity. This is associated with a significant displacement reduction.

The propagation of waves on the surface of the foundation and the half-space is presented in Fig. 6, where the magnitude of the displacement field at the nodes is shown. Clearly, the two wave fronts of the compression and the shear wave are observed. The wave front of the compression

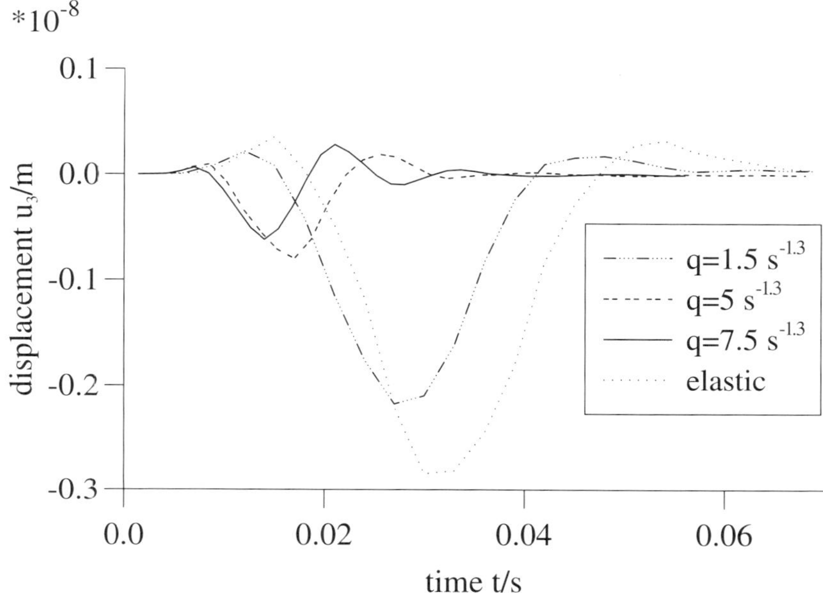

Figure 5. Displacement u_3 perpendicular to the surface on point A for different values q

wave can be observed on the transition from the foundation slab into the half-space. The wave front of the slower wave belongs to the shear wave.

References

Ahmad, S., and Manolis, G. (1987). Dynamic analysis of 3-D structures by a transformed boundary element method. *Computational Mechanics* 2:185–196.

Bagley, R., and Torvik, P. (1986). On the Fractional Calculus Model of Viscoelastic Behaviour. *Journal of Rheology* 30(1):133–155.

Beskos, D. (1987). Boundary element methods in dynamic analysis. *Applied Mechanics Review* 40(1):1–23.

Beskos, D. E. (1997). Boundary element methods in dynamic analysis: Part II (1986-1996). *Applied Mechanics Review* 50(3):149–197.

Carini, A., Diligenti, M., and Maier, G. (1991). Boundary integral equation analysis in linear viscoelasticity: Variational and saddle point formulations. *Computational Mechanics* 8:87–98.

Christensen, R. (1971). *Theory of Viscoelasticity*. New York: Academic Press.

Crump, K. S. (1976). Numerical Inversion of Laplace Transforms Using a Fourier Series Approximation. *Journal of the Association for Computing Machinery* 23(1):89–96.

Cruse, T., and Rizzo, F. (1968). A direct formulation and numerical solution of the general transient elastodynamic problem, I. *Journal of Mathematical Analysis and Applications* 22:244–259.

Durbin, F. (1974). Numerical inversion of Laplace transforms: an efficient improvement to Dubner and Abate's method. *The Computer Journal* 17(4):371–376.

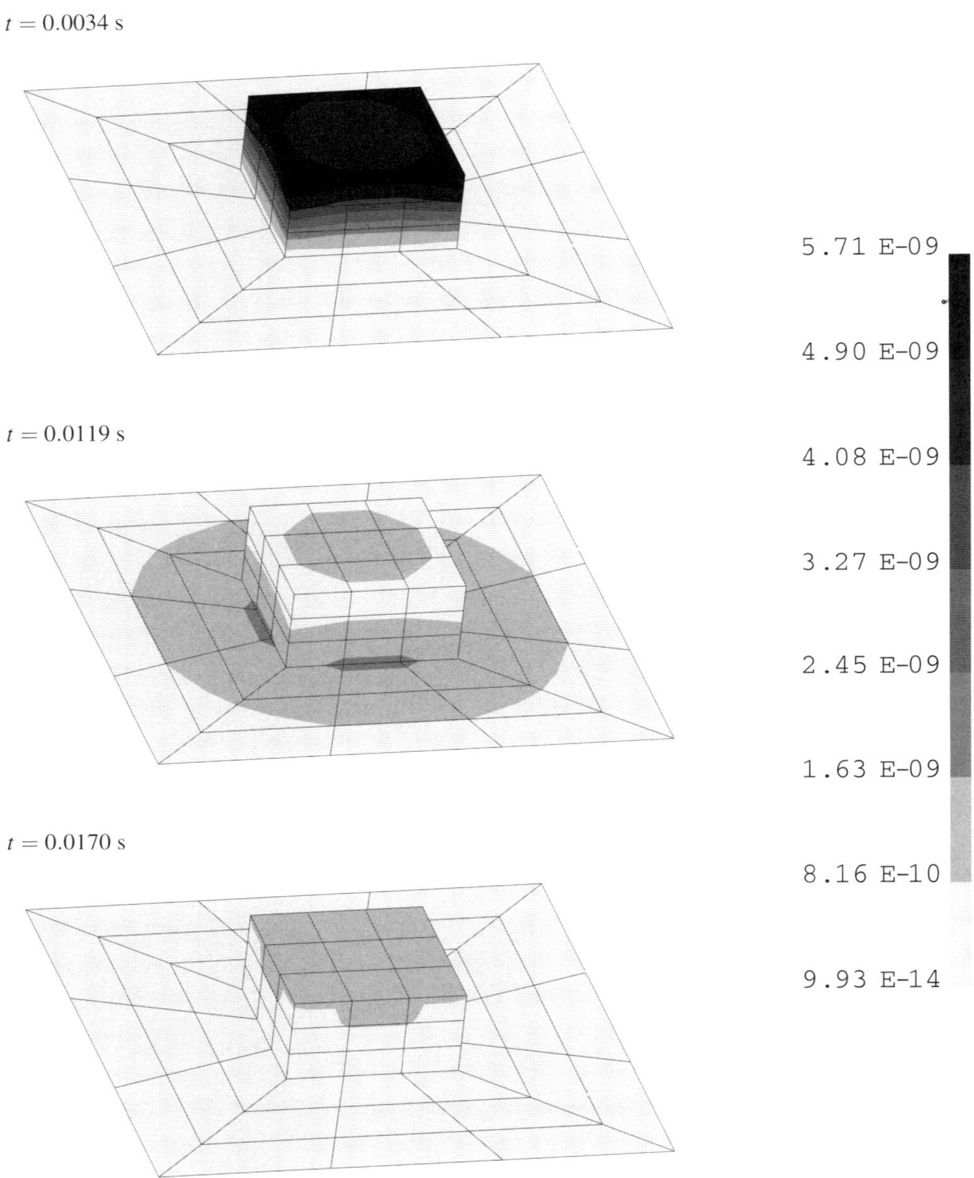

Figure 6. Displacement field of propagating waves on the surfaces of foundation slab and half-space

Flügge, W. (1975). *Viscoelasticity*. New York, Heidelberg, Berlin: Springer–Verlag.

Gaul, L., and Fiedler, C. (1997). *Boundary Element Method in Statics and Dynamics (in German)*. Braunschweig, Wiesbaden: Vieweg.

Gaul, L., and Schanz, M. (1992). BEM formulation in time domain for viscoelastic media based on analytical time integration. In Brebbia, C., Dominguez, J., and Paris, F., eds., *Boundary Elements XIV*, volume II, 223–234. Southampton: Computational Mechanics Publications.

Gaul, L., and Schanz, M. (1994a). Dynamics of viscoelastic solids treated by boundary element approaches in time domain. *European Journal of Mechanics A/Solids* 13(4–suppl.):43–59.

Gaul, L., and Schanz, M. (1994b). A viscoelastic boundary element formulation in time domain. *Archives of Mechanics* 46(4):583–594.

Gaul, L., and Schanz, M. (1999). A comparative study of three boundary element approaches to calculate the transient response of viscoelastic solids with unbounded domains. *Computer Methods in Applied Mechanics and Engineering* 179(1-2):111–123.

Gaul, L., Klein, P., and Plenge, M. (1988). Dynamic boundary element analysis of foundation slabs on layered soil. In Brebbia, C., ed., *Boundary Elements X*, volume 4, 29–44. Southampton, Boston: Computational Mechanics Publications.

Gaul, L., Klein, P., and Kempfle, S. (1991). Damping Description Involving Fractional Operators. *Mechanical Systems and Signal Processing* 5(2):81–88.

Graff, D. (1954). Über den Reziprozitätssatz in der Dynamik elastischer Körper. *Ingenieur Archiv* 22:45–46.

Grünwald, A. K. (1867). Ueber "begrenzte" Derivationen und deren Anwendung. *Zeitschrift für Mathematik und Physik* 12:441–480.

Guiggiani, M., and Gigante, A. (1990). A general algorithm for multidimensional cauchy principal value integrals in the boundary element method. *Journal of Applied Mechanics, ASME* 57:906–915.

Kobayashi, S., and Kawakami, T. (1985). Application of BE-FE combined method to analysis of dynamic interactions between structures and viscoelastic soil. In Brebbia, C., and Maier, G., eds., *Boundary Elements VII*, volume I, 6–3 – 6–12. Berlin, Heidelberg: Springer.

Lubich, C. (1988a). Convolution quadrature and discretized operational calculus. I. *Numerische Mathematik* 52:129–145.

Lubich, C. (1988b). Convolution quadrature and discretized operational calculus. II. *Numerische Mathematik* 52:413–425.

Manolis, G., and Beskos, D. (1981). Dynamic stress concentration studies by boundary integrals and Laplace transform. *International Journal for Numerical Methods in Engineering* 17:573–599.

Narayanan, G., and Beskos, D. (1982). Numerical pperational methods for time-dependent linear problems. *International Journal for Numerical Methods in Engineering* 18:1829–1854.

Oldham, K. B., and Spanier, J. (1974). *The Fractional Calculus*, volume 111 of *Mathematics in Science and Engineering*. New York, London: Academic Press.

Rossikhin, Y., and Shitikova, M. (1997). Applications of fractional calculus to dynamic problems of linear and nonlinear hereditary mechanics of solids. *Applied Mechanics Review* 50(1):15–67.

Schanz, M., and Antes, H. (1997a). Application of 'operational quadrature methods' in time domain boundary element methods. *Meccanica* 32(3):179–186.

Schanz, M., and Antes, H. (1997b). A new visco- and elastodynamic time domain boundary element formulation. *Computational Mechanics* 20(5):452–459.

Schanz, M., and Gaul, L. (1993). Implementation of viscoelastic behaviour in a time domain boundary element formulation. *Applied Mechanics Review* 46(11, Part 2):S41–S46.

Schanz, M. (1994). Eine Randelementformulierung im Zeitbereich mit verallgemeinerten viskoelastischen Stoffgesetzen. Bericht aus dem Institut A für Mechanik Heft 1, Universität Stuttgart.

Schanz, M. (1999). A boundary element formulation in time domain for viscoelastic solids. *Communications in Numerical Methods in Engineering* 15:799–809.

Schanz, M. (2001). *Wave Propagation in Viscoelastic and Poroelastic Continua: A Boundary Element Approach*. Lecture Notes in Applied Mechanics. Berlin, Heidelberg, New York: Springer-Verlag.

Sim, W., and Kwak, B. (1988). Linear viscoelastic analysis in time domain by boundary element method. *Computers & Structures* 29(4):531–539.

Talbot, A. (1979). The Accurate Numerical Inversion of Laplace Transforms. *Journal of the Institute of Mathematics and its Applications* 23:97–120.

Tanaka, M. (1987). *New Integral Equation Approach to Viscoelastic Problems*, volume 3: Computational Aspects of *Topics in Boundary Element Research*. Berlin, Heidelberg: Springer. chapter 2, 25–35.

Wolf, J., and Dabre, G. (1986). Time-domain boundary element method in viscoelasticity with application to a spherical cavity. *Soil Dynamics and Earthquake Engineering* 5:138–148.

CHAPTER 10

A Hybrid Boundary Element Approach without Singular Boundary Integrals

Lothar Gaul and Friedrich Moser *

Institute A of Mechanics, University of Stuttgart, Germany

Abstract. A boundary element formulation for 3D-elastostatics and 3D-elastodynamics is presented which avoids singular boundary integrals. The proposed method is based on a generalized variational principle. A weighted superposition of static fundamental solutions is used for the field approximation in the domain, whereas the displacement and stress field on the boundary are interpolated by well-known polynomial shape functions. By separating time- and space-dependence a symmetric equation of motion is derived with time-independent mass and stiffness matrix. The domain integral over inertia terms, leading to the mass matrix, is analytically transformed to the boundary. Thus, a boundary only formulation is derived. Comparing numerical results with analytical solutions clearly shows that the obtained system of equations is well-suited for dynamic problems.

1 Introduction

The combination of different discretization methods leads to reliable numerical solutions even for challenging tasks in engineering, because the advantages of the involved methods are combined. The finite element method (FEM) is widely used since powerful user-friendly computer codes are available. In order to efficiently couple the symmetric FEM with the boundary element method (BEM) a symmetric BEM formulation is needed – the hybrid displacement boundary element method (HDBEM) fulfills this requirement. In elastostatics as well as in elastodynamics the method is based on generalized variational principles (DeFigueiredo and Brebbia, 1989; Gaul and Fiedler, 1996). The independent field variables are the displacements in the domain as well as the displacements and the tractions on the boundary. The domain field is approximated by a weighted superposition of fundamental solutions, where the weighting factors are unknowns. The displacement and traction field on the boundary are interpolated by shape functions multiplied by nodal data. Using these approximations a symmetric equation system for unknown boundary displacements is obtained by integration over the discretized boundary. Besides regular, weakly and strongly singular integrals also hypersingular integrals have to be numerically evaluated. Latter integrals appear when the load point of the weakly singular displacement fundamental solution coincides with the load point of the strongly singular traction fundamental solution. These integrals are indirectly calculated by considering a stress-free rigid body motion (DeFigueiredo and

* Support by the Deutsche Forschungsgemeinschaft DFG of the Graduate Collegium 'Modelling and discretization methods for continua and fluids' at the University of Stuttgart is gratefully acknowledged.

Brebbia, 1989; Dumont, 1999), similar to the rigid body motion technique used in collocation BEM. This technique leads to an overdetermined system of equations for the unknown elements of main diagonal submatrices. The overdetermined system is solved using a least squares procedure. Thus, the rigid body motion is stress-free only in a least squares sense. In addition, for a general load case the numerical results obtained by this rigid body motion technique are not satisfying (Moser, 2001). Therefore, a non-singular formulation is developed. The collocation of discretization nodes and load points is circumvented by moving the load points away from the boundary outside of the domain, similar to the approach presented in (Davi, 1992). All boundary integrals become regular and therefore can be evaluated directly by a standard quadrature formula.

Based on the non-singular concept an equation of motion is derived in section 2 for linear elastodynamics. Due to time-invariant system matrices this equation can be used for forced and transient vibration analyses as well as for eigenvalue problems. The calculation of the mass matrix is discussed in detail in section 3 since the domain integral over inertia effects is analytically transformed to the boundary. Numerical examples validate the proposed non-singular HDBEM formulation in section 4.

2 Non-Singular HDBEM Formulation for Elastodynamics

For a linear-elastic continuum Hamilton's principle requires stationarity of the functional (Washizu, 1975)

$$\Pi_{\mathrm{H}}(u_i) = \int_{t_1}^{t_2} \left[\int_{\Omega} \left(-\frac{1}{2}\rho\,\dot{u}_i\,\dot{u}_i + \frac{1}{2}\sigma_{ij}\,\varepsilon_{ij} - b_i\,u_i\,\mathrm{d}\Omega \right) \mathrm{d}\Omega - \int_{\Gamma_t} \bar{t}_i\,u_i\,\mathrm{d}\Gamma \right] \mathrm{d}t \stackrel{!}{=} \text{stat.} \tag{1}$$

with the displacement boundary condition $u_i = \bar{u}_i$ on Γ_u and initial conditions. On Γ_t traction boundary conditions $\bar{t}_i$ are prescribed. For the boundary $\Gamma = \Gamma_u \cup \Gamma_t$ holds. The functional contains the kinetic energy of the domain Ω with mass density ρ, the strain energy and the work of body forces b_i and tractions $\bar{t}_i$ on Γ_t. The strain tensor ε_{ij} and the stress tensor σ_{ij} are both functions of the displacement vector u_i. Therefore, Hamilton's principle is a single-field functional of the independent variable u_i. This principle is generalized by decoupling the domain displacements u_i and the displacements $\tilde{u}_i$ on the boundary Γ. Adding $\int_\Gamma \lambda_i(\tilde{u}_i - u_i)\,\mathrm{d}\Gamma = 0$ to (1), compatibility of the independent variables u_i and $\tilde{u}_i$ is enforced in a weak sense by Lagrange multipliers λ_i. Stationarity of this multi-field functional leads to Euler equations for a vanishing first variation with independent variables u_i, $\tilde{u}_i$, λ_i. The Lagrange multiplier λ_i can be physically interpreted as the traction vector $\tilde{t}_i$ on the boundary, leading to a 3-field principle (Gaul and Fiedler, 1996)

$$\Pi_{\mathrm{HD}}(u_i, \tilde{u}_i, \tilde{t}_i) = \int_{t_1}^{t_2} \left[\int_{\Omega} \left(-\frac{1}{2}\rho\,\dot{u}_i\,\dot{u}_i + \frac{1}{2}\sigma_{ij}\,\varepsilon_{ij} - b_i\,u_i \right) \mathrm{d}\Omega \right.$$
$$\left. - \int_{\Gamma_t} \bar{t}_i\,u_i\,\mathrm{d}\Gamma + \int_{\Gamma} \tilde{t}_i \left(\tilde{u}_i - u_i \right) \mathrm{d}\Gamma \right] \mathrm{d}t \stackrel{!}{=} \text{stat.}\,. \tag{2}$$

Stationarity requires that the first variation vanishes, $\delta\Pi_{\mathrm{HD}} = 0$. The first variation is reformulated by neglecting body forces b_i, by integration by parts of the inertia term with respect to time $(\delta u_i(t_1) = \delta u_i(t_2) = 0)$ and by applying the divergence theorem to the strain energy term (Gaul and Fiedler, 1996)

$$\delta\Pi_{\mathrm{HD}} = \int_{t_1}^{t_2} \left[+ \int_{\Omega} \left\{ \rho\, \ddot{u}_i - \sigma_{ij,j} \right\} \delta u_i \, \mathrm{d}\Omega + \int_{\Gamma} \left\{ \underbrace{\sigma_{ij}\, n_j}_{t_i} - \tilde{t}_i \right\} \delta u_i \, \mathrm{d}\Gamma \right.$$
$$\left. + \int_{\Gamma} \left\{ \tilde{u}_i - u_i \right\} \delta \tilde{t}_i \, \mathrm{d}\Gamma + \int_{\Gamma_t} \left\{ \tilde{t}_i - \bar{t}_i \right\} \delta \tilde{u}_i \, \mathrm{d}\Gamma \right] \mathrm{d}t = 0\,. \quad (3)$$

This is the basic equation for the non-singular HDBEM formulation.

For the field approximation in the domain fundamental solutions are used. But contrary to regular BEM formulations the load points of fundamental solutions are not collocated with the nodes of boundary discretization. Instead, the load points are moved outside the domain. This approach leads to several advantages: no domain modification has to be used in order to completely include or exclude the singular points. Thus, the geometry-dependent evaluation of so-called integral free terms is avoided. Even though singular fundamental solutions are applied, the integration over the discretized boundary only deals with regular integral kernels. This is why the presented method is called non-singular. The generation of the load points is based on the boundary discretization. For each discretization node one load point is generated. In (Moser, 2001) it is shown that the load point placement according to

$$\xi_i = x_i + \eta\, D\, d'_i \quad (4)$$

gives very good results, where ξ_i is the coordinate vector of the load point related to the node x_i. η is a global scaling factor, D is the diameter of the largest element adjacent to the particular node and d'_i is a normalized vector giving the direction into which the load point is moved with respect to the particular node

$$d'_i = \frac{d_i}{|d_i|}\,, \qquad \text{with} \qquad d_i = \sum_{e=1}^{E} n_i^e\,. \quad (5)$$

n_i^e is the normal vector of the element e evaluated at the node of interest. E is the number of local elements adjacent to the selected node. Further discussion on the load point generation can be found in (Moser, 2001), where it is also shown, that the scaling factor η should be chosen from the interval $1.5 \leq \eta \leq 2.5$ in order to get optimal numerical results.

For both, boundary and domain variables, the space-dependence is separated from the time-dependence. The following approach is already successfully used in acoustics (Gaul and Wenzel, 1998). The domain field is approximated by superposition of static fundamental solutions

$$u_i(x_i, t) = \boldsymbol{u}^{*\mathrm{T}}(x_i)\, \boldsymbol{\gamma}(t)\,, \quad (6)$$
$$t_i(x_i, t) = \boldsymbol{t}^{*\mathrm{T}}(x_i)\, \boldsymbol{\gamma}(t)\,, \quad (7)$$

where x_i is a domain point, $\boldsymbol{u}^{*\mathrm{T}}$ and $\boldsymbol{t}^{*\mathrm{T}}$ are fundamental solution matrices containing the well-known Kelvin fundamental solutions, weighted by time-dependent fictitious loads $\boldsymbol{\gamma}(t)$. Differentiation of (6) with respect to time gives

$$\ddot{u}_i(x_i,t) = \boldsymbol{u}^{*\mathrm{T}}(x_i)\,\ddot{\boldsymbol{\gamma}}(t)\ . \tag{8}$$

The fundamental solutions do not change in a variational sense. Therefore, the first variation of (6) is given by

$$\delta u_i(x_i,t) = \boldsymbol{u}^{*\mathrm{T}}(x_i)\,\delta\boldsymbol{\gamma}(t)\ . \tag{9}$$

On the boundary space-dependent polynomial shape-functions Φ are multiplied by time-dependent nodal data. Thus, the boundary field approximation is formally equal to the domain approximation

$$\tilde{u}_i(x_i,t) = \boldsymbol{\Phi}^{\mathrm{T}}(x_i)\,\tilde{\boldsymbol{u}}(t)\ , \tag{10}$$

$$\tilde{t}_i(x_i,t) = \boldsymbol{\Phi}^{\mathrm{T}}(x_i)\,\tilde{\boldsymbol{t}}(t)\ . \tag{11}$$

The first variations are

$$\delta\tilde{u}_i(x_i,t) = \boldsymbol{\Phi}^{\mathrm{T}}(x_i)\,\delta\tilde{\boldsymbol{u}}(t)\ , \tag{12}$$

$$\delta\tilde{t}_i(x_i,t) = \boldsymbol{\Phi}^{\mathrm{T}}(x_i)\,\delta\tilde{\boldsymbol{t}}(t)\ . \tag{13}$$

Traction boundary conditions $\bar{t}_i$ are approximated similar to (11). Substituting (6)–(13) in (3), the space-independent variables $\tilde{\boldsymbol{u}}(t)$, $\tilde{\boldsymbol{t}}(t)$ and $\boldsymbol{\gamma}(t)$ can be extracted from the spatial integrals. Using static fundamental solutions and positioning the load points outside the domain, the domain integral over the stress divergence $\sigma_{ij,j}$ vanishes, thus leading to

$$\begin{aligned}\delta\Pi_{\mathrm{HD}} = \int_{t_1}^{t_2} \Bigg[&\left(\ddot{\boldsymbol{\gamma}}^{\mathrm{T}} \int_{\Omega} \rho\,\boldsymbol{u}^*\,\boldsymbol{u}^{*\mathrm{T}}\,\mathrm{d}\Omega + \boldsymbol{\gamma}^{\mathrm{T}} \int_{\Gamma} \boldsymbol{t}^*\,\boldsymbol{u}^{*\mathrm{T}}\,\mathrm{d}\Gamma - \tilde{\boldsymbol{t}}^{\mathrm{T}} \int_{\Gamma} \boldsymbol{\Phi}\,\boldsymbol{u}^*\,\mathrm{d}\Gamma \right) \delta\boldsymbol{\gamma} \\ &+ \left(\tilde{\boldsymbol{t}}^{\mathrm{T}} \int_{\Gamma_t} \boldsymbol{\Phi}\boldsymbol{\Phi}^{\mathrm{T}}\,\mathrm{d}\Gamma - \bar{\boldsymbol{t}}^{\mathrm{T}} \int_{\Gamma_t} \boldsymbol{\Phi}\boldsymbol{\Phi}^{\mathrm{T}}\,\mathrm{d}\Gamma \right) \delta\tilde{\boldsymbol{u}} \\ &+ \left(\tilde{\boldsymbol{u}}^{\mathrm{T}} \int_{\Gamma} \boldsymbol{\Phi}\boldsymbol{\Phi}^{\mathrm{T}}\,\mathrm{d}\Gamma - \boldsymbol{\gamma}^{\mathrm{T}} \int_{\Gamma} \boldsymbol{u}^*\,\boldsymbol{\Phi}^{\mathrm{T}}\,\mathrm{d}\Gamma \right) \delta\tilde{\boldsymbol{t}} \Bigg] \mathrm{d}t = 0\ . \end{aligned} \tag{14}$$

The definition of system matrices used in static HDBEM analyses (DeFigueiredo and Brebbia, 1989) is extended to elastodynamics

$$\begin{aligned} \boldsymbol{N} &= \int_{\Omega} \rho\,\boldsymbol{u}^*\,\boldsymbol{u}^{*\mathrm{T}}\,\mathrm{d}\Omega\ , & \boldsymbol{F} &= \int_{\Gamma} \boldsymbol{t}^*\,\boldsymbol{u}^{*\mathrm{T}}\,\mathrm{d}\Gamma\ , \\ \boldsymbol{G} &= \int_{\Gamma} \boldsymbol{u}^*\,\boldsymbol{\Phi}^{\mathrm{T}}\,\mathrm{d}\Gamma\ , & \boldsymbol{L} &= \int_{\Gamma} \boldsymbol{\Phi}\boldsymbol{\Phi}^{\mathrm{T}}\,\mathrm{d}\Gamma\ . \end{aligned} \tag{15}$$

Using (15) the vanishing first variation of (14) leads to three Euler equations

$$\ddot{\gamma}^{\mathrm{T}}\boldsymbol{N} + \gamma^{\mathrm{T}}\boldsymbol{F} - \tilde{\boldsymbol{t}}^{\mathrm{T}}\boldsymbol{G}^{\mathrm{T}} = \boldsymbol{0}\,, \tag{16}$$

$$\gamma^{\mathrm{T}}\boldsymbol{G} - \tilde{\boldsymbol{u}}^{\mathrm{T}}\boldsymbol{L} = \boldsymbol{0}\,, \tag{17}$$

$$\bar{\boldsymbol{t}}^{\mathrm{T}}\boldsymbol{L} - \tilde{\boldsymbol{t}}^{\mathrm{T}}\boldsymbol{L} = \boldsymbol{0}\,. \tag{18}$$

Contrary to singular BEM formulations the calculation of the system matrices according to (15) requires no limiting process since all integral kernels are regular. The matrices $\boldsymbol{N}$, $\boldsymbol{F}$ and $\boldsymbol{L}$ are symmetric. Using (18) the vector of equivalent nodal forces is obtained

$$\boldsymbol{q} = \boldsymbol{L}^{\mathrm{T}}\bar{\boldsymbol{t}} = \boldsymbol{L}^{\mathrm{T}}\tilde{\boldsymbol{t}}\,. \tag{19}$$

(17) is solved for the fictitious loads. Differentiation with respect to time gives also

$$\gamma = \left(\boldsymbol{G}^{\mathrm{T}}\right)^{-1}\boldsymbol{L}\tilde{\boldsymbol{u}} = \boldsymbol{R}\tilde{\boldsymbol{u}} \qquad \Longrightarrow \qquad \ddot{\gamma} = \boldsymbol{R}\ddot{\tilde{\boldsymbol{u}}}\,, \tag{20}$$

with the definition $\boldsymbol{R}{=}\left(\boldsymbol{G}^{\mathrm{T}}\right)^{-1}\boldsymbol{L}$. Solving (16) for boundary tractions $\tilde{\boldsymbol{t}}$ and substituting (19) and (20), leads to a symmetric equation of motion

$$\boldsymbol{M}\ddot{\tilde{\boldsymbol{u}}} + \boldsymbol{K}\,\tilde{\boldsymbol{u}} = \boldsymbol{q}\,, \qquad \text{with} \qquad \begin{aligned}\boldsymbol{M} &= \boldsymbol{R}^{\mathrm{T}}\boldsymbol{N}\boldsymbol{R}\,,\\ \boldsymbol{K} &= \boldsymbol{R}^{\mathrm{T}}\boldsymbol{F}\boldsymbol{R}\,.\end{aligned} \tag{21}$$

The mass and stiffness matrix are time-independent. Therefore, this equation system can be used for forced and transient vibration analyses as well as for solving eigenvalue problems. Due to symmetry properties, FEM solvers can be adopted. After solving (21) for boundary displacements $\tilde{\boldsymbol{u}}(t)$, the fictitious loads are calculated with (20). (6) and (7) lead to displacements and tractions within the domain. Besides symmetry this efficient field point evaluation without additional boundary integration is the most important advantage of HDBEM compared to CBEM.

3 Mass Matrix

Inertia terms are obviously domain loads and therefore, the mass matrix $\boldsymbol{M}$ requires a domain integration, which also can be seen in definition (15) of the kernel matrix $\boldsymbol{N}$. To keep a boundary only formulation a transformation to the boundary is needed. A general applicable method for this task is the dual reciprocity method (DRM) (Nardini and Brebbia, 1985; Partridge et al., 1992), which also can be successfully adapted to the HDBEM mass matrix (Moser, 2001). A more accurate approach is an analytical transformation to the boundary using a so-called higher order fundamental solution (Gaul and Fiedler, 1997; Gaul et al., 2000). This method is presented next.

For each load point combination (p, q) a submatrix $\boldsymbol{n}^{pq}$ is defined by

$$\boldsymbol{n}^{pq} = n_{ij}^{pq} = \int\limits_{\Omega} \rho\, u_{li}^{*p}\, u_{lj}^{*q}\; \mathrm{d}\Omega\,. \tag{22}$$

For the sake of clarity the load point indices p and q are omitted in the following discussion. The domain integral is analytically transformed to the boundary by using an auxiliary stress field $\sigma^{\mathrm{a}}_{lki}$ (higher order fundamental solution) with the property $\sigma^{\mathrm{a}}_{lki,k} \equiv \rho\, u^*_{li}$. The higher order fundamental solution $\sigma^{\mathrm{a}}_{lki}$ and the corresponding displacement field u^{a}_{ij} are given by (Moser, 2001)

$$u^{\mathrm{a}}_{ij} = \frac{-r\,\rho}{128\pi\,\mu^2\,(1-\nu)^2}\Big[(3-4\nu)\,r_{,i}\,r_{,j} - (13-28\nu+16\nu^2)\,\delta_{ij}\Big]\,, \tag{23}$$

$$\begin{aligned}\sigma^{\mathrm{a}}_{ijk} = \frac{\rho}{64\pi\,\mu\,(1-\nu)^2}\Big[&(3-4\nu)\,r_{,i}\,r_{,j}\,r_{,k} + (5-12\nu+8\nu^2)(r_{,j}\,\delta_{ik} + r_{,i}\,\delta_{jk})\\ &-(3-8\nu+8\nu^2)\,r_{,k}\,\delta_{ij}\Big]\,,\end{aligned} \tag{24}$$

$$\begin{aligned}t^{\mathrm{a}}_{ij} = \sigma^{\mathrm{a}}_{ikj}\,n_k = \frac{\rho}{64\pi\,\mu\,(1-\nu)^2}\Big[&(3-4\nu)\frac{\partial r}{\partial n}\,r_{,i}\,r_{,j} + (5-12\nu+8\nu^2)(\frac{\partial r}{\partial n}\,\delta_{ij} + r_{,i}\,n_j)\\ &-(3-8\nu+8\nu^2)\,r_{,j}\,n_i\Big]\,.\end{aligned} \tag{25}$$

Using $\sigma^{\mathrm{a}}_{lki,k} \equiv \rho\, u^*_{li}$, applying the differentiation rule $(a\,b)_{,i} = a_{,i}\,b + a\,b_{,i}$ and Gauss' theorem to the obtained first domain integral, leads to

$$\begin{aligned}n_{ij} &= \int_\Omega \sigma^{\mathrm{a}}_{lki,k}\,u^*_{lj}\,\mathrm{d}\Omega = \int_\Omega \left(\sigma^{\mathrm{a}}_{lki}\,u^*_{lj}\right)_{,k}\,\mathrm{d}\Omega - \int_\Omega \sigma^{\mathrm{a}}_{lki}\,u^*_{lj,k}\,\mathrm{d}\Omega\\ &= \int_\Gamma u^*_{lj}\,\sigma^{\mathrm{a}}_{lki}\,n_k\,\mathrm{d}\Gamma - \int_\Omega u^*_{lj,k}\,C_{lkmn}\,u^{\mathrm{a}}_{mi,n}\,\mathrm{d}\Omega\,.\end{aligned} \tag{26}$$

The stress tensor is replaced by $\sigma^{\mathrm{a}}_{lki} = C_{lkmn}\,u^{\mathrm{a}}_{mi,n}$, where C_{lkmn} is the symmetric elasticity tensor of a linear elastic material. Since the material properties of the auxiliary field are equal to those of the fundamental solution field, one obtains $\sigma^*_{mnj} = C_{lkmn}\,u^*_{lj,k}$. Therefore the remaining domain integral has a similar structure to the first integral in (26). Again, using the differentiation rule and Gauss' theorem leads to

$$n_{ij} = \int_\Gamma u^*_{lj}\,\sigma^{\mathrm{a}}_{lki}\,n_k\,\mathrm{d}\Gamma - \int_\Gamma u^{\mathrm{a}}_{mi}\,\sigma^*_{mnj}\,n_n\,\mathrm{d}\Gamma + \int_\Omega \sigma^*_{mnj,n}\,u^{\mathrm{a}}_{mi}\,\mathrm{d}\Omega\,. \tag{27}$$

The still remaining domain integral vanishes since the load points of the fundamental solution field are positioned outside the domain Ω. Using Cauchy's theorem, (27) is reformulated leading to two regular boundary integrals

$$n^{pq}_{ij} = \int_\Gamma u^{*p}_{lj}\,(t^{\mathrm{a}}_{li})^q\,\mathrm{d}\Gamma - \int_\Gamma (u^{\mathrm{a}}_{li})^q\,t^{*p}_{lj}\,\mathrm{d}\Gamma\,, \tag{28}$$

where the load point indices are included again. Thus, the presented transformation makes (21) a boundary only formulation.

4 Numerical Examples

To demonstrate the accuracy of the proposed method a static and two dynamic numerical examples are presented and compared with the corresponding analytical solutions.

First, an infinite plate with a circular hole under a static tensile load is investigated. In the static case (21) simplifies to $\boldsymbol{K}\,\tilde{\boldsymbol{u}} = \boldsymbol{q}$. Due to symmetry just a quarter of the plate has to be modeled, see Fig. 1 ($l_{x_1} = l_{x_2} = 10a,\ h = a$). Assuming plane stress conditions and using

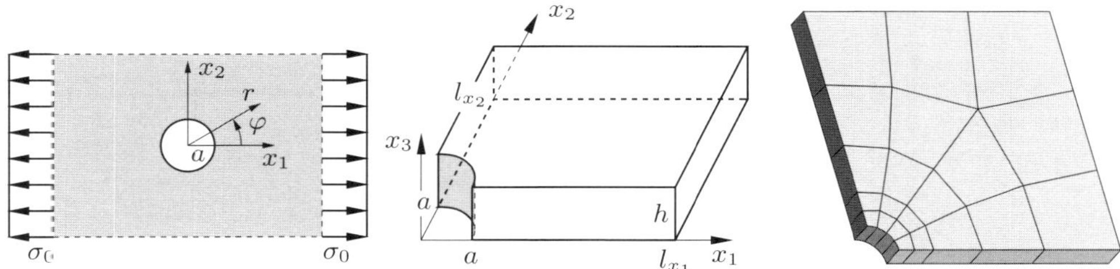

Figure 1. Plate with circular hole under tensile load: load case, coordinate system and 3D boundary discretization with quadratic shape functions.

Airy's stress function an analytical solution for radial and tangential stresses can be obtained

$$\sigma_r = \frac{\sigma_0}{2}\left[1 - \frac{a^2}{r^2} + \left(1 - 4\frac{a^2}{r^2} + 3\frac{a^4}{r^4}\right)\cos(2\varphi)\right] , \tag{29}$$

$$\sigma_\varphi = \frac{\sigma_0}{2}\left[1 + \frac{a^2}{r^2} + \left(1 + 3\frac{a^4}{r^4}\right)\cos(2\varphi)\right] . \tag{30}$$

This example is chosen because of the stress concentration at the hole. For three angles φ the normalized stresses are depicted in Fig. 2. For the two extrem angles (0°, 90°) the domain solution is evaluated on the boundary of the modeled domain. The influence of particular fundamental solutions is obvious. But the results are much better compared to a singular HDBEM formulation, where the singular fundamental solutions dominate the approximation close to the boundary. For other angles, e.g. $\varphi = 45^\circ$, the numerical results are almost identical with the analytical solution. Using singular fundamental solutions for the domain approximation explains the very good numerical results at stress concentrations, even though a relative coarse mesh is used.

For a beam with rectangular cross-section, see Fig. 3 ($h = 10$, $b = 5$, $l = 150$), and one fixed end as displacement boundary condition, an eigenvalue analysis is performed. The obtained eigenfrequencies are listed in Tab. 1, where also the relative errors of these frequencies are specified. The frequencies of the first few modes are very accurate. As expected, solutions for higher eigenmodes get worse due to the finite boundary discretization.

Next, a transient solution with the standard Newmark time-integration scheme (Bathe and Wilson, 1976) (time increment $\Delta t = 1\mu$s) is calculated with the above mentioned beam model. Again, one end of the model is fixed. On the free end a constant normal stress is applied, the time dependence is the Heaviside step function at $t = 0$. For this 1D-problem the analytical solution is known (Graff, 1975). In Fig. 4 (left) the nodal displacements $\tilde{u}_3$ are depicted as a function of time t and longitudinal coordinate x_3. The numerical results excellently coincide with the analytical solution. In an equivalent plot, the corresponding stresses σ_{33} within the domain are

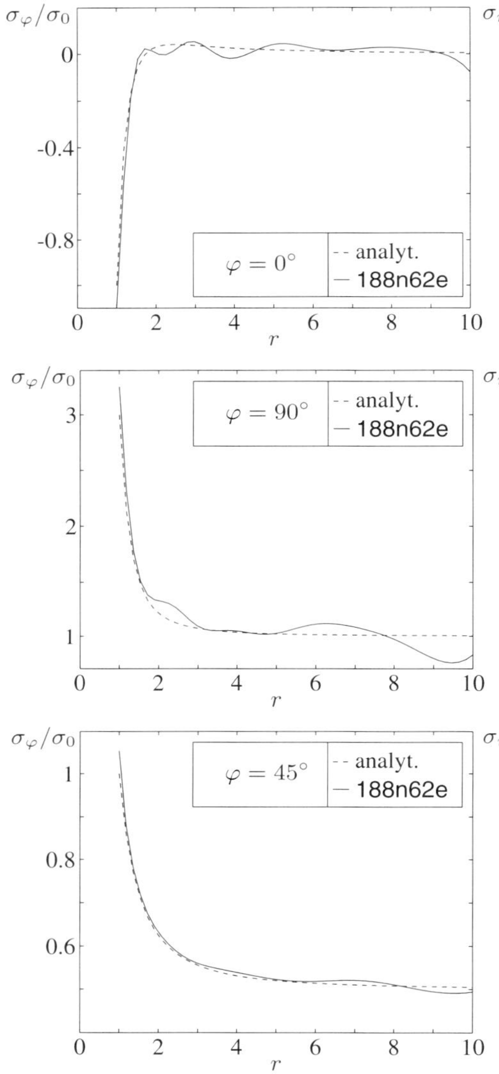

Figure 2. Domain solutions for radial and tangential stresses σ_r and σ_φ for a 3D HDBEM model with 62 quadratic elements and 188 nodes.

Figure 3. 3D HDBEM model of a beam with 62 quadratic elements and 188 nodes.

Table 1. Eigenfrequencies ω [kHz] und relative error [%].

order	1.	2.	3.	4.	5.	6.	7.
axial mode	52.829 (-0.1)	158.680 (0.1)	265.118 (0.3)	372.504 (0.7)	481.329 (1.2)	591.495 (1.7)	703.862 (2.4)
bending mode	1.167 (2.5)	7.106 (-0.4)	19.740 (-1.2)	38.295 (-2.2)	62.533 (-3.4)	92.088 (-4.7)	126.624 (-6.2)

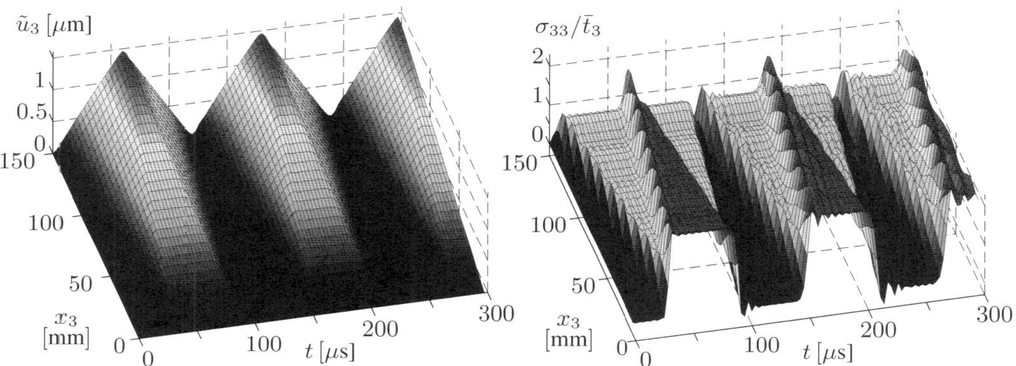

Figure 4. Transient solution: nodal displacements $\tilde{u}_3$ and internal stress σ_{33}.

illustrated in Fig. 4 (right). To calculate this internal solution no additional boundary integration is necessary, which is an advantage compared to the CBEM. In both diagrams the propagating nature of the stress wave can be clearly seen. Because of finite discretization the theoretically propagating stress wave (Heaviside function) can not be approximated exactly. Considering the used mesh the infinite time derivative of the stress function is reproduced well. In spite of the obtained overshooting the stress levels turn out satisfactory, see Fig. 5. In this diagram the numerical stress result at an internal point is compared to the analytical solution, which is developed as a series expansion with 40 members.

5 Conclusions

The paper presents a new symmetric non-singular boundary element formulation for elastodynamics based on a mixed variational principle. Earlier published HDBEM formulations discretization nodes and load points are collocated. This collocation is not mandatory. Therefore, the load points are moved outside the domain. Thus, singular integration is completely avoided. The mass matrix is calculated by an analytical transformation of the domain integral to the boundary. A time-invariant symmetric equation of motion is obtained. The applicability of the proposed method to elastodynamic problems is validated by different numerical examples.

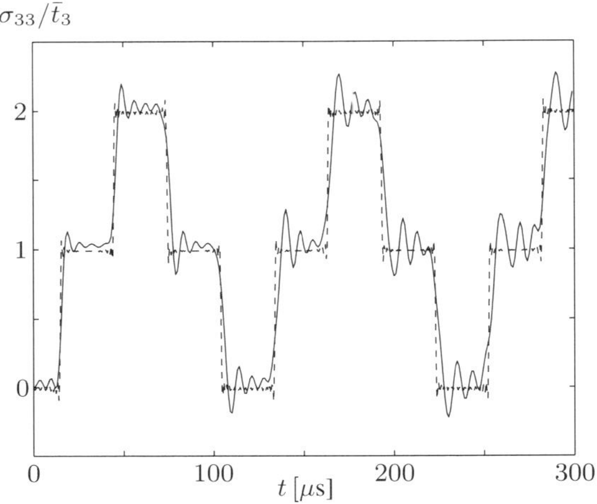

Figure 5. Transient internal stress solution σ_{33}.

References

Bathe, K. J., and Wilson, E. L. (1976). *Numerical Methods in Finite Element Analysis.* Englwood Cliffs: Prentice-Hall, Inc.

Davi, G. (1992). A hybrid displacement variational formulation of bem for elastostatics. *Engineering Analysis with Boundary Elements* 10:219–224.

DeFigueiredo, T. G. B., and Brebbia, C. A. (1989). A new hybrid displacement variational formulation of bem for elastostatics. volume 1, 47–57. Springer-Verlag.

Dumont, N. A. (1999). An assessment of the spectral properties of the matrices obtained in the boundary element methods. In *Proc. 15th Brazilian Congress of Mechanical Engineering.*

Gaul, L., and Fiedler, C. (1996). Fundamentals of the dynamic hybrid boundary element method. *Zeitschrift für angewandte Mathematik und Mechanik* 74(4):539–542.

Gaul, L., and Fiedler, C. (1997). *Methode der Randelemente in Statik und Dynamik.* Braunschweig: Vieweg Verlag.

Gaul, L., and Wenzel, W. (1998). A symmetric boundary element formulation for time-domain analyses of acoustic problems. In *Proc. of Euro Noise 98*, volume 1, 123–128.

Gaul, L., Wagner, M., and Wenzel, M. (2000). Hybrid boundary element methods in frequency and time domain. In von Estorff, O., ed., *Boundary Elements in Acoustics, Advances and Applications*, 121–164. Southampton: WIT Press.

Graff, K. F. (1975). *Wave Motion in Elastic Solids.* London: Oxford University Press.

Moser, F. (2001). Nicht-singuläre räumliche Randelementformulierung der Elastodynamik. Bericht aus dem Institut A für Mechanik (to be published), Universität Stuttgart.

Nardini, D., and Brebbia, C. A. (1985). Boundary element formulations of mass matrices for dynamic analysis. *Topics in boundary element research* Vol. 2: Time-dependent and vibration problems:191–208.

Partridge, P. W., Brebbia, C. A., and Wrobel, L. C. (1992). *The Dual Reciprocity Boundary Element Method.* Southampton: Computational Mechanics Publications.

Washizu, K. (1975). *Variational methods in elasticity and plasticity.* 2. Aufl. Oxford: Pergamon Press Ldt.

Chapter 11
The Boundary Contour Method

Subrata Mukherjee

Department of Theoretical and Applied Mechanics, Kimball Hall,
Cornell University, Ithaca, NY 14853, USA
e-mail: sm85@cornell.edu

Abstract

This chapter presents applications of the boundary contour method (BCM) in three dimensional (3-D) linear elasticity. Following a brief introduction, BCM and hypersingular BCM (HBCM) formulations, for this class of problems, are presented in Section 2. This is followed by sections describing shape sensitivity analysis, shape optimization and error analysis and adaptivity with the BCM. Numerical results for selected examples are included throughout the chapter.

1 Introduction

A brief introduction to the BCM, HBCM, and their applications in 3-D linear elasticity, appears in this section.

1.1 The Boundary Contour Method

The usual boundary element method (BEM), for three-dimensional (3-D) linear elasticity, requires numerical evaluations of surface integrals on boundary elements on the surface of a body (see, for example, [Mukh1982]). [NagaLM94] (for 2-D linear elasticity) and [NagaML96] (for 3-D linear elasticity) have recently proposed a novel approach, called the boundary contour method (BCM), that achieves a further reduction in dimension! The BCM, for 3-D linear elasticity problems, only requires numerical evaluation of line integrals over the closed bounding contours of the usual (surface) boundary elements.

The central idea of the BCM is the exploitation of the divergence-free property of the usual BEM integrand and a very useful application of Stokes' theorem, to analytically convert surface integrals on boundary elements to line integrals on closed contours that bound these elements. [Lutz91] first proposed an application of this idea for the Laplace equation and Nagarajan et al. generalized this idea to linear elasticity. Numerical results for two-dimensional (2-D) problems, with linear boundary elements, are presented in [NagaLM94], while results with quadratic boundary elements appear in [PhanMM97]. Three-dimensional elasticity problems, with quadratic boundary elements, is the subject of [NagaML96] and [MukhMSN97]. Hypersingular boundary contour formulations, for two-dimensional [PhanMM98b] and three-dimensional [MukhM98] linear elasticity, have been proposed recently. A symmetric Galerkin BCM for 2-D linear elasticity appears in [NovaS99].

1.2 The Hypersingular Boundary Element Method

Hypersingular boundary element equations (HBIEs) are derived from a differentiated version of the usual boundary integral equations (BIEs). HBIEs have diverse important applications and are the subject of considerable current research (see, for example, [KrisRR92], [TanaSS94], [Paul95] and [ChenH99] for recent surveys of the field). HBIEs, for example, have been employed for the evaluation of boundary stresses (e.g. [Guig94], [WildA98], [ZhaoL99], [ChatM99]), in wave scattering (e.g. [KrisSRR90]), in fracture mechanics (e.g. [GrayMI90], [LutzIG92], [Paul95], [GrayP98], [Mukh01]), in order to obtain symmetric Galerkin boundary element formulations (e.g. [Bonn95], [GrayBK95], [GrayP97a], [GrayP97b]), to obtain the hypersingular boundary contour method (HBCM) [PhanMM98b], [MukhM98]), to obtain the hypersingular boundary node method (HBNM) ([ChatMP01a]), and for error analysis ([PaulGZ96], [Meno96], [MenoPM99], [PaulMM01] [ChatPM01b]) and adaptivity [ChatPM01b]. The relationship between finite parts of strongly singular and hypersingular integrals, and the HBIE, is discussed in [TohM94], [Mukh00a] and [Mukh00b]. A lively debate (e.g. [MartR96], [CrusR96]), on smoothness requirements on boundary variables for collocating an HBIE on the boundary of a body, has apparently been concluded recently [MartRC98].

1.3 Shape Sensitivity Analysis with the BCM and the HBCM

Design sensitivity coefficients (DSCs), which are defined as rates of change of physical response quantities with respect to changes in design variables, are useful for various applications such as in judging the robustness of a given design, in reliability analysis and in solving inverse and design optimization problems. There are three methods for design sensitivity analysis (e.g. [HaugCK86]), namely, the finite difference approach (FDA), the adjoint structure approach (ASA) and the direct differentiation approach (DDA). The DDA is of interest in this work.

The goal of obtaining BCM sensitivity equations can be achieved in two equivalent ways. In the 2-D work by [PhanMM98a], design sensitivities are obtained by first converting the discretized BIEs into their boundary contour version, and then applying the DDA (using the concept of the material derivative) to this BCM version. This approach, while relatively straightforward in principle, becomes extremely algebraically intensive for 3-D elasticity problems. [MukhSM99] offers a novel alternative derivation, using the opposite process, in which the DDA is first applied to the regularized BIE and then the resulting equations are converted to their boundary contour version. It is important to point out that this process of converting the sensitivity BIE into a BCM form is quite challenging. This new derivation, for sensitivities of surface variables [MukhSM99], as well as for internal variables [MukhSM00], for 3-D elasticity problems, is presented in this chapter. The reader is referred to [PhanM99] for a corresponding derivation for 2-D elasticity

1.4 Shape Optimization with the BCM

Shape optimization refers to the optimal design of the shape of structural components and is of great importance in mechanical engineering design. A typical gradient based shape optimization procedure is an iterative process in which iterative improvements are carried out over successive designs until an optimal design is accepted . A domain based method such as the finite element method (FEM) typically requires discretization of the entire domain of a body many times during this iterative process. The BEM, however, only requires surface discretization, so that mesh generation and remeshing procedures can be carried out much more easily for the BEM than for the FEM. Also, surface stresses are typically obtained very accurately in the BEM. As a result, the BEM has been a popular method for shape optimization in linear mechanics. Some examples are references [ChoiK88], [SandW88], [Yang90], [SaigK90], [WeiCLM94], [YamaSK94], [TafF95] and the book [Zhao91].

In addition to having the same meshing advantages as the usual BEM, the BCM, as explained above, offers a further reduction in dimension. Also, surface stresses can be obtained very easily and accurately by the BCM without the need for additional shape function differentiation as is commonly required with the BEM. These properties make the BCM very attractive as the computational engine for stress analysis for use in shape optimization. Shape optimization in 2-D linear elasticity, with the BCM, has been presented by [PhanMM98c]. The corresponding 3-D problem is presented in [ShiM99] and is discussed in this chapter.

1.5 Error Analysis and Adaptivity with the BCM and the HBCM

[Paul95] and [PaulGZ96] first proposed the idea of obtaining a hypersingular residual by substituting the BEM solution of a problem into the hypersingular BEM (HBEM) for the same problem; and then using this residual as an element error estimator in the BEM. It has been proved that ([Meno96], [MenoPM99], [PaulMM01]), under certain conditions, this residual is related to a measure of the local error on a boundary element, and has been used to postulate local error estimates on that element. This idea has been applied to the collocation BEM ([PaulGZ96], [MenoPM99], [PaulMM01]) and to the symmetric Galerkin BEM ([PaulG99]). Very recently, residuals have been obtained in the context of the boundary node method (BNM) [ChatPM01b] and used to obtain local error estimates (at the element level) and then to drive an h-adaptive mesh refinement process. An analogous approach for error estimation and h-adaptivity, in the context of the BCM, is described in [MukhM01] and is described in this chapter.

2 The BCM and the HBCM

Derivations of the BCM and HBCM, for 3-D linear elasticity, together with representative numerical results for selected problems, are presented in this section.

2.1 Boundary Surface and Boundary Contour Integral Equations

A regularized form of the standard boundary integral equation [Rizzo67], for 3-D linear elasticity, can be written as :

$$0 = \int_{\partial B} [\, U_{ik}(\mathbf{x},\mathbf{y})\, \sigma_{ij}(\mathbf{y}) \; - \Sigma_{ijk}(\mathbf{x},\mathbf{y})\, \{u_i(\mathbf{y}) \; - u_i(\mathbf{x})\, \}\,]\, \mathbf{e}_j \cdot d\mathbf{S}(\mathbf{y}) \; \equiv \int_{\partial B} \mathbf{F}_k \cdot d\mathbf{S}(\mathbf{y}) \tag{1}$$

Here, ∂B is the bounding surface of a body B (B is an open set) with infinitesimal surface area $d\mathbf{S} = dS\mathbf{n}$, where $\mathbf{n}$ is the unit outward normal to ∂B at a point on it. The stress tensor is $\boldsymbol{\sigma}$, the displacement vector is $\mathbf{u}$ and $\mathbf{e}_j (j = 1, 2, 3)$ are global Cartesian unit vectors. The BEM Kelvin kernels are written in terms of (boundary) source and field points. These are :

$$\Sigma_{ijk} = -\frac{1}{8\pi(1-\nu)r^2} [\, (1-2\nu)(r_{,i}\delta_{jk} + r_{,j}\delta_{ik} - r_{,k}\delta_{ij}) + 3r_{,i}r_{,j}r_{,k}\,] \tag{2}$$

$$U_{ik} = \frac{1}{16\pi(1-\nu)Gr} [\, (3-4\nu)\delta_{ik} + r_{,i}r_{,k}\,] \tag{3}$$

in terms of r, the Euclidean distance between the source and field points $\mathbf{x}$ and $\mathbf{y}$, and the shear modulus G and Poisson's ratio ν of the isotropic elastic solid. (Source and field points are also referred to as p and q (for internal points) and as P and Q (for boundary points), respectively, in this work). Also, $\boldsymbol{\delta}$ is the Kronecker delta and $_{,i} = \partial/\partial y_i$. The range of indices in these and all other equations in this paper is 1,2,3, unless specified otherwise.

It has been shown in ([NagaLM94], [NagaML96]) and [MukhMSN97] that the integrand vector $\mathbf{F}_k$ in equation(1) is divergence free (except at the point of singularity P=Q) and that the surface integral in it, over an open surface patch $S \in \partial B$, can be converted to a contour integral around the bounding curve C of S, by applying Stokes' theorem. Therefore, vectors $\mathbf{V}_k$ exist such that :

$$\int_S \mathbf{F}_k \cdot d\mathbf{S} = \oint_C \mathbf{V}_k \cdot d\mathbf{r} \tag{4}$$

Since the vectors $\mathbf{F}_k$ contain the unknown fields $\mathbf{u}$ and $\boldsymbol{\sigma}$, shape functions must be chosen for these variables, and potential functions derived for each linearly independent shape function, in order to determine the vectors $\mathbf{V}_k$. Also, since the kernels in equation(1) are functions only of $z_k = y_k - x_k$ (and not of the source and field coordinates separately), these shape functions must also be written in the coordinates z_k in order to determine the potential vectors $\mathbf{V}_k$. Finally, these shape functions are global in nature and are chosen to satisfy, a priori, the Navier-Cauchy equations of equilibrium. The weights, in linear combinations of these shape functions, however, are defined piecewise on boundary elements.

Quadratic shape functions are used in this work. With :

$$z_k = y_k - x_k \tag{5}$$

one has, on a boundary element :

$$u_i = \sum_{\alpha=1}^{27} \beta_\alpha \overline{u}_{\alpha i}(y_1, y_2, y_3) = \sum_{\alpha=1}^{27} \hat{\beta}_\alpha(x_1, x_2, x_3) \overline{u}_{\alpha i}(z_1, z_2, z_3) \tag{6}$$

$$\sigma_{ij} = \sum_{\alpha=1}^{27} \beta_\alpha \overline{\sigma}_{\alpha ij}(y_1, y_2, y_3) = \sum_{\alpha=1}^{27} \hat{\beta}_\alpha(x_1, x_2, x_3)\overline{\sigma}_{\alpha ij}(z_1, z_2, z_3) \tag{7}$$

where $\overline{u}_{\alpha i}$, $\overline{\sigma}_{\alpha ij}$ (with $i = 1, 2, 3$ and $\alpha = 1, 2, .., 27$) are the shape functions and β_α are the weights in the linear combinations of the shape functions. Each boundary element has, associated with it, 27 constants β_α which will be related to physical variables on that element. This set of β 's differ from one element to the next.

The displacement shape functions for $\alpha = 1, 2, 3$ are constants, those for $\alpha = 4, ..., 12$ are of first degree and those for $\alpha = 13, ..., 27$ are of second degree. There are a total of 27 linearly independent (vector) shape functions $\overline{\mathbf{u}}_\alpha$. The shape functions for the stresses are obtained from those for the displacements through the use of Hooke's law. The shape functions $\overline{u}_{\alpha i}$ and $\overline{\sigma}_{\alpha ij}$ are given in [MukhMSN97]

It is easy to show that the coordinate transformation (5) results in the constants $\hat{\beta}_j$ being related to the β_α's as follows :

$$\hat{\beta}_i = \sum_{\alpha=1}^{27} S_{i\alpha}(x_1, x_2, x_3)\beta_\alpha \ , \quad i = 1, 2, 3 \tag{8}$$

$$\hat{\beta}_k = \sum_{\alpha=1}^{27} R_{n\alpha}(x_1, x_2, x_3)\beta_\alpha \ , \ k = 4, 5, ...12, \quad n = k - 3 \tag{9}$$

$$\hat{\beta}_\alpha = \beta_\alpha \ , \qquad \alpha = 13, 14, .., 27 \tag{10}$$

where

$$\begin{aligned} S_{i\alpha} &= \overline{u}_{\alpha i}(x_1, x_2, x_3) \ , \qquad i = 1, 2, 3, \quad \alpha = 1, 2, ..., 27 \\ R_{k\alpha} &= \left.\frac{\partial \overline{u}_{\alpha \ell}(y_1, y_2, y_3)}{\partial y_j}\right|_{(x_1, x_2, x_3)} \ , \ k = 1, 2, .., 9, \quad \alpha = 1, 2, ..., 27 \end{aligned}$$

with $j = 1 + \lfloor \frac{k-1}{3} \rfloor$ and $\ell = k - 3j + 3$. Here, the symbol $\lfloor n \rfloor$, called the floor of n, denotes the largest integer less than or equal to n.
It is useful to note that the matrices $\mathbf{S}$ and $\mathbf{R}$ are functions of only the source point coordinates (x_1, x_2, x_3).

The procedure for designing boundary elements in the 3-D BCM is discussed in detail in [NagaML96] and [MukhMSN97]. A set of primary physical variables a_k, whose number must match the number (here 27) of artificial variables β_k on a boundary element, are chosen first. The first step in the BCM solution procedure is to determine the unspecified primary physical variables in terms of those prescribed from the boundary conditions. Later, secondary physical variables as well as stresses, at boundary points, are obtained from a simple post-processing procedure. Unlike in the standard BEM, it is particularly easy to obtain surface variables, such as stresses and curvatures, in the BCM. Surface stresses are discussed in Section 2.2.

A square invertible transformation matrix $\mathbf{T}$ relates the vectors $\mathbf{a}$ and $\boldsymbol{\beta}$ on element m according to the equation :

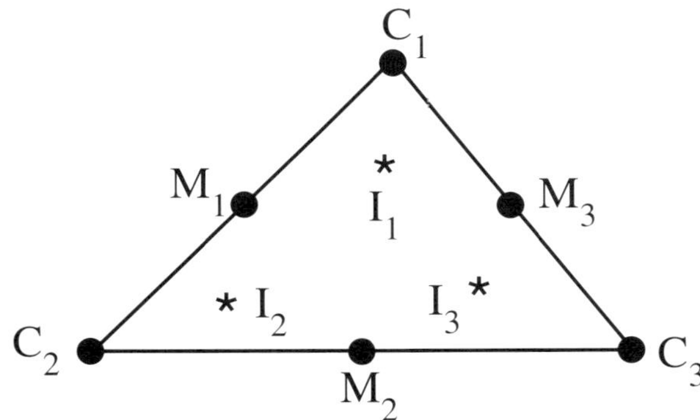

Figure 1: The CIM9 boundary element (from [MukhMSN97])

$$\overset{m}{a} = \overset{m}{T}\overset{m}{\beta} \tag{11}$$

The CIM9 boundary element, shown in Figure 1, is used in the present work. The displacement $\mathbf{u}$ is the primary physical variable at the three corner nodes C_i and the three midside nodes M_i, while tractions are primary variables at the internal nodes I_i. Thus, there are a total of 27 primary variables. The BCM equations are collocated at the 6 peripheral nodes as well as at the centroid of the element. In a typical discretization procedure, some of the peripheral nodes may lie on corners or edges, while the internal nodes are always located at regular points where the boundary ∂B is locally smooth. It is of obvious advantage to have to deal only with displacement components, that are always continuous, on edges and corners, while having traction components only at regular boundary points. It is important to restate that the BCM is versatile enough to handle any well-posed problem in linear elasticity - all the secondary variables can be easily determined by simple post-processing once the primary BCM equations are solved.

Details of the shape functions and intrinsic coordinates, that are used to define the geometry of the boundary elements, are available in [MukhMSN97]. Also, the procedure for obtaining the vector potentials $\mathbf{V_k}$, for nonsingular as well as singular integrands, are available in [NagaML96] and [MukhMSN97]. Finally, the regularized BIE, equation (1), is converted into a regularized boundary contour equation (BCE) that can be collocated (as in the usual BEM) at any boundary (surface) point - including those on edges and corners, as long as the displacement is continuous there. This equation is :

$$\begin{aligned} 0 &= \frac{1}{2}\sum_{m=1}^{M}\sum_{\alpha=13}^{27}\left[\oint_{L_m}(\overline{\sigma}_{\alpha ij}U_{ik} - \overline{u}_{\alpha i}\Sigma_{ijk})\,\epsilon_{jnt}z_n dz_t\right]\left[\overset{m}{T}{}^{-1}\overset{m}{a}\right]_{\alpha} \\ &+ \sum_{m=1}^{M}\sum_{\alpha=4}^{12}\left[\oint_{L_m}(\overline{\sigma}_{\alpha ij}U_{ik} - \overline{u}_{\alpha i}\Sigma_{ijk})\,\epsilon_{jnt}z_n dz_t\right]\left[R\,\overset{m}{T}{}^{-1}\overset{m}{a}\right]_{\alpha-3} \end{aligned}$$

$$+\sum_{\substack{m=1 \\ m\notin\mathcal{S}}}^{M}\sum_{\alpha=1}^{3}\left[\oint_{L_m} D_{\alpha jk}dz_j\right]\left[S\left(\overset{m}{T}{}^{-1}\overset{m}{a}-\overset{P}{T}{}^{-1}\overset{P}{a}\right)\right]_\alpha \tag{12}$$

with

$$\begin{aligned}\oint_{L_m} D_{\alpha jk}dz_j &= -\int_{S_m}\Sigma_{\alpha jk}\mathbf{e}_j\cdot d\mathbf{S}\\ &= \frac{1}{8\pi(1-\nu)}\oint_{L_m}\epsilon_{kij}\frac{r_{,\alpha}r_{,i}}{r}dz_j+\frac{1-2\nu}{8\pi(1-\nu)}\oint_{L_m}\epsilon_{\alpha kj}\frac{1}{r}dz_j+\frac{\Theta}{4\pi}\delta_{\alpha k}\end{aligned} \tag{13}$$

Here L_m is the bounding contour of the surface element S_m. In the above, Θ is the solid angle (subtended by a surface element m at a collocation point $\mathbf{x}$), which is defined as :

$$\Theta=\int_{S_m}\frac{\mathbf{r}\cdot d\mathbf{S}}{r^3} \tag{14}$$

Also, $\overset{m}{T}$ and $\overset{m}{a}$ are the transformation matrix and primary physical variable vectors on element *m*, $\overset{P}{T}$ and $\overset{P}{a}$ are the same quantities evaluated on any element that belongs to the set $\mathcal{S}$ of elements that contain the source point $\mathbf{x}$, and ϵ_{ijk} is the usual alternating symbol.

The procedure for obtaining an assembled discretized form of equation (12) is described in [MukhMSN97] and [MukhM98]. The final result is :

$$Ka = 0 \tag{15}$$

which is written as

$$Ax = By \tag{16}$$

where $\mathbf{x}$ contains the unknown and $\mathbf{y}$ the known (from the boundary conditions) values of the primary physical variables on the surface of the body. Once these equations are solved, the vector $\mathbf{a}$ is completely known. Now, at a post processing step, $\overset{m}{\beta}_\alpha$ can be easily obtained on each boundary element from equation (11).

2.2 Boundary Displacements and Stresses

A very useful consequence of using global shape functions is that, once the standard BCM is solved, it is very easy to obtain the displacements, stresses and curvatures at a regular off-contour boundary point (ROCBP) on the bounding surface of a body. Here, a point at an edge or corner is called an irregular point while at a regular point the boundary is locally smooth. Also, a regular boundary point can lie on or away from a boundary contour. The former is called a regular contour point (RCP), the latter a regular off-contour boundary point. A point inside a body is called an internal point.

First, one obtains $\overset{m}{\beta}_\alpha$ from equation (11), then uses equations (8 and 9) to get $\hat{\beta}^m_\alpha$, $\alpha = 1, 2, .., 12$. The curvatures, which are piecewise constant on each boundary element, are obtained by direct differentiation of equation (6). (Curvatures are given in [MukhSM99]. Finally, one has the following results.

2.2.1 Surface displacements.

$$[u_i(P)] = \begin{bmatrix} \hat{\beta}_1 \\ \hat{\beta}_2 \\ \hat{\beta}_3 \end{bmatrix}_P \tag{17}$$

2.2.2 Surface stresses.

$$[u_{i,j}(P)] = \begin{bmatrix} \hat{\beta}_4 & \hat{\beta}_7 & \hat{\beta}_{10} \\ \hat{\beta}_5 & \hat{\beta}_8 & \hat{\beta}_{11} \\ \hat{\beta}_6 & \hat{\beta}_9 & \hat{\beta}_{12} \end{bmatrix}_P \tag{18}$$

Equation (18) can be used to find the displacement gradients at P. Hooke's law would then give the stress $\sigma_{ij}(P)$. An alternative approach is to use equation (7) together with all the β_α on an element.

It should be noted that the simple approach, described above, cannot be used to find internal stresses since the constants $\overset{m}{\beta}_\alpha$ are only meaningful on the boundary of a body. Therefore, an internal point representation of the differentiated BCE, for the internal displacement gradients $u_{i,j}(p)$, is necessary (see Section 2.4). It is also of interest to examine the limiting process of a differentiated BCE as an internal point p approaches a boundary point P. This issue is of great current interest in the BEM community in the context of the standard and hypersingular BIE (HBIE - see, for example, [MartR96], [CrusR96], [MartRC98], [MukhM01]). Further, the hypersingular BCE (HBCE) must be understood if one wishes to collocate the HBCE as the primary integral equation, as may be necessary, for example, in applications such as fracture mechanics, symmetric Galerkin formulations or adaptive analysis (please see Section 1.2 of this chapter). This topic is the subject of Section 2.3 of this chapter.

2.3 Hypersingular Boundary Surface and Boundary Contour Integral Equations

A hypersingular boundary integral equation (HBIE) can be obtained by differentiating the standard BIE *at an internal point*, with respect to the coordinates of this internal source point. A regularized version of this equation, containing, at most, weakly singular integrals, is (see, for example, [CrusR96] for a detailed discussion) :

$$\begin{aligned} 0 &= \int_{\partial B} U_{ik,n}(\mathbf{x},\mathbf{y})\ [\sigma_{ij}(\mathbf{y}) \ - \sigma_{ij}(\mathbf{x})\,]\, n_j(\mathbf{y})\, dS(\mathbf{y}) \\ &- \int_{\partial B} \Sigma_{ijk,n}(\mathbf{x},\mathbf{y})\ [u_i(\mathbf{y}) \ - u_i(\mathbf{x}) \ - u_{i,\ell}(\mathbf{x})\ (y_\ell - x_\ell)]\, n_j(\mathbf{y})\, dS(\mathbf{y}) \end{aligned} \tag{19}$$

[MartRC98] - Appendix II2, p. 905, (see, also, [MukhM01]) have proved that (19) can be collocated even at an edge or corner point $\mathbf{x}$ on the surface of a 3-D body, provided that the displacement and stress fields in (19) satisfy certain smoothness requirements.

The regularized HBIE (19) can be converted to a regularized hypersingular boundary contour equation (HBCE). Details are available in [MukhM98]. The result is:

$$
\begin{aligned}
0 = & -\sum_{m=1}^{M}\sum_{\alpha=13}^{27}\left[\oint_{L_m}(\overline{\sigma}_{\alpha ij}U_{ik}-\overline{u}_{\alpha i}\Sigma_{ijk})\,\epsilon_{jnt}dz_t\right]\left[\overset{m}{T}{}^{-1}\overset{m}{a}\right]_{\alpha} \\
& +\sum_{m=1}^{M}\sum_{\alpha=4}^{12}\left[\oint_{L_m}(\overline{\sigma}_{\alpha ij}U_{ik}-\overline{u}_{\alpha i}\Sigma_{ijk})\,\epsilon_{jst}z_s dz_t\right]\left[R_{,N}\,\overset{m}{T}{}^{-1}\overset{m}{a}\right]_{\alpha-3} \\
& -\sum_{\substack{m=1\\ m\notin S}}^{M}\sum_{\alpha=4}^{12}\left[\oint_{L_m}(\overline{\sigma}_{\alpha ij}U_{ik}-\overline{u}_{\alpha i}\Sigma_{ijk})\,\epsilon_{jnt}dz_t\right]\left[R\left(\overset{m}{T}{}^{-1}\overset{m}{a}-\overset{P}{T}{}^{-1}\overset{P}{a}\right)\right]_{\alpha-3} \\
& +\sum_{\substack{m=1\\ m\notin S}}^{M}\sum_{\alpha=1}^{3}\left[\oint_{L_m}D_{\alpha jk}dz_j\right]\left[S_{,N}\left(\overset{m}{T}{}^{-1}\overset{m}{a}-\overset{P}{T}{}^{-1}\overset{P}{a}\right)\right]_{\alpha} \\
& +\sum_{\substack{m=1\\ m\notin S}}^{M}\sum_{\alpha=1}^{3}\left[\oint_{L_m}\Sigma_{\alpha jk}\epsilon_{jnt}dz_t\right]\left[S\left(\overset{m}{T}{}^{-1}\overset{m}{a}-\overset{P}{T}{}^{-1}\overset{P}{a}\right)\right]_{\alpha}
\end{aligned}
\tag{20}
$$

where, as before, S is the set of boundary elements that contains the source point $\mathbf{x}$. The derivatives $R_{,N}$ and $S_{,N}$ in (20) are taken with respect to the source point coordinates x_n. In equation (20), the integrands in the first two terms are regular $(O(1))$. The third and fourth (potentially strongly singular, $O(1/r)$) as well as the fifth (potentially hypersingular, $O(1/r^2)$) need to be evaluated only on nonsingular elements.

2.4 Internal Displacements and Stresses

At this stage, it is a simple matter to derive equations for displacements and stresses at a point inside a body. These equations are derived in [MukhSM00]. They are given below.

2.4.1 Internal displacements.

$$
\begin{aligned}
u_k(p) = & \frac{1}{2}\sum_{m=1}^{M}\sum_{\alpha=13}^{27}\left[\oint_{L_m}(\overline{\sigma}_{\alpha ij}U_{ik}-\overline{u}_{\alpha i}\Sigma_{ijk})\,\epsilon_{jnt}z_n dz_t\right]\left[\overset{m}{T}{}^{-1}\overset{m}{a}\right]_{\alpha} \\
& +\sum_{m=1}^{M}\sum_{\alpha=4}^{12}\left[\oint_{L_m}(\overline{\sigma}_{\alpha ij}U_{ik}-\overline{u}_{\alpha i}\Sigma_{ijk})\,\epsilon_{jnt}z_n dz_t\right]\left[R\,\overset{m}{T}{}^{-1}\overset{m}{a}\right]_{\alpha-3} \\
& +\sum_{m=1}^{M}\sum_{\alpha=1}^{3}\left[\oint_{L_m}D_{\alpha jk}dz_j\right]\left[S\,\overset{m}{T}{}^{-1}\overset{m}{a}\right]_{\alpha}
\end{aligned}
\tag{21}
$$

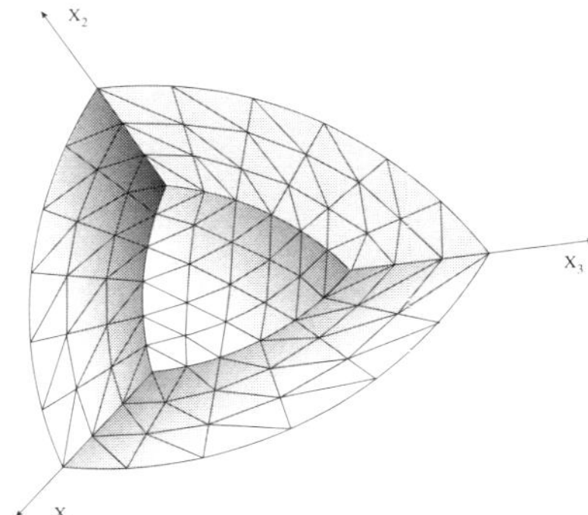

Figure 2: A typical mesh on the surface of a one-eighth sphere (from [MukhSM99])

2.4.2 Internal stresses.

$$
\begin{aligned}
u_{k,n}(p) \quad = \quad & -\sum_{m=1}^{M}\sum_{\alpha=13}^{27}\left[\oint_{L_m}(\overline{\sigma}_{\alpha ij}U_{ik}-\overline{u}_{\alpha\imath}\Sigma_{ijk})\,\epsilon_{jnt}dz_t\right]\left[\overset{m}{T}{}^{-1}\overset{m}{a}\right]_{\alpha} \\
& +\sum_{m=1}^{M}\sum_{\alpha=4}^{12}\left[\oint_{L_m}(\overline{\sigma}_{\alpha ij}U_{ik}-\overline{u}_{\alpha i}\Sigma_{ijk})\,\epsilon_{jst}z_s dz_t\right]\left[R_{,N}\,\overset{m}{T}{}^{-1}\overset{m}{a}\right]_{\alpha-3} \\
& -\sum_{m=1}^{M}\sum_{\alpha=4}^{12}\left[\oint_{L_m}(\overline{\sigma}_{\alpha ij}U_{ik}-\overline{u}_{\alpha i}\Sigma_{ijk})\,\epsilon_{jnt}dz_t\right]\left[R\,\overset{m}{T}{}^{-1}\overset{m}{a}\right]_{\alpha-3} \\
& +\sum_{m=1}^{M}\sum_{\alpha=1}^{3}\left[\oint_{L_m}D_{\alpha jk}dz_j\right]\left[S_{,N}\,\overset{m}{T}{}^{-1}\overset{m}{a}\right]_{\alpha} \\
& +\sum_{m=1}^{M}\sum_{\alpha=1}^{3}\left[\oint_{L_m}\Sigma_{\alpha jk}\epsilon_{jnt}dz_t\right]\left[S\,\overset{m}{T}{}^{-1}\overset{m}{a}\right]_{\alpha} \qquad (22)
\end{aligned}
$$

Hooke's law is now used to obtain the internal stress from the internal displacement gradient. Curvatures at an internal point are given in [MukhSM00].

2.5 Numerical Results

Numerical results from the BCM, for selected 3-D examples, are available in [NagaML96], [MukhMSN97] and, from the HBCM, in [MukhM98]. Typical results, for a thick hollow sphere under internal pressure, are given below. The inner and outer radii of the sphere are 1 and 2 units, respectively, the shear modulus $G = 1$ and Poisson's ration $\nu = 0.3$. The internal pressure is 1. A generic surface mesh on a one-eighth sphere is shown in Figure 2. Three levels of discretization - coarse, medium and fine, are used in this work. Mesh statistics are shown in Table 1.

mesh	number of elements		
	on each flat plane	on each curved surface	total
coarse	12	9	54
medium	36	36	180
fine	64	64	320

Table 1: Mesh statistics on a one-eighth sphere.

2.5.1 Surface displacements from the BCM and the HBCM.

First, please note that (20) has two free indices, k and n, so that it represents nine equations. These equations arise from $u_{k,n}$ (please see the detailed derivation of equations (19) and (20) in [MukhM98]). Different strategies are possible for collocating (20) at a boundary point. The first is to use all nine equations. The second in to use six corresponding to $\epsilon_{kn} = (1/2)(u_{k,n} + u_{n,k})$. The six equation strategy amounts to replacing E_{kn}, the right hand side of (20), by $(1/2)(E_{kn} + E_{nk})$. Both the nine and six equation strategies lead to overdetermined systems, but are convenient for collocating at irregular boundary points since the source point normal is not involved in these cases. A third, the three equation strategy, suitable for collocation at regular points, corresponds to the traction components τ_n. In this case, the right hand side (E_{kn}) of (20) is replaced by $[\lambda E_{mm}\delta_{kn} + \mu(E_{kn} + E_{nk})]n_k(P)$, where λ and μ are Lamé constants, δ_{ij} are components of the Kronecker delta and Hooke's law is used. The three equation strategy involving the traction components is not convenient for collocating the HBCE at a point on an edge or a corner of a body where the normal to the body surface has a jump discontinuity. In view of the assumed continuity of the stress tensor at such a point, this situation leads to a jump in traction at that point, unless the stress tensor is zero there. One would, therefore, need to use multiple source points, each belonging to a smooth surface meeting at that irregular point, and collocate separately at these points. Since the primary purpose here is to demonstrate collocation of (20) at irregular boundary points, only the nine and six equation strategies are used below.

It should be mentioned here that, for the HBCM in 2-D elasticity, a strategy corresponding to the first one above has been successfully employed by [PhanMM98b] and a strategy corresponding the third one above has been implemented by [ZhouCS99].

The overdetermined system of linear algebraic equations, resulting from the nine and six equation strategies mentioned above, have been solved by using a subroutine based on QR decomposition of the system matrix. This subroutine has been obtained form the IMSL software package.

It is seen from Figure 2 that many of the collocation points lie on edges and six of them lie on corners of the surface of the one-eighth sphere. These results, displayed in Figure 3, show a comparison of the BCM (from equation (12)) and HBCM (from equation (20)) results with the exact solution of the problem [TimoG70]. The first and last points along the axis lie on corners, the rest lie along an edge. The agreement between the exact, BCM and 9 equation HBCM solution is seen to be excellent.

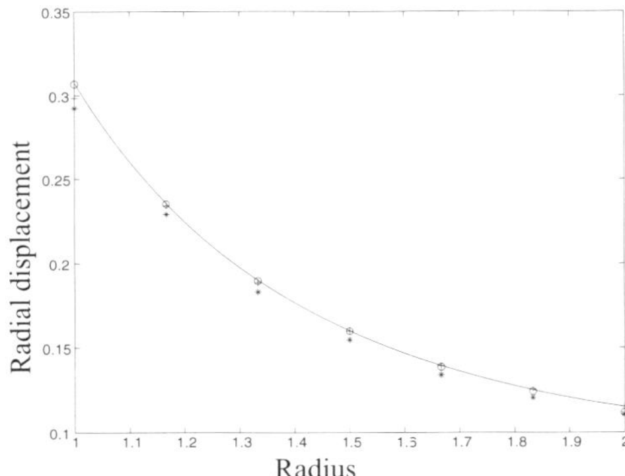

Figure 3: Hollow sphere under internal pressure. Radial displacement as a function of radius along the x_1 axis. Numerical solutions are obtained from the medium mesh. Exact solution: —, BCM solution: $\circ\circ\circ\circ$, 6 equation HBCM solution: $****$, 9 equation HBCM solution: $++++$ (from [MukhM01])

2.5.2 Surface stresses.

Stresses on the outer surface $(R = b)$ of the sphere, obtained from equation (18) and Hooke's law, are shown in Figure 4. The nodes are chosen at the centroids of the boundary elements. The agreement between the numerical and analytical solutions is seen to be excellent.

2.5.3 Internal stresses.

Internal stresses along the line $x_1 = x_2 = x_3$, obtained from equation (22) and Hooke's law, appear in Figure 5. Excellent agreement is observed between the numerical and analytical solutions.

3 Shape Sensitivity Analysis with the BCM

Shape sensitivity analysis of surface and internal displacements and stresses, in 3-D linear elasticity, is the subject of this section. Further details are available in [MukhSM99] and [MukhSM00].

3.1 Sensitivities of Surface Variables

Sensitivities of surface displacements and tractions are discussed below. Further details are available in [MukhSM99].

3.1.1 Shape sensitivity analysis.

One starts with the usual regularized boundary integral equation (1) for linear elasticity which is valid at both an internal point $\mathbf{x} \in B$ as well as at a boundary point $\mathbf{x} \in \partial B$. An internal point $\mathbf{x} \in B$ is considered first. Equation (1) is rewritten as :

mesh	number of elements		
	on each flat plane	on each curved surface	total
coarse	12	9	54
medium	36	36	180
fine	64	64	320

Table 1: Mesh statistics on a one-eighth sphere.

2.5.1 Surface displacements from the BCM and the HBCM.

First, please note that (20) has two free indices, k and n, so that it represents nine equations. These equations arise from $u_{k,n}$ (please see the detailed derivation of equations (19) and (20) in [MukhM98]). Different strategies are possible for collocating (20) at a boundary point. The first is to use all nine equations. The second in to use six corresponding to $\epsilon_{kn} = (1/2)(u_{k,n} + u_{n,k})$. The six equation strategy amounts to replacing E_{kn}, the right hand side of (20), by $(1/2)(E_{kn} + E_{nk})$. Both the nine and six equation strategies lead to overdetermined systems, but are convenient for collocating at irregular boundary points since the source point normal is not involved in these cases. A third, the three equation strategy, suitable for collocation at regular points, corresponds to the traction components τ_n. In this case, the right hand side (E_{kn}) of (20) is replaced by $[\lambda E_{mm}\delta_{kn} + \mu(E_{kn} + E_{nk})]n_k(P)$, where λ and μ are Lamé constants, δ_{ij} are components of the Kronecker delta and Hooke's law is used. The three equation strategy involving the traction components is not convenient for collocating the HBCE at a point on an edge or a corner of a body where the normal to the body surface has a jump discontinuity. In view of the assumed continuity of the stress tensor at such a point, this situation leads to a jump in traction at that point, unless the stress tensor is zero there. One would, therefore, need to use multiple source points, each belonging to a smooth surface meeting at that irregular point, and collocate separately at these points. Since the primary purpose here is to demonstrate collocation of (20) at irregular boundary points, only the nine and six equation strategies are used below.

It should be mentioned here that, for the HBCM in 2-D elasticity, a strategy corresponding to the first one above has been successfully employed by [PhanMM98b] and a strategy corresponding the third one above has been implemented by [ZhouCS99].

The overdetermined system of linear algebraic equations, resulting from the nine and six equation strategies mentioned above, have been solved by using a subroutine based on QR decomposition of the system matrix. This subroutine has been obtained form the IMSL software package.

It is seen from Figure 2 that many of the collocation points lie on edges and six of them lie on corners of the surface of the one-eighth sphere. These results, displayed in Figure 3, show a comparison of the BCM (from equation (12)) and HBCM (from equation (20)) results with the exact solution of the problem [TimoG70]. The first and last points along the axis lie on corners, the rest lie along an edge. The agreement between the exact, BCM and 9 equation HBCM solution is seen to be excellent.

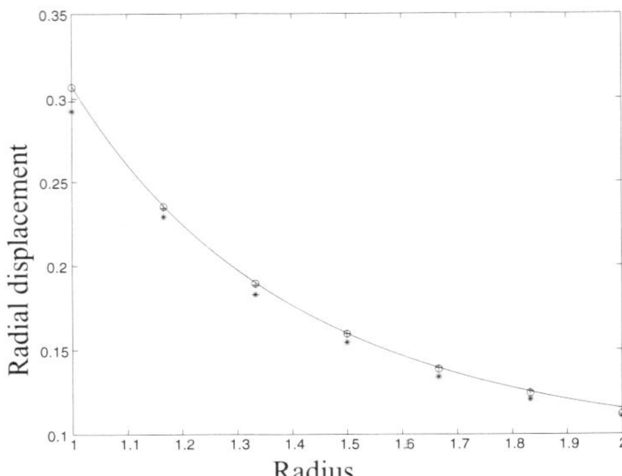

Figure 3: Hollow sphere under internal pressure. Radial displacement as a function of radius along the x_1 axis. Numerical solutions are obtained from the medium mesh. Exact solution: —, BCM solution: $\circ\circ\circ\circ$, 6 equation HBCM solution: $****$, 9 equation HBCM solution: $++++$ (from [MukhM01])

2.5.2 Surface stresses.

Stresses on the outer surface $(R = b)$ of the sphere, obtained from equation (18) and Hooke's law, are shown in Figure 4. The nodes are chosen at the centroids of the boundary elements. The agreement between the numerical and analytical solutions is seen to be excellent.

2.5.3 Internal stresses.

Internal stresses along the line $x_1 = x_2 = x_3$, obtained from equation (22) and Hooke's law, appear in Figure 5. Excellent agreement is observed between the numerical and analytical solutions.

3 Shape Sensitivity Analysis with the BCM

Shape sensitivity analysis of surface and internal displacements and stresses, in 3-D linear elasticity, is the subject of this section. Further details are available in [MukhSM99] and [MukhSM00].

3.1 Sensitivities of Surface Variables

Sensitivities of surface displacements and tractions are discussed below. Further details are available in [MukhSM99].

3.1.1 Shape sensitivity analysis.

One starts with the usual regularized boundary integral equation (1) for linear elasticity which is valid at both an internal point $\mathbf{x} \in B$ as well as at a boundary point $\mathbf{x} \in \partial B$. An internal point $\mathbf{x} \in B$ is considered first. Equation (1) is rewritten as :

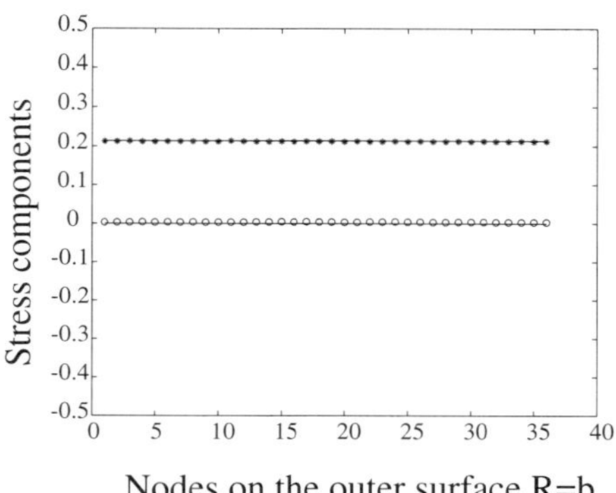

Figure 4: Hollow sphere under internal pressure. Stresses on the outer surface $R = b$. Exact solutions —. Numerical solutions from the medium mesh: $\sigma_{\theta\theta} = \sigma_{\phi\phi}$ $****$, σ_{RR} $\circ\circ\circ\circ$ (from [MukhSM99])

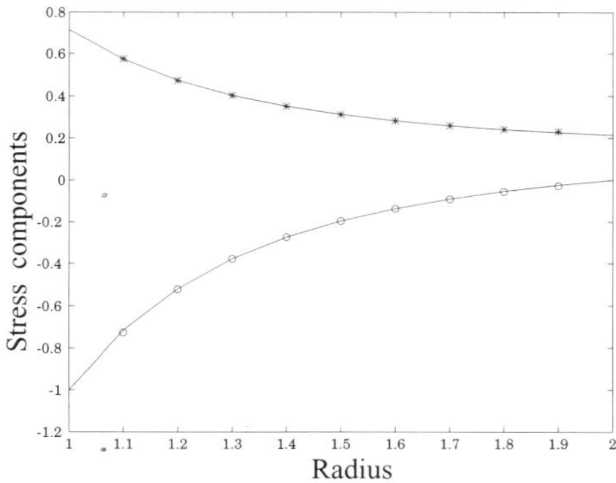

Figure 5: Hollow sphere under internal pressure. Stresses as functions of radius along the line $x_1 = x_2 = x_3$. Exact solutions —. Numerical solutions from the fine mesh: $\sigma_{\theta\theta} = \sigma_{\phi\phi}$ $* * * *$, σ_{RR} $\circ \circ \circ \circ$ (from [MukhSM00])

$$0 = \int_{\partial B} \left[U_{ik}(\mathbf{x},\mathbf{y})\sigma_{ij}(\mathbf{y},b) - \Sigma_{ijk}(\mathbf{x},\mathbf{y})(u_i(\mathbf{y},b) - u_i(\mathbf{x},b)) \right] n_j(\mathbf{y}) dS(\mathbf{y}) \tag{23}$$

As mentioned above, b is a shape design variable and the spatial coordinates of the source and field points depend on b , i.e. $\mathbf{x}(b), \mathbf{y}(b)$. Define, as in equation (1),

$$F_{jk}(\mathbf{x},\mathbf{y},b) = U_{ik}(\mathbf{x},\mathbf{y})\sigma_{ij}(\mathbf{y},b) - \Sigma_{ijk}(\mathbf{x},\mathbf{y})[u_i(\mathbf{y},b) - u_i(\mathbf{x},b)] \tag{24}$$

Now the (total) sensitivity of a function $f(\mathbf{x}(b), \mathbf{y}(b), b)$, in a materials derivative sense, is defined as :

$$\overset{*}{f} \equiv \frac{df}{db} = \frac{\partial f}{\partial x_k} \overset{*}{x}_k + \frac{\partial f}{\partial y_k} \overset{*}{y}_k + \frac{\partial f}{\partial b} \tag{25}$$

while partial sensitivities of displacements and stresses are defined as:

$$\overset{\triangle}{u}_i \equiv \frac{\partial u_i}{\partial b}, \quad \overset{\triangle}{\sigma}_{ij} \equiv \frac{\partial \sigma_{ij}}{\partial b} \tag{26}$$

It should be noted that:

$$\overset{\triangle}{u}_i = \overset{*}{u}_i - u_{i,k} \overset{*}{y}_k \ , \quad \overset{\triangle}{\sigma}_{ij} = \overset{*}{\sigma}_{ij} - \sigma_{ij,k} \overset{*}{y}_k \tag{27}$$

where $, k \equiv \frac{\partial}{\partial y_k}$

Taking the sensitivity (total derivative) of equation (23) with respect to b, one gets:

$$\begin{aligned} 0 \quad = \quad & \overset{*}{x}_r \int_{\partial B} \frac{\partial F_{jk}(\mathbf{x},\mathbf{y},b)}{\partial x_r} n_j(\mathbf{y}) dS(\mathbf{y}) \\ & + \int_{\partial B} \frac{\partial F_{jk}(\mathbf{x},\mathbf{y},b)}{\partial y_r} \overset{*}{y}_r \, n_j(\mathbf{y}) dS(\mathbf{y}) + \int_{\partial B} F_{jk}(\mathbf{x},\mathbf{y},b) \frac{d}{db} [n_j(\mathbf{y}) dS(\mathbf{y})] \\ & + \int_{\partial B} \left[U_{ik}(\mathbf{x},\mathbf{y}) \overset{\triangle}{\sigma}_{ij} (\mathbf{y},b) - \Sigma_{ijk}(\mathbf{x},\mathbf{y})(\overset{\triangle}{u}_i (\mathbf{y},b) - \overset{\triangle}{u}_i (\mathbf{x},b)) \right] n_j(\mathbf{y}) dS(\mathbf{y}) \end{aligned} \tag{28}$$

It should be noted that the last integrand above is $\frac{\partial F_{jk}(\mathbf{x},\mathbf{y},b)}{\partial b}$.

The first integral on the right hand side of equation (28) is zero because the integral in equation (23) vanishes for all values of $\mathbf{x} \in B$. Let the second and third integrals together be called I_k and the last integral J_k. Thus:

$$I_k + J_k = 0 \tag{29}$$

Each of these surface integrals can be converted to line integrals. Details are available in [MukhSM99]. The final result is:

$$
\begin{aligned}
0 \;=\; & \sum_{m=1}^{M}\sum_{\alpha=13}^{27}\left[\oint_{L_m}(\overline{\sigma}_{\alpha ij}U_{ik}-\overline{u}_{\alpha i}\Sigma_{ijk})\,\epsilon_{jnt}\,\overset{*}{z}_n\,dz_t\right]\left[\overset{m}{T}{}^{-1}\overset{m}{a}\right]_\alpha \\
&+\sum_{m=1}^{M}\sum_{\alpha=4}^{12}\left[\oint_{L_m}(\overline{\sigma}_{\alpha ij}U_{ik}-\overline{u}_{\alpha i}\Sigma_{ijk})\,\epsilon_{jnt}\,\overset{*}{z}_n\,dz_t\right]\left[R\,\overset{m}{T}{}^{-1}\overset{m}{a}\right]_{\alpha-3} \\
&-\sum_{\substack{m=1\\ m\notin\mathcal{S}}}^{M}\sum_{\alpha=1}^{3}\left[\oint_{L_m}\Sigma_{\alpha jk}\epsilon_{jnt}\,\overset{*}{z}_n\,dz_t\right]\left[S\left(\overset{m}{T}{}^{-1}\overset{m}{a}-\overset{P}{T}{}^{-1}\overset{P}{a}\right)\right]_\alpha \\
&+\frac{1}{2}\sum_{m=1}^{M}\sum_{\alpha=13}^{27}\left[\oint_{L_m}(\overline{\sigma}_{\alpha ij}U_{ik}-\overline{u}_{\alpha i}\Sigma_{ijk})\,\epsilon_{jnt}z_n dz_t\right]\left[\overset{m}{T}{}^{-1}\overset{*m}{a}+\left(\overset{m}{T}{}^{-1}\right)^{*}\overset{m}{a}\right]_\alpha \\
&+\sum_{m=1}^{M}\sum_{\alpha=4}^{12}\left[\oint_{L_m}(\overline{\sigma}_{\alpha ij}U_{ik}-\overline{u}_{\alpha i}\Sigma_{ijk})\,\epsilon_{jnt}z_n dz_t\right]\left[R\left(\overset{m}{T}{}^{-1}\overset{*m}{a}+\left(\overset{m}{T}{}^{-1}\right)^{*}\overset{m}{a}\right)\right]_{\alpha-3} \\
&+\sum_{\substack{m=1\\ m\notin\mathcal{S}}}^{M}\sum_{\alpha=1}^{3}\left[\oint_{L_m}D_{\alpha jk}dz_j\right]\times \\
&\left[S\left(\overset{m}{T}{}^{-1}\overset{*m}{a}-\overset{P}{T}{}^{-1}\overset{*P}{a}+\left(\overset{m}{T}{}^{-1}\right)^{*}\overset{m}{a}-\left(\overset{P}{T}{}^{-1}\right)^{*}\overset{P}{a}\right)\right]_\alpha
\end{aligned}
\tag{30}
$$

Comparison of the above sensitivity equation (30) with the standard BCE (12) reveals that the integrals in its last three terms are identical to those in the standard BCE. Therefore, its discretized form can be written with the same coefficient matrix A as for the standard BCM, i.e.:

$$K\,\overset{*}{a}= h \tag{31}$$

where the right hand side vector $h = -\overset{*}{K}\,a$ can be computed from equation (30) by using the boundary values of the primary variables a that are known at this stage.

Finally, the usual switching of columns leads to:

$$A\,\overset{*}{x}= B\,\overset{*}{y} + h \tag{32}$$

where $\overset{*}{x}$ contains the unknown and $\overset{*}{y}$ the known values of boundary sensitivities. In many applications, $\overset{*}{y}= 0$

3.1.2 Sensitivities of surface stresses.

On any boundary element:

$$\overset{*}{\beta}= T^{-1}\,\overset{*}{a} + (T^{-1})^{*}a \tag{33}$$

in which it is convenient to evaluate the sensitivity of T^{-1} from the formula:

$$(T^{-1})^* = -T^{-1} \overset{*}{T} T^{-1} \tag{34}$$

The first step is to use equation (33) to find $\overset{*}{\beta}$ on each element.

Sensitivities of displacement gradients $v_{ij} \equiv u_{i,j}$ are computed by differentiating equation (18). The result is:

$$\left[\overset{*}{v}_{ij}(P)\right] = \begin{bmatrix} \overset{*}{\hat{\beta}}_4 & \overset{*}{\hat{\beta}}_7 & \overset{*}{\hat{\beta}}_{10} \\ \overset{*}{\hat{\beta}}_5 & \overset{*}{\hat{\beta}}_8 & \overset{*}{\hat{\beta}}_{11} \\ \overset{*}{\hat{\beta}}_6 & \overset{*}{\hat{\beta}}_9 & \overset{*}{\hat{\beta}}_{12} \end{bmatrix}_P \tag{35}$$

where, by differentiating equation (9), one has:

$$\overset{*}{\hat{\beta}}_k = \sum_{\alpha=1}^{27} R_{n\alpha}(x_1, x_2, x_3)\, \overset{*}{\beta}_\alpha + \sum_{\alpha=1}^{27} \overset{*}{R}_{n\alpha}(\overset{*}{x}_1, \overset{*}{x}_2, \overset{*}{x}_3)\beta_\alpha\ , \ k = 4, 5, ...12, \ \ n = k - 3 \tag{36}$$

Note that $\overset{*}{R}_{n\alpha}(\overset{*}{x}_1, \overset{*}{x}_2, \overset{*}{x}_3)$ involves sensitivities of the source point coordinates (x_1, x_2, x_3)

Finally, Hooke's law is used to determine the stress sensitivities from the sensitivities of the displacement gradients.

3.2 Sensitivities of Internal Variables

Sensitivities of internal displacements and stresses are discussed below. Further details are available in [MukhSM00].

3.2.1 Sensitivities of displacements.

One starts with the usual boundary integral equation for linear elasticity at an internal point $\mathbf{x} \in B$. This equation is (see (1)):

$$u_k(\mathbf{x}\,, b) = \int_{\partial B} [\, U_{ik}(\mathbf{x}, \mathbf{y})\, \sigma_{ij}(\mathbf{y}\,, b) - \Sigma_{ijk}(\mathbf{x}, \mathbf{y})\, u_i(\mathbf{y}\,, b)\,]\, n_j(\mathbf{y})\, dS(\mathbf{y}) \tag{37}$$

The corresponding BCM equation is (21).

Define:

$$G_{jk}(\mathbf{x}, \mathbf{y}, b) = U_{ik}(\mathbf{x}, \mathbf{y})\sigma_{ij}(\mathbf{y}, b) - \Sigma_{ijk}(\mathbf{x}, \mathbf{y})u_i(\mathbf{y}, b) \tag{38}$$

Taking the sensitivity of equation (37), one gets:

$$\begin{aligned} \overset{*}{u}_k(\mathbf{x}) &= \overset{*}{x}_r \int_{\partial B} \frac{\partial G_{jk}(\mathbf{x}, \mathbf{y}, b)}{\partial x_r} n_j(\mathbf{y}) dS(\mathbf{y}) \\ &+ \int_{\partial B} \frac{\partial G_{jk}(\mathbf{x}, \mathbf{y}, b)}{\partial y_r}\, \overset{*}{y}_r\, n_j(\mathbf{y}) dS(\mathbf{y}) + \int_{\partial B} G_{jk}(\mathbf{x}, \mathbf{y}, b) \frac{d}{db} [n_j(\mathbf{y}) dS(\mathbf{y})] \end{aligned}$$

$$+\int_{\partial B}\left[U_{ik}(\mathbf{x},\mathbf{y})\overset{\triangle}{\sigma}_{ij}(\mathbf{y},b)-\Sigma_{ijk}(\mathbf{x},\mathbf{y})\overset{\triangle}{u}_i(\mathbf{y},b)\right]n_j(\mathbf{y})dS(\mathbf{y}) \tag{39}$$

Let the first term on the right hand side of equation (39) be called J_{1k} , the second and third integrals together be called J_{2k} and the last integral be J_{3k}. Therefore, one has:

$$\overset{*}{u}_k(\mathbf{x})=J_{1k}+J_{2k}+J_{3k} \tag{40}$$

It is obvious that:

$$J_{1k}=u_{k,r}(\mathbf{x})\overset{*}{x}_r \tag{41}$$

Using exactly the same procedure described in [MukhSM99], one gets:

$$J_{2k}=\sum_{m=1}^{M}\oint_{L_m}\epsilon_{jnt}G_{jk}\overset{*}{z}_n\,dz_t \tag{42}$$

Finally,

$$J_{3k}=\overset{\triangle}{u}_k \tag{43}$$

Substituting equations (41, 42, 43 and 38) into (40), one obtains the expression:

$$\overset{*}{u}_k(\mathbf{x})=u_{k,p}(\mathbf{x})\overset{*}{x}_p+\sum_{m=1}^{M}\oint_{L_m}\left[U_{ik}(\mathbf{x},\mathbf{y})\sigma_{ij}(\mathbf{y},b)-\Sigma_{ijk}(\mathbf{x},\mathbf{y})u_i(\mathbf{y},b)\right]\epsilon_{jnt}\overset{*}{z}_n\,dz_t+\overset{\triangle}{u}_k \tag{44}$$

An explicit form of equation (44) is obtained by writing the displacements and stresses in terms of their shape functions from equations (6 and 7). Please refer to [MukhSM99] for details of the treatment of the partial sensitivity $\overset{\triangle}{u}_k$. Finally, the boundary contour integral form of the displacement sensitivity equation is:

$$\begin{aligned}
\overset{*}{u}_k(\mathbf{x}) &= u_{k,r}(\mathbf{x})\overset{*}{x}_r+\sum_{m=1}^{M}\sum_{\alpha=13}^{27}\left[\oint_{L_m}(\overline{\sigma}_{\alpha ij}U_{ik}-\overline{u}_{\alpha i}\Sigma_{ijk})\epsilon_{jnt}\overset{*}{z}_n\,dz_t\right]\left[\overset{m}{T}{}^{-1}\overset{m}{a}\right]_\alpha \\
&+\sum_{m=1}^{M}\sum_{\alpha=4}^{12}\left[\oint_{L_m}(\overline{\sigma}_{\alpha ij}U_{ik}-\overline{u}_{\alpha i}\Sigma_{ijk})\epsilon_{jnt}\overset{*}{z}_n\,dz_t\right]\left[R\,\overset{m}{T}{}^{-1}\overset{m}{a}\right]_{\alpha-3} \\
&-\sum_{m=1}^{M}\sum_{\alpha=1}^{3}\left[\oint_{L_m}\Sigma_{\alpha jk}\epsilon_{jnt}\overset{*}{z}_n\,dz_t\right]\left[S\;\overset{m}{T}{}^{-1}\overset{m}{a}\right]_\alpha \\
&+\frac{1}{2}\sum_{m=1}^{M}\sum_{\alpha=13}^{27}\left[\oint_{L_m}(\overline{\sigma}_{\alpha ij}U_{ik}-\overline{u}_{\alpha i}\Sigma_{ijk})\epsilon_{jnt}z_n dz_t\right]\left[\overset{m}{T}{}^{-1}\overset{*m}{a}+\left(\overset{m}{T}{}^{-1}\right)^{*}\overset{m}{a}\right]_\alpha
\end{aligned}$$

$$
\begin{aligned}
&+\sum_{m=1}^{M}\sum_{\alpha=4}^{12}\left[\oint_{L_m}(\overline{\sigma}_{\alpha ij}U_{ik}-\overline{u}_{\alpha i}\Sigma_{ijk})\,\epsilon_{jnt}z_n dz_t\right]\times \\
&\left[R\left(\overset{m}{T}{}^{-1}\overset{*m}{a}+\left(\overset{m}{T}{}^{-1}\right)^{*}\overset{m}{a}\right)\right]_{\alpha-3} \\
&+\sum_{m=1}^{M}\sum_{\alpha=1}^{3}\left[\oint_{L_m}D_{\alpha jk}dz_j\right]\left[S\left(\overset{m}{T}{}^{-1}\overset{*m}{a}+\left(\overset{m}{T}{}^{-1}\right)^{*}\overset{m}{a}\right)\right]_{\alpha}
\end{aligned}
\tag{45}
$$

3.2.2 Sensitivities of stresses.

This time, the starting point is the displacement gradient BIE. This well known equation is:

$$
v_{kr}(\mathbf{x},b)\equiv u_{k,r}(\mathbf{x},b)=-\int_{\partial B}\left[U_{ik,r}(\mathbf{x},\mathbf{y})\sigma_{ij}(\mathbf{y},b)-\Sigma_{ijk,r}(\mathbf{x},\mathbf{y})u_i(\mathbf{y},b)\right]n_j(\mathbf{y})dS(\mathbf{y})
\tag{46}
$$

Now, one defines:

$$
H_{jkr}(\mathbf{x},\mathbf{y})=U_{ik,r}(\mathbf{x},\mathbf{y})\sigma_{ij}(\mathbf{y},b)-\Sigma_{ijk,r}(\mathbf{x},\mathbf{y})u_i(\mathbf{y},b)
\tag{47}
$$

One has $H_{jkr,j}=0$ at an internal point since:

$$
H_{jkr}(\mathbf{x},\mathbf{y})=-\frac{\partial G_{jk}(\mathbf{x},\mathbf{y})}{\partial x_r}
\tag{48}
$$

and $G_{jk,j}=0$.

The exact same reasoning as in Section 3.2.1 leads to an equation for the sensitivities of displacement gradients at an internal point. This is:

$$
\begin{aligned}
\overset{*}{v}_{kr}(\mathbf{x}) &= u_{k,rp}(\mathbf{x})\,\overset{*}{x}_p-\sum_{m=1}^{M}\oint_{L_m}\left[U_{ik,r}(\mathbf{x},\mathbf{y})\sigma_{ij}(\mathbf{y},b)-\Sigma_{ijk,r}(\mathbf{x},\mathbf{y})u_i(\mathbf{y},b)\right]\epsilon_{jnt}\,\overset{*}{z}_n\,dz_t \\
&\quad-\int_{\partial B}\left[U_{ik,r}(\mathbf{x},\mathbf{y})\,\overset{\triangle}{\sigma}_{ij}(\mathbf{y},b)-\Sigma_{ijk,r}(\mathbf{x},\mathbf{y})\,\overset{\triangle}{u}_i(\mathbf{y},b)\right]n_j(\mathbf{y})dS(\mathbf{y})
\end{aligned}
\tag{49}
$$

Finally, one writes the displacements and stresses in terms of their shape functions in order to obtain an explicit form of equation (49). It should be noted that the second term in the right hand side of equation (49) is analogous to the integral on the right hand side of equation (44) (with $\mathbf{G}$ replaced by $\mathbf{H}$), while the last term in equation (49), $\overset{\triangle}{v}_{kr}$, is analogous to an expression for the displacement gradient at an internal point. The displacement gradient BCE (22) is very useful for obtaining an explicit expression for this integral. The final result is:

$$\begin{aligned}
\overset{*}{v}_{kr}(\mathbf{x}) &= u_{k,rp}(\mathbf{x})\,\overset{*}{x}_p - \sum_{m=1}^{M}\sum_{\alpha=13}^{27}\left[\oint_{L_m}(\overline{\sigma}_{\alpha ij}U_{ik,r} - \overline{u}_{\alpha i}\Sigma_{ijk,r})\,\epsilon_{jnt}\,\overset{*}{z}_n\,dz_t\right]\left[\overset{m}{T}{}^{-1}\overset{m}{a}\right]_{\alpha} \\
&- \sum_{m=1}^{M}\sum_{\alpha=4}^{12}\left[\oint_{L_m}(\overline{\sigma}_{\alpha ij}U_{ik,r} - \overline{u}_{\alpha i}\Sigma_{ijk,r})\,\epsilon_{jnt}\,\overset{*}{z}_n\,dz_t\right]\left[R\,\overset{m}{T}{}^{-1}\overset{m}{a}\right]_{\alpha-3} \\
&+ \sum_{m=1}^{M}\sum_{\alpha=1}^{3}\left[\oint_{L_m}\Sigma_{\alpha jk,r}\epsilon_{jnt}\,\overset{*}{z}_n\,dz_t\right]\left[S\;\overset{m}{T}{}^{-1}\overset{m}{a}\right]_{\alpha} \\
&- \sum_{m=1}^{M}\sum_{\alpha=13}^{27}\left[\oint_{L_m}(\overline{\sigma}_{\alpha ij}U_{ik} - \overline{u}_{\alpha i}\Sigma_{ijk})\,\epsilon_{jrt}dz_t\right]\left[\overset{m}{T}{}^{-1}\overset{*m}{a} + \left(\overset{m}{T}{}^{-1}\right)^{*}\overset{m}{a}\right]_{\alpha} \\
&+ \sum_{m=1}^{M}\sum_{\alpha=4}^{12}\left[\oint_{L_m}(\overline{\sigma}_{\alpha ij}U_{ik} - \overline{u}_{\alpha i}\Sigma_{ijk})\,\epsilon_{jnt}z_n dz_t\right]\times \\
&\left[R_{,r}\left(\overset{m}{T}{}^{-1}\overset{*m}{a} + \left(\overset{m}{T}{}^{-1}\right)^{*}\overset{m}{a}\right)\right]_{\alpha-3} \\
&- \sum_{m=1}^{M}\sum_{\alpha=4}^{12}\left[\oint_{L_m}(\overline{\sigma}_{\alpha ij}U_{ik} - \overline{u}_{\alpha i}\Sigma_{ijk})\,\epsilon_{jrt}dz_t\right]\left[R\left(\overset{m}{T}{}^{-1}\overset{*m}{a} + \left(\overset{m}{T}{}^{-1}\right)^{*}\overset{m}{a}\right)\right]_{\alpha-3} \\
&+ \sum_{m=1}^{M}\sum_{\alpha=1}^{3}\left[\oint_{L_m}D_{\alpha jk}dz_j\right]\left[S_{,r}\left(\overset{m}{T}{}^{-1}\overset{*m}{a} + \left(\overset{m}{T}{}^{-1}\right)^{*}\overset{m}{a}\right)\right]_{\alpha} \\
&+ \sum_{m=1}^{M}\sum_{\alpha=1}^{3}\left[\oint_{L_m}\Sigma_{\alpha jk}\epsilon_{jrt}dz_t\right]\left[S\left(\overset{m}{T}{}^{-1}\overset{*m}{a} + \left(\overset{m}{T}{}^{-1}\right)^{*}\overset{m}{a}\right)\right]_{\alpha}
\end{aligned} \tag{50}$$

The curvatures $u_{k,rp}$ in the above equation can be obtained from equation (30) in [MukhSM00]. Stress sensitivities can be easily obtained from $\overset{*}{v}_{kr}$ by using Hooke's law.

3.3 Numerical Results

Numerical results, for sensitivities of the displacement and stresses in a hollow sphere under internal pressure, appear below. These results correspond to the displacement and stress distributions displayed in Figures 3 - 5. In all cases, the design variable is a, the inside radius of the sphere, and a linear design velocity profile is used. This is:

$$\overset{*}{R} = \frac{b-R}{b-a} \tag{51}$$

3.3.1 Sensitivities of surface displacement.

Displacement sensitivities at surface points are calculated from equation (30). The sensitivity of the radial displacement profile in Figure 3 is shown in Figure 6. It is seen that the numerical

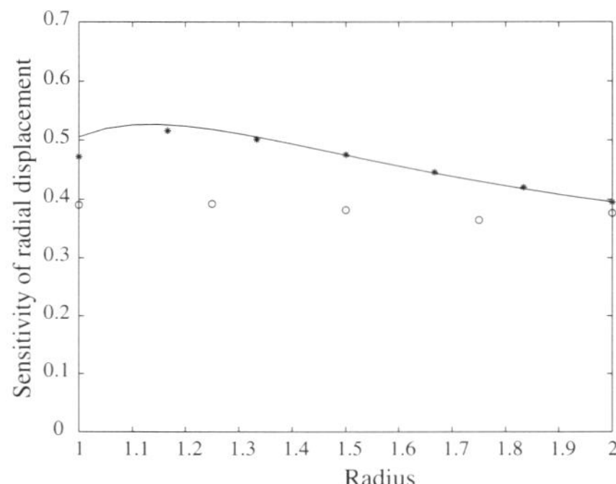

Figure 6: Hollow sphere under internal pressure. Sensitivity of radial displacement along the x_1 axis. Exact solution —. Numerical solutions: coarse mesh $\circ\circ\circ\circ$, medium mesh $*\,*\,**$ (from [MukhSM99])

results for the coarse mesh exhibit large errors; however, they do appear to converge to the exact solution with increasing mesh density. Please see [ChanM97] for a discussion of analytical solutions for design sensitivities for various examples.

3.3.2 Sensitivities of surface stresses.

Stress sensitivities on the surface of the sphere are calculated from equations (35 - 36). As shown in Figure 7, numerical solutions from the medium mesh, for stress sensitivities on the outer surface $R = b$, agree well with the exact solution. The corresponding stress solution is shown in Figure 4.

3.3.3 Sensitivities of internal stresses.

Sensitivities of internal stresses, shown in Figure 5, appear in Figure 8. These sensitivities are calculated from equations (50) and Hooke's law. Very good agreement is observed between the analytical and numerical solutions.

4 Shape Optimization with the BCM

Shape optimization of 3-D elasticity problems, with the BCM, is the subject of this section. Further details are available in [ShiM99].

4.1 Shape Optimization

An optimal shape design problem can be stated as a minimization problem under certain constraints. Its general form can be stated as follows:

$$\text{Minimize} \quad f(\mathbf{b}) \tag{52}$$

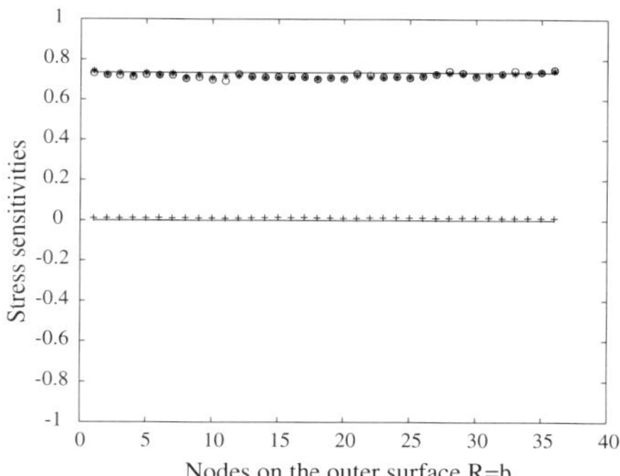

Figure 7: Hollow sphere under internal pressure. Stress sensitivities on the outer surface $R = b$. Exact solutions —. Numerical solutions from the medium mesh: $\overset{*}{\sigma}_{RR}$ $+ + + +$, $\overset{*}{\sigma}_{\theta\theta}$ $\circ\circ\circ\circ$, $\overset{*}{\sigma}_{\phi\phi}$ $* * **$ (from [MukhSM99])

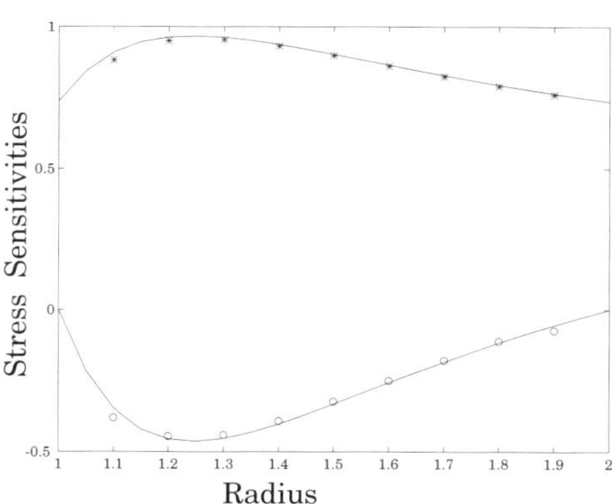

Figure 8: Hollow sphere under internal pressure. Stress sensitivities along the line $x_1 = x_2 = x_3$. Exact solutions —. Numerical solutions from the fine mesh: $\overset{*}{\sigma}_{\theta\theta}$ $* * **$, $\overset{*}{\sigma}_{RR}$ $\circ \circ \circ\circ$ (from [MukhSM00])

$$\text{Subject to} \quad g_i(\mathbf{b}) \geq 0, \qquad i = 1, ..., N_g \tag{53}$$

$$h_j(\mathbf{b}) = 0, \qquad j = 1, ..., N_h \tag{54}$$

$$b_k^\ell \leq b_k \leq b_k^u, \quad k = 1, ..., N \tag{55}$$

in which $\mathbf{b} = \langle b_1, b_2, ..., b_N \rangle^T$ are the design variables, $f(\mathbf{b})$ is the objective function, and $g_i(\mathbf{b})$ and $h_j(\mathbf{b})$ are the inequality and equality constraints, respectively. Equation (55) gives side constraints that are used to limit the search region of an optimization problem. Here, the parameters b_k^ℓ and b_k^u denote the lower and upper bounds, respectively, of the design variable b_k.

The most common mathematical programming approaches, used in gradient based optimization algorithms, are the successive linear programming (SLP) and successive quadratic programming (SQP) methods. In the SQP method, the optimization problem is approximated by expanding the objective function in a second order Taylor series about the current values of the design variables, while the constraints are expanded in a first order Taylor series.

The subroutine DNCONF from the IMSL library is coupled with a 3-D BCM code for elastic stress analysis in order to carry out the shape optimization examples that are described in Section 4.2. This subroutine, that uses the SQP method, is based on the FORTRAN subroutine NLPQL developed by [Schi86]. The required gradients of the objective functions, constraints etc. are calculated internally by the optimization subroutine by the finite difference method.

4.2 Numerical Results

An illustrative shape optimization example in 3-D linear elasticity is solved using the BCM coupled with the IMSL optimization subroutine mentioned above. The example is that of optimizing the shape of a fillet in a tension bar whose volume is selected as the optimization function. An optimal shape is sought that minimizes the volume (and therefore the weight when the bar material has constant density) without causing yielding anywhere in the bar. This is the axisymmetric version of the planar problem described in [PhanMM98c] - section 5.3. Of course, the full 3-D BCM code is used here.

The initial cross-section of the axisymmetric bar is shown in Figure 9. The Young's modulus, Poisson's ratio and allowable von Mises stress are taken as $E = 10^7 psi$, $\nu = 0.3$ and $\hat{\sigma}^{VM} = 120$ psi, respectively. The bar is loaded by a uniform axial tensile traction of 100 psi. The design surface is the surface of revolution (initially a cone) obtained by revolving the curve ED about the symmetry axis AB. The end circles of this surface of revolution are kept fixed. The variable boundary ED is modeled as a cubic spline (using the IMSL subroutine DCSDEC) which is defined by the fixed end points E and D and by the variable points $C_k, k = 1, 2, 3$. The points E, $C_k, k = 1, 2, 3$ and D have equally spaced projections on the axis of symmetry while the radii $r_k, k = 1, 2, 3$ are the design variables.

Quadratic CIM9 elements (see Figure 1) are used in this numerical example. A total of 160 elements are used to discretize the surface of the bar. These are distributed as follows. There are 16 elements on each of the end circles (on the planes z = 0 and z = 20, respectively) of the bar. The surface of revolution described by the edge FE has 32 elements, the one described by DC has 32 elements, and the design surface has 64 elements, respectively.

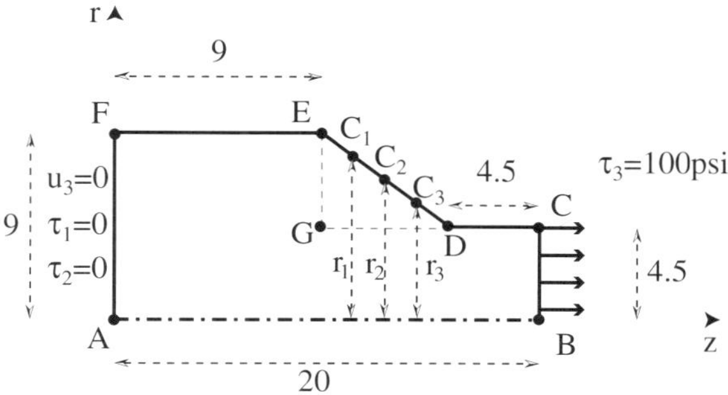

Figure 9: Modeling of a bar with a fillet (from [ShiM99])

The objective function is the volume of the axisymmetric object bounded by the plane $z = 9$, the cylindrical surface $r = 4.5$ and the design surface. This is:

$$\phi(r_1, r_2, r_3) = \int_{z_G}^{z_D} \pi \left[r^2(r_1, r_2, r_3, z) - (4.5)^2 \right] dz \tag{56}$$

The axisymmetric shape of the body is maintained during the optimization process and the design nodes $C_k,\ k = 1, 2, 3$ are constrained to lie within the triangle EGD. In addition, one has:

$$\sigma_i^{VM} / \hat{\sigma}^{VM} \leq 1.0, \qquad 1 \leq i \leq n_s \tag{57}$$

where σ_i^{VM} are the values of the von Mises stress at the centroids of certain elements on the curved surface of the bar. Here, since the physical problem is axisymmetric, the elements chosen are the ones whose centroids lie along FEDC, so that $n_s = 16$.

The usual definition of the von Mises stress is used. This is:

$$[\sigma^{VM}]^2 = (3/2) s_{ij} s_{ij} \tag{58}$$

in terms of s_{ij}, the deviatoric components of the stress σ_{ij}.

The final converged solution, which is obtained after just 1 iteration, is shown in Figure 10 The initial value of the objective function is 551.35 with one stress constraint being violated. The final (converged) value is 150.39 when 2 stress constraints are active. Thus, the final volume of the design portion of the bar is about 27.28 % of its initial value. In the corresponding 2-D problem [PhanMM98c], the design area reduces to 49.85% of its initial value.

5 Error Analysis and Adaptivity with the BCM and the HBCM

Error analysis and adaptivity for 3-D elasticity problems, with the BCM and the HBCM, is presented in this section. Further details are available in [MukhM01].

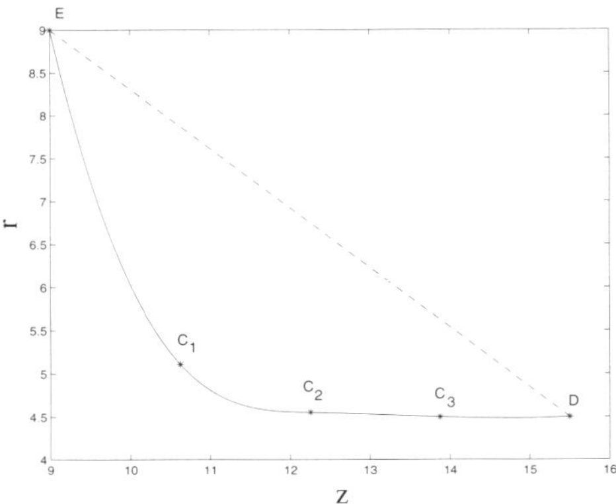

Figure 10: Optimal shape of the fillet (from [ShiM99])

5.1 Hypersingular Residuals as Local Error Estimators

The idea of using hypersingular residuals, to obtain local error estimates for the BIE, was first proposed by [Paul95] and [PaulGZ96]. This idea has been applied to the collocation BEM ([PaulGZ96], [MenoPM99], [PaulMM01]), to the symmetric Galerkin BEM ([PaulG99]) and to the BNM ([ChatPM01b]). The main idea, applied to the BCM, is as follows.

The usual BCM equation (12) is solved first for the boundary variables (tractions and displacements) $\mathbf{a}$. Next, this value of $\mathbf{a}$ is input into the right hand side of equation (20) in order to obtain the hypersingular residuals v_{kn} in the displacement gradients $u_{k,n}$. (Equation (20) yields the hypersingular residuals in the displacement gradients - see Section 2.5.1). Next, the stress residuals are obtained from Hooke's law:

$$s_{kn} = \lambda v_{mm}\delta_{kn} + \mu(v_{kn} + v_{nk}) \tag{59}$$

Finally, a scalar measure r of the residual, evaluated at the centroid of a triangular surface element, is postulated based on the idea of energy. This is:

$$r = s_{kn}v_{kn} \tag{60}$$

It has been proved in [MenoPM99], [PaulMM01] (for the BIE) that, under certain favorable conditions, real positive constants c_1 and c_2 exist such that:

$$c_1 r \leq \epsilon \leq c_2 r \tag{61}$$

where r is some scalar measure of a hypersingular residual and ϵ is a scalar measure of the exact local error. Thus, a hypersingular residual is expected to provide a good estimate of the local error on a boundary element. It should be mentioned here that the definitions of the residuals used in [MenoPM99] and [PaulMM01] are analogous to, but different in detail from, the ones proposed in this work.

In the rest of this chapter, $e = r$, where r, defined in equation (60) (and evaluated at an element centroid), is the hypersingular residual, and e is the local element error estimator that is used to drive an h-adaptive procedure with the BCM.

5.2 Adaptive Meshing Strategy

The flow chart for adaptive meshing is shown in Figure 11.

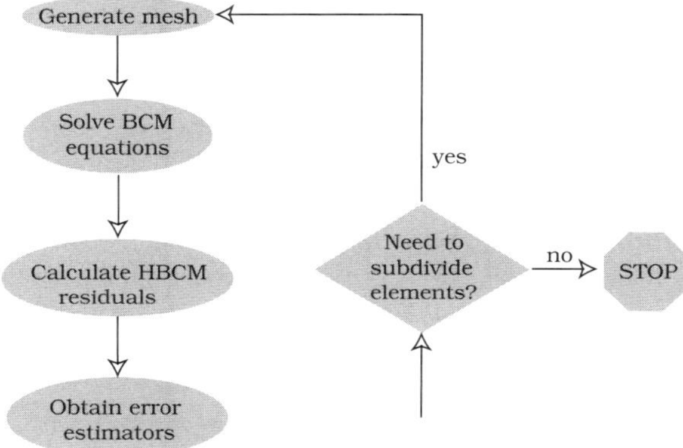

Figure 11: Flow chart for adaptive meshing (from [MukhM01])

The remeshing strategy is based on the values of the error estimator e at each element centroid. This strategy is shown in Figure 12 in which $\bar{e}$ is the average value of the error estimator e over all the boundary elements.

$\gamma = e/\bar{e}$	$1 \leqslant \gamma < 2$	$2 \leqslant \gamma < 3$	$3 \leqslant \gamma$
	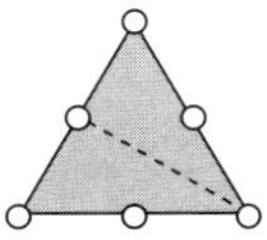	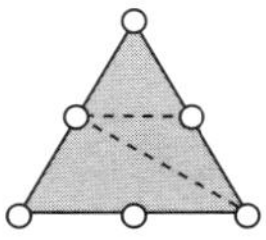	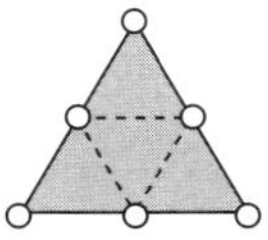
One element is split into	2 elements	3 elements	4 elements

Figure 12: Remeshing strategy (from [MukhM01])

A possible criterion for stopping cell refinement can be:

$$\bar{e} \leq e_{global} \tag{62}$$

where e_{global} has a preset value that depends on the level of overall desired accuracy.

5.3 Numerical Results

This example is concerned with a short cylinder which is clamped at the bottom and subjected to unit tensile traction on the top surface (Figure 13(a)). The radius and length of the cylinder are each 2 units, the shear modulus of the cylinder material is 1.0 and the Poisson's ratio is 0.3 (in consistent units). The initial mesh on the top (loaded) and bottom (clamped) faces of the cylinder are identical and are shown in Figure 13 (b).

It is known ([Crus69], [Pick44]) that, for this problem, the normal stress component σ_{33} varies slowly over much of the clamped face, but exhibits sharp gradients near its boundary. This stress component becomes singular on the boundary of the clamped face. The behavior of the shearing stress component σ_{zr} (here $r, \theta, z \equiv 3$ are the usual polar coordinates) on the clamped face is qualitatively similar to that of σ_{33}. The stresses are uniform on the loaded face.

It is seen from Figure 13 that this behavior is captured well by the adaptive scheme. Element error estimators are obtained from equations (20), (59) and (60) after first averaging the traction results from (12) within each element and then using these averaged traction values. Figure 13 (b) shows that these element error estimators are largest on the elements near the boundary of the clamped face. As a consequence (Figures 13 (b), (c), (d)) the region near the boundary of the clamped face is refined most while the mesh on the loaded face of the cylinder is left unaltered. Also, Figures 7 (c), (d) in [MukhM01] show that some mesh refinement takes place on the bottom layer of the curved surface of the cylinder, which is nearest to the clamped face, while the rest of the mesh on it remains unaltered.

Finally, the mesh statistics, together with $\bar{e}$, the average value of the error estimator e over the entire surface of the cylinder, appear in Table 2. As expected, $\bar{e}$ is seen to decrease with mesh refinement.

6 Acknowledgements

This research has been supported by NSF SBIR Phase I and Phase II grants to Dehan Engineering Numerics, Ithaca, NY. Some of the computing has been performed at the Cornell Theory Center. The contributions of Y.X. Mukherjee, X. Shi, E. D. Lutz, A. Nagarajan and A-V Phan, to this research, are gratefully acknowledged.

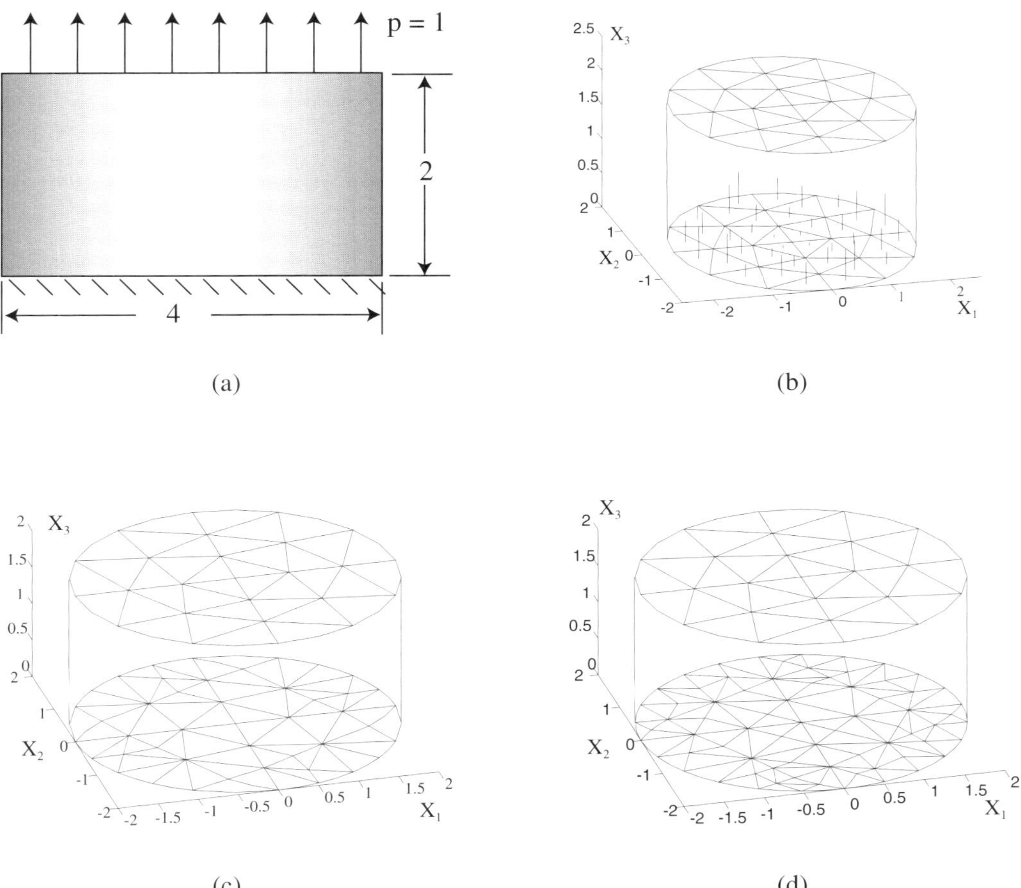

Figure 13: Adaptive meshing of the top and bottom faces of a short clamped cylinder under tension (analysis with the BCM). (a) geometry and loading (b) initial mesh with element error estimators (c) mesh at the end of the first adaptive step (d) mesh at the end of the second adaptive step (from [MukhM01])

mesh	# of elements	# of nodes	$\bar{e}$
initial	144	290	0.0086799
after 1st adaptive step	192	386	0.0048994
after 2nd adaptive step	246	494	0.0042723

Table 2: Mesh statistics and $\bar{e}$ for the clamped cylinder under tension

References

[Bonn95] Bonnet, M. (1995). Regularized direct and indirect symmetric variational BIE formulations for three-dimensional elasticity. *Engineering Analysis with Boundary Elements* 15:93-102.

[ChanM97] Chandra, A., and Mukherjee, S. (1997). *Boundary Element Methods in Manufacturing* New York:Oxford University Press.

[ChatM99] Chati, M.K., and Mukherjee, S. (1999). Evaluation of gradients on the boundary using fully regularized hypersingular boundary integral equations. *Acta Mechanica* 135:41-45.

[ChatMP01a] Chati, M.K., Mukherjee, S., and Paulino, G.H. (2001). The meshless hypersingular boundary node method for three-dimensional potential theory and linear elasticity problems. *Engineering Analysis with Boundary Elements* 25:639-653.

[ChatPM01b] Chati, M.K., Paulino, G.H., and Mukherjee, S. (2001). The meshless standard and hypersingular boundary node methods - applications to error estimation and adaptivity in three-dimensional problems. *International Journal for Numerical methods in Engineering* 50:2233-2269.

[ChenH99] Chen, J.T., and Hong, H-K. (1999). Review of dual boundary element methods with emphasis on hypersingular integrals and divergent series. *ASME Applied Mechanics Reviews* 52:17-33.

[ChoiK88] Choi, J.H., and Kwak, B.M. (1988). Boundary integral equation method for shape optimization of elastic structures. *International Journal for Numerical Methods in Engineering* 26:1579-1595.

[Crus69] Cruse, T.A. (1969). Numerical solutions in three-dimensional elastostatics. *International Journal of Solids and Structures* 5:1259-1274.

[CrusR96] Cruse, T.A., and Richardson, J.D. (1996). Non-singular Somigliana stress identities in elasticity. *International Journal for Numerical methods in Engineering* 39:3273-3304.

[GrayMI90] Gray, L.J., Martha, L.F., and Ingraffea A.R. (1990). Hypersingular integrals in boundary element fracture analysis. *International Journal for Numerical methods in Engineering* 29:1135-1158.

[GrayBK95] Gray, L.J., Balakrishna, C., and Kane, J.H. (1995). Symmetric Galerkin fracture analysis. *Engineering Analysis with Boundary Elements* 15:103-109.

[GrayP97a] Gray, L.J., and Paulino, G.H. (1997). Symmetric Galerkin boundary integral formulation for interface and multizone problems. *International Journal for Numerical methods in Engineering* 40:3085-3101.

[GrayP97b] Gray, L.J., and Paulino, G.H. (1997). Symmetric Galerkin boundary integral fracture analysis for plane orthotropic elasticity. *Computational Mechanics* 20:26-33.

[GrayP98] Gray, L.J., and Paulino, G.H. (1998). Crack tip interpolation revisited. *SIAM Journal of Applied Mathematics* 58:428-455.

[Guig94] Guiggiani, M. (1994). Hypersingular formulation for boundary stress evaluation. *Engineering Analysis with Boundary Elements* 13:169-179.

[HaugCK86] Haug, E.J., Choi, K.K., and Komkov, V. (1986). *Design Sensitivity Analysis of Structural Systems.* New York:Academic Press.

[KrisSRR90] Krishnasamy, G., Schmerr, L.W., Rudolphi, T.J., and Rizzo, F.J. (1990). Hypersingular boundary integral equations : some applications in acoustic and elastic wave scattering. *ASME Journal of Applied Mechanics* 57:404-414.

[KrisRR92] Krishnasamy, G., Rizzo, F.J. and Rudolphi, T.J. (1992). Hypersingular boundary integral equations: their occurrence, interpretation, regularization and computation. In Banerjee, P.K., and Kobayashi, S., eds., *Developments in Boundary Element Methods-7*, Elsevier Applied Science: London. 207-252.

[Lutz91] Lutz, E.D. (1991). *Numerical Methods for Hypersingular and Near-singular Boundary Integrals in Fracture Mechanics.* Ph.D. dissertation, Cornell University, Ithaca, NY.

[LutzIG92] Lutz, E.D., Ingraffea, A.R., and Gray, L.J. (1992). Use of 'simple solutions' for boundary integral methods in elasticity and fracture analysis. *International Journal for Numerical methods in Engineering* 35:1737-1751.

[MartR96] Martin, P.A., and Rizzo, F.J. (1996). Hypersingular integrals : how smooth must the density be ? *International Journal for Numerical methods in Engineering* 39:687-704.

[MartRC98] Martin, P.A., Rizzo, F.J., and Cruse, T.A. (1998). Smoothness-relaxation strategies for singular and hypersingular integral equations. *International Journal for Numerical methods in Engineering* 42:885-906.

[Meno96] Menon, G. (1996). *Hypersingular error estimates in boundary element methods.* M.S. Thesis, Cornell University, Ithaca, NY

[MenoPM99] Menon, G., Paulino, G.H., and Mukherjee, S. (1999). Analysis of hypersingular residual error estimates in boundary element methods for potential problems. *Computer Methods in Applied Mechanics and Engineering* 173:449-473.

[Mukh1982] Mukherjee, S. (1982). *Boundary Element Methods in Crep and Fracture.* London:Elsevier.

[Mukh00a] Mukherjee, S. (2000). CPV and HFP integrals and their applications in the boundary element method. *International Journal of Solids and Structures* 37:6623-6634.

[Mukh00b] Mukherjee, S. (2000). Finite parts of singular and hypersingular integrals with irregular boundary source points. *Engineering Analysis with Boundary Elements* 24:767-776.

[Mukh01] Mukherjee, S. (2001). On boundary integral equations for cracked and for thin bodies. *Mathematics and Mechanics of Solids* 6:47-64.

[MukhMSN97] Mukherjee, Y.X., Mukherjee, S., Shi, X., and Nagarajan, A. (1997). The boundary contour method for three-dimensional linear elasticity with a new quadratic boundary element. *Engineering Analysis with Boundary Elements* 20:35-44.

[MukhM98] Mukherjee, S., and Mukherjee, Y. X. (1998). The hypersingular boundary contour method for three-dimensional linear elasticity. *ASME Journal of Applied Mechanics* 65:300-309.

[MukhSM99] Mukherjee, S., Shi, X., and Mukherjee, Y.X. (1999). Surface variables and their sensitivities in three-dimensional linear elasticity by the boundary contour method. *Computer Methods in Applied Mechanics and Engineering* 173:387-402.

[MukhSM00] Mukherjee, S., Shi, X., and Mukherjee, Y.X. (2000). Internal variables and their sensitivities in three-dimensional linear elasticity by the boundary contour method. *Computer Methods in Applied Mechanics and Engineering* 187:289-306.

[MukhM01] Mukherjee, Y.X., and Mukherjee, S. (2001). Error analysis and adaptivity in three-dimensional linear elasticity by the usual and hypersingular boundary contour method. *International Journal of Solids and Structures* 38:161-178.

[NagaLM94] Nagarajan, A., Lutz, E.D., and Mukherjee, S. (1994). A novel boundary element method for linear elasticity with no numerical integration for 2-D and line integrals for 3-D problems. *ASME Journal of Applied Mechanics* 61:264-269.

[NagaML96] Nagarajan, A., Mukherjee, S., and Lutz, E.D. (1996). The boundary contour method for three-dimensional linear elasticity. *ASME Journal of Applied Mechanics* 63:278-286.

[NovaS99] Novati, G., and Springhetti, R. (1999). A Galerkin boundary contour method for two-dimensional linear elasticity. *Computational Mechanics* 23:53-62.

[Paul95] Paulino, G.H. (1995). *Novel formulations of the boundary element method for fracture mechanics and error estimation.* Ph.D. dissertation, Cornell University, Ithaca, NY

[PaulGZ96] Paulino, G.H., Gray, L.J., and Zarkian, V. (1996). Hypersingular residuals - a new approach for error estimation in the boundary element method. *International Journal for Numerical methods in Engineering* 39:2005-2029.

[PaulG99] Paulino, G.H., and Gray, L.J. (1999). Galerkin residuals for adaptive symmetric-Galerkin boundary element methods. *ASCE Journal of Engineering Mechanics* 125:575-585.

[PaulMM01] Paulino, G.H., Menon, G., and Mukherjee S. (2001). Error estimation using hypersingular integrals in boundary element methods for linear elasticity. *Engineering Analysis with Boundary Elements* Manuscript in print.

[PhanMM97] Phan, A-V., Mukherjee, S., and Mayer J.R.R. (1997). The boundary contour method for two-dimensional linear elasticity with quadratic boundary elements. *Computational Mechanics* 20:310-319.

[PhanMM98a] Phan, A-V., Mukherjee, S., and Mayer J.R.R. (1998). A boundary contour formulation for design sensitivity analysis in two-dimensional linear elasticity. *International Journal of Solids and Structures* 35:1981-1999.

[PhanMM98b] Phan, A-V., Mukherjee, S., and Mayer J.R.R. (1998). The hypersingular boundary contour method for two-dimensional linear elasticity. *Acta Mechanica* 130:209-225.

[PhanMM98c] Phan, A-V., Mukherjee, S., and Mayer J.R.R. (1998). Stresses, stress sensitivities and shape optimization in two-dimensional linear elasticity by the boundary contour method. *International Journal for Numerical methods in Engineering* 42:1391-1407.

[PhanM99] Phan, A-V., and Mukherjee, S. (1999). On design sensitivity analysis in linear elasticity by the boundary contour method. *Engineering Analysis with Boundary Elements* 23:195-199.

[Pick44] Pickett, G. (1944). Application of the Fourier method to the solution of certain boundary value problems in the theory of elasticity. *ASME Journal of Applied Mechanics* 66:A176-A182.

[Rizzo67] Rizzo, F.J. (1967). An integral equation approach to boundary value problems of classical elastostatics *Quarterly of Applied Mathematics* 25:83-95.

[SandW88] Sandgren, E., and Wu, S.J. (1988). Shape optimization using the boundary element method with substructuring. *International Journal for Numerical methods in Engineering* 26:1913-1924.

[SaigK90] Saigal S., and Kane, J.H. (1990). Boundary-element shape optimization system for aircraft structural components. *AIAA Journal* 28:1203-1204.

[Schi86] Schittkowski, K. (1986). NLPQL: a FORTRAN subroutine solving constrained nonlinear programming problems. *Annals of Operations Research* 5:485-500.

[ShiM99] Shi, X., and Mukherjee, S. (1999). Shape optimization in three-dimensional linear elasticity by the boundary contour method. *Engineering Analysis with Boundary Elements* 23:627-637.

[TafF95] Tafreshi, A., and Fenner, R.T. (1995). General-purpose computer program for shape optimization of engineering structures using the boundary element method. *Computers and Structures* 56:713-720.

[TanaSS94] Tanaka, M., Sladek, V., and Sladek, J. (1994). Regularization techniques applied to boundary element methods. *ASME Applied Mechanics Reviews* 47:457-499.

[TohM94] Toh, K-C., and Mukherjee, S. (1994). Hypersingular and finite part integrals in the boundary element method. *International Journal of Solids and Structures* 31:2299-2312.

[TimoG70] Timoshenko, S.P., and Goodier, J.N. (1970) *Theory of Elasticity*, 3rd. ed. New York:McGraw Hill.

[WeiCLM94] Wei, X., Chandra, A., Leu, L.J., and Mukherjee, S. (1994). Shape optimization in elasticity and elasto-viscoplasticity by the boundary element method. *International Journal of Solids and Structures* 31:533-550.

[WildA98] Wilde, A.J., and Aliabadi, M.H. (1998). Direct evaluation of boundary stresses in the 3D BEM of elastostatics. *Communications in Numerical Methods in Engineering* 14:505-517.

[YamaSK94] Yamazaki, K., Sakamoto, J., and Kitano, M. (1994). Three-dimensional shape optimization using the boundary element method. *AIAA Journal* 32:1295-1301.

[Yang90] Yang, R.J. (1990). Component shape optimization using BEM. *Computers and Structures* 37:561-568.

[Zhao91] Zhao, Z. (1991). *Shape design sensitivity analysis and optimization using the boundary element method.* New York:Springer.

[ZhaoL99] Zhao, Z.Y., and Lan, S.R. (1999). Boundary stress calculation - a comparison study. *Computers and Structures* 71:77-85.

[ZhouCS99] Zhou, S., Cao, Z., and Sun, S. (1999). The traction boundary contour method for linear elasticity. *International Journal for Numerical Methods in Engineering* 46:1883-1895.

Chapter 12
The Boundary Node Method

Subrata Mukherjee

Department of Theoretical and Applied Mechanics, Kimball Hall,
Cornell University, Ithaca, NY 14853, USA
e-mail: sm85@cornell.edu

Abstract

This chapter presents applications of the boundary node method (BNM) in three dimensional (3-D) linear elasticity. Following a brief introduction, and a section on surface approximants, derivations of the BNM and the hypersingular BNM (HBNM) are presented in Section 3. This is followed by a section describing error estimation and adaptivity with the BNM and the HBNM. Numerical results for selected examples are included throughout the chapter.

1 Introduction

A brief introduction to the BNM, HBNM, and their applications in 3-D linear elasticity, appears in this section.

Conventional computational engines such as the Finite Difference Method (FDM), Finite Element Method (FEM), and the Boundary Element Method (BEM) require meshing of either the domain (FEM and FDM) or the surface (BEM) of a solid body. Although significant progress has been made in three-dimensional (3-D) meshing algorithms (see Mackerle [Mack93]), the task of meshing a 3-D object with complicated geometry can be arduous, time consuming, computationally cumbersome and expensive. Such shortcomings are greatly amplified when one considers problems with changing geometry such as crack propagation, finite deformation, phase change or shape optimization. The main difficulty in these problems is the task of re-meshing a 3-D object many times during a solution process.

1.1 Meshfree Methods

In recent years, novel computational algorithms have been proposed that circumvent some of the problems associated with 3-D meshing. These methods have been collectively referred to as "meshfree" or "meshless" methods. Nayroles et al. [NayrTV92] proposed a method called the Diffuse Element Method (DEM). The main idea of their work is to replace the usual FEM interpolation by a "diffuse approximation". Their strategy consists of using a least-squares approximation scheme to interpolate the field variables - these are called MLS interpolants in Ref.

[NayrTV92]. The idea of MLS interpolants, for curve and surface fitting, is described in a book by Lancaster and Salkauskas [LancS90]. Nayroles et al. [NayrTV92] have applied the DEM to two-dimensional (2-D) problems in potential theory and linear elasticity.

During the relatively short span of less than a decade, great progress has been made in solid mechanics applications of meshfree methods. Meshfree methods proposed to date include the element-free Galerkin (EFG) method ([BelyLG94]), the reproducing kernel particle method (RKPM) ([LiuJZ95]), $h-p$ clouds ([DuarO96a], [DuarO96b], [OdenDZ98]), the meshless local Petrov-Galerkin (MLPG) approach ([AtluZ98a], [AtluZ98b]), the local boundary integral equation (LBIE) method ([ZhuZA98], [SladSA00]), the natural element method (NEM) ([SukuMB98], [SukuMSB01]), the generalized finite element method (GFEM) ([StroBC00]), the extended finite element method (X-FEM) ([MoesDB99], [DolbMB00], [SukuMMB00]), the finite point method (FPM) ([OnatIZT96, OnatIZTS96]), the finite cloud method (FCM) ([AlurL01]), the boundary cloud method (BCLM) ([LiA01]) and the boundary node method (BNM) ([MukhM97a, KothMM99, ChatM00, ChatMM99, ChatMP01a, ChatPM01b, GowrM01].

1.1.1 The element-free Galerkin method (EFG).

The EFG method deserves special mention because of its considerable success during recent years. The main idea in the EFG method is to use Moving Least Squares (MLS) approximants to construct the trial functions used in the Galerkin weak form. Integration is typically carried out using cells - although attempts have been made ([BeisB96, ChenWYY01]) to use nodal integration and eliminate cells altogether. An early problem with application of essential boundary conditions in the EFG has been solved ([ChenPWL96, KaljS97, MukhM97b]). A wide variety of problems have been solved using the EFG method. In the introductory paper [BelyLG94], the EFG method was applied to 2-D problems in linear elasticity and heat conduction. Since then, the method has been applied, for example, to solve problems in elasto-plasticity ([BarrS99]), fracture mechanics ([SukuMBB97]), crack growth ([BelyLG95], [XuS98a]), dynamic fracture ([KrysB99], [BelyLGT95], [BelyT96]), elasto-plastic fracture mechanics ([XuS98b], [XuS99]), plate bending ([KrysB95]), thin shells ([KrysB96]), damage mechanics ([AskePd00]) and sensitivity analysis and shape optimization ([BobaM01a], [BobaM01b]).

1.1.2 Boundary only meshfree methods.

S. Mukherjee, together with his research collaborators, has recently pioneered a breakthrough computational approach called the Boundary Node Method (BNM) [MukhM97a, KothMM99, ChatM00, ChatMM99, ChatMP01a, ChatPM01b]. Another example of a boundary-based meshless method is the local BIE (LBIE) [ZhuZA98] approach. The LBIE, however, is not strictly a boundary method since it requires evaluation of integrals over certain surfaces (called L_s in [ZhuZA98]) that can be regarded as "closure surfaces" of boundary elements. Li and Aluru [LiA01] have very recently proposed a boundary only method called the Boundary Cloud Method (BCM). This method is very similar to the BNM in that scattered points are used for constructing interpolation functions and these interpolations are used with the appropriate Boundary Integral Equations (BIEs) for the problem. However, a key feature of this very nice paper is that, unlike the BNM where surface curvilinear coordinates must be employed, the usual Cartesian coordinates can be used in the BCM.

The BNM is a combination of the MLS interpolation scheme and the standard Boundary Integral Equation (BIE) method. The method divorces the traditional coupling between spatial discretization (meshing) and interpolation (as commonly practiced in the FEM or in the BEM). Instead, a "diffuse" interpolation, based on MLS interpolants, is used to represent the unknown functions; and surface cells, with a very flexible structure (e.g. any cell can be arbitrarily subdivided without affecting its neighbors) are used for integration (see Figure 1). Thus, the BNM retains the *meshless attribute of the EFG method and the dimensionality advantage of the BEM*. As a consequence, the BNM only requires the specification of *points on the 2-D bounding surface* of a 3-D body (including crack faces in fracture mechanics problems), together with unstructured surface cells, thereby practically eliminating the meshing problem. In contrast, the FEM needs volume meshing, the BEM needs surface meshing, and the EFG needs points throughout the domain of a body.

1.2 Hypersingular Boundary Element Method

Hypersingular boundary element equations (HBIEs) are derived from a differentiated version of the usual boundary integral equations (BIEs). HBIEs have diverse important applications and are the subject of considerable current research (see, for example, [KrisRR92], [TanaSS94], [Paul95] and [ChenH99] for recent surveys of the field). HBIEs, for example, have been employed for the evaluation of boundary stresses (e.g. [Guig94], [WildA98], [ZhaoL99], [ChatM99]), in wave scattering (e.g. [KrisSRR90]), in fracture mechanics (e.g. [GrayMI90], [LutzIG92], [Paul95], [GrayP98], [Mukh01]), in order to obtain symmetric Galerkin boundary element formulations (e.g. [Bonn95], [GrayBK95], [GrayP97a], [GrayP97b]), to obtain the hypersingular boundary contour method (HBCM) [PhanMM98b], [MukhM98]), to obtain the hypersingular boundary node method (HBNM) ([ChatMP01a]), and for error analysis ([PaulGZ96], [Meno96], [MenoPM99], [PaulMM01] [ChatPM01b]) and adaptivity [ChatPM01b]. The relationship between finite parts of strongly singular and hypersingular integrals, and the HBIE, is discussed in [TohM94], [Mukh00a] and [Mukh00b]. A lively debate (e.g. [MartR96], [CrusR96]), on smoothness requirements on boundary variables for collocating an HBIE on the boundary of a body, has apparently been concluded recently [MartRC98].

1.3 Boundary Node Method

Mukherjee and Mukherjee [MukhM97a] have recently proposed a meshfree method called the boundary node method (BNM). The BNM has been used for solving two-dimensional (2-D) problems in potential theory ([MukhM97a]) and linear elasticity ([KothMM99]), for 3-D problems in potential theory ([ChatM00]) and linear elasticity ([ChatMM99]), and for error analysis and adaptivity ([ChatPM01b]). The hypersingular BNM (HBNM) has also been employed to solve 3-D problems in potential theory and linear elasticity ([ChatMP01a]). This method is a combination of the moving least squares (MLS) approximation scheme and the standard boundary integral equation (BIE) method. The method divorces the traditional coupling between spatial discretization (meshing) and interpolation as commonly practiced in the FEM or in the BEM. Instead, a "diffuse" approximation, based on MLS approximants, is used to represent the unknown functions; and surface cells, with a very flexible structure (e.g. any cell can be arbitrarily subdivided without affecting its neighbors) are used for integration (Figure 1). Thus, the BNM retains the *meshless attribute of the EFG method and the dimensionality advantage of the BEM.* As a consequence, the BNM only requires the specification of *points on the 2-D bounding surface* of a 3-D body (including crack faces in fracture mechanics problems), together with unstructured surface cells, thereby practically eliminating the meshing problem that has been referred to previously. In contrast, the FEM needs volume meshing, the BEM needs surface meshing, and the EFG needs points throughout the domain of a body.

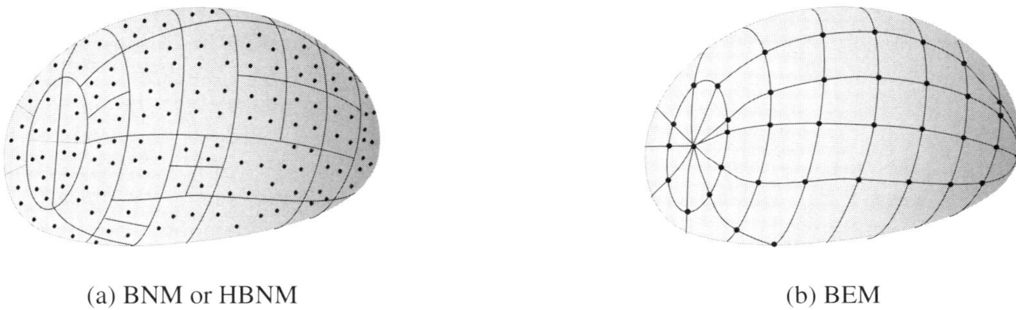

(a) BNM or HBNM (b) BEM

Figure 1: Comparison of BNM or HBNM and BEM input data structure: (a) BNM or HBNM - Cells and collocation nodes; (b) BEM - mesh with elements and collocation nodes (from [ChatPM01b])

1.4 Error Analysis and Adaptivity with the BNM and the HBNM

[Paul95] and [PaulGZ96] first proposed the idea of obtaining a hypersingular residual by substituting the BEM solution of a problem into the hypersingular BEM (HBEM) for the same problem; and then using this residual as an element error estimator in the BEM. It has been proved that ([Meno96], [MenoPM99], [PaulMM01]), under certain conditions, this residual is related to a measure of the local error on a boundary element, and has been used to postulate local error estimates on that element. This idea has been applied to the collocation BEM ([PaulGZ96], [MenoPM99], [PaulMM01]) and to the symmetric Galerkin BEM ([PaulG99]). Very recently, residuals have been obtained in the context of the boundary node method (BNM) [ChatPM01b] and used to obtain local error estimates (at the element level) and then to drive an h-adaptive mesh refinement process. This work is described in this chapter. An analogous approach for error estimation and h-adaptivity, in the context of the BCM, is described in [MukhM01].

2 Surface Approximants

A moving least-squares (MLS) approximation scheme, using curvilinear coordinates on the surface of a 3-D solid body, is suitable for the BNM. Such a scheme ([ChatM00] for problems in potential theory and [ChatMM99] for linear elasticity) is briefly described below.

2.1 Moving Least Squares (MLS) Approximants

It is assumed that, for 3-D problems, the bounding surface ∂B of a solid body is the union of piecewise smooth segments called panels. On each panel, one defines surface curvilinear coordinates (s_1, s_2). For 3-D problems in potential theory, let u be the potential function and $\tau \equiv \partial u/\partial n$ the flux. (Here $\mathbf{n}$ is an unit outward normal to ∂B at a point on it). For 3-D linear elasticity, let u denote a component of the displacement vector $\mathbf{u}$ and τ be a component of the traction vector $\boldsymbol{\tau}$ on ∂B. One defines :

$$u(\mathbf{s}) = \sum_{i=1}^{m} p_i(\mathbf{s} - \mathbf{s}^E)a_i = \mathbf{p}^T(\mathbf{s} - \mathbf{s}^E)\mathbf{a}\,, \quad \tau(\mathbf{s}) = \sum_{i=1}^{m} p_i(\mathbf{s} - \mathbf{s}^E)b_i = \mathbf{p}^T(\mathbf{s} - \mathbf{s}^E)\mathbf{b} \tag{1}$$

The monomials p_i (see below) are evaluated in local coordinates $(s_1 - s_1^E, s_2 - s_2^E)$ where (s_1^E, s_2^E) are the global coordinates of an evaluation point E. It is important to state here that a_i and b_i are not constants. Their functional dependencies are determined later. The name "moving least squares" arises from the fact that the quantities a_i and b_i are not constants. The integer m is the number of monomials in the basis used for u and τ. Quadratic approximants, for example, are of the form:

$$\mathbf{p}^T(\tilde{s}_1, \tilde{s}_2) = [1, \tilde{s}_1, \tilde{s}_2, \tilde{s}_1^2, \tilde{s}_2^2, \tilde{s}_1\tilde{s}_2], \quad m = 6, \quad \tilde{s}_i = s_i - s_i^E \;\; ; \quad i = 1, 2 \tag{2}$$

The coefficients a_i and b_i are obtained by minimizing the weighted discrete L_2 norms:

$$R_u = \sum_{I=1}^{n} w_I(d)\left[\mathbf{p}^T(\mathbf{s}^I - \mathbf{s}^E)\mathbf{a} - \hat{u}_I\right]^2\,, \quad R_\tau = \sum_{I=1}^{n} w_I(d)\left[\mathbf{p}^T(\mathbf{s}^I - \mathbf{s}^E)\mathbf{b} - \hat{\tau}_I\right]^2 \tag{3}$$

where the summation is carried out over the n boundary nodes for which the weight function satisfies the inequality $w_I(d) \neq 0$. (Weight functions are defined in Section 2.3). The quantity $d = g(\mathbf{s}, \mathbf{s}_I)$ is the length of the geodesic on ∂B between $\mathbf{s}$ and $\mathbf{s}^I$. These n nodes are said to be within the domain of dependence of a point $\mathbf{s}$ (evaluation point E in Figure 2(a)). Also, $(s_1^I - s_1^E, s_2^I - s_2^E)$ are the local surface coordinates of the boundary nodes with respect to the evaluation point $\mathbf{s}^E = (s_1^E, s_2^E)$ and $\hat{u}_I$ and $\hat{\tau}_I$ are the approximations to the nodal values u_I and τ_I. These equations above can rewritten in compact form as:

$$R_u = (\mathbf{P}(\mathbf{s}^I - \mathbf{s}^E)\mathbf{a} - \hat{\mathbf{u}})^T\mathbf{W}(\mathbf{s}, \mathbf{s}^I)(\mathbf{P}(\mathbf{s}^I - \mathbf{s}^E)\mathbf{a} - \hat{\mathbf{u}}) \tag{4}$$

$$R_\tau = (\mathbf{P}(\mathbf{s}^I - \mathbf{s}^E)\mathbf{b} - \hat{\boldsymbol{\tau}})^T\mathbf{W}(\mathbf{s}, \mathbf{s}^I)(\mathbf{P}(\mathbf{s}^I - \mathbf{s}^E)\mathbf{b} - \hat{\boldsymbol{\tau}}) \tag{5}$$

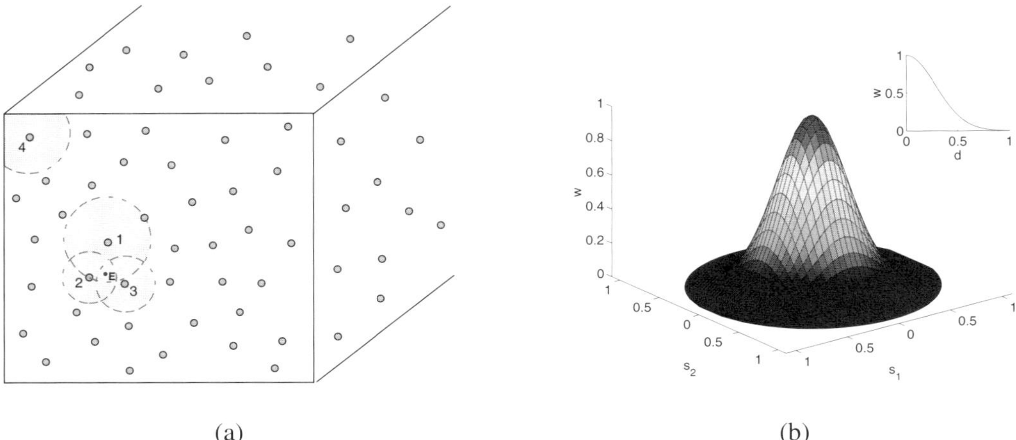

(a) (b)

Figure 2: Domain of dependence and range of influence. (a) The nodes 1, 2 and 3 lie within the domain of dependence of the evaluation point E. The ranges of influence of nodes 1, 2, 3 and 4 are shown as gray circles. The range of influence of node 4 is truncated at the edges of the body. (b) Gaussian weight function defined on the range of influence of a node (from [ChatPM01b])

where $\hat{\mathbf{u}}^T = (\hat{u}_1, \hat{u}_2, \cdots, \hat{u}_n)$, $\hat{\boldsymbol{\tau}}^T = (\hat{\tau}_1, \hat{\tau}_2, \cdots, \hat{\tau}_n)$, $\mathbf{P}(\mathbf{s}^I)$ is an $n \times m$ matrix whose k_{th} row is:

$$[1, p_2(s_1^k, s_2^k), ..., p_m(s_1^k, s_2^k)]$$

and $\mathbf{W}(\mathbf{s}, \mathbf{s}^I)$ is an $n \times n$ diagonal matrix with $w_{kk} = w_k(d)$ (no sum over k). A typical weight function $w_I(d)$ is illustrated in Figure 2(b) and is discussed in Section 2.3.

The stationarity of R_u and R_t, with respect to $\mathbf{a}$ and $\mathbf{b}$, respectively, leads to the equations:

$$\mathbf{a}(\mathbf{s}) = \mathbf{A}^{-1}(\mathbf{s})\mathbf{B}(\mathbf{s})\hat{\mathbf{u}}\,, \quad \mathbf{b}(\mathbf{s}) = \mathbf{A}^{-1}(\mathbf{s})\mathbf{B}(\mathbf{s})\hat{\boldsymbol{\tau}} \tag{6}$$

where

$$\mathbf{A}(\mathbf{s}) = \mathbf{P}^T(\mathbf{s}^I - \mathbf{s}^E)\mathbf{W}(\mathbf{s}, \mathbf{s}^I)\mathbf{P}(\mathbf{s}^I - \mathbf{s}^E), \;\; \mathbf{B}(\mathbf{s}) = \mathbf{P}^T(\mathbf{s}^I - \mathbf{s}^E)\mathbf{W}(\mathbf{s}, \mathbf{s}^I) \tag{7}$$

It is noted from above that the coefficients a_i and b_i turn out to be functions of $\mathbf{s}$. Substitution of equations (6) into equations (1) leads to:

$$u(\mathbf{s}) = \sum_{I=1}^{n} \Phi_I(\mathbf{s})\hat{u}_I, \;\; \tau(\mathbf{s}) = \sum_{I=1}^{n} \Phi_I(\mathbf{s})\hat{\tau}_I \tag{8}$$

where the approximating functions Φ_I are:

$$\Phi_I(\mathbf{s}) = \sum_{j=1}^{m} p_j(\mathbf{s} - \mathbf{s}^E)(\mathbf{A}^{-1}\mathbf{B})_{jI}(\mathbf{s}) \tag{9}$$

As mentioned previously, $\hat{\mathbf{u}}$ and $\hat{\boldsymbol{\tau}}$ are approximations to the actual nodal values $\mathbf{u}$ and $\boldsymbol{\tau}$. The two sets of values are related by equations (8). Discretized versions of (8) can be written as:

$$[\mathbf{H}]\{\hat{\mathbf{u}}_k\} = \{\mathbf{u}_k\}\,, \quad [\mathbf{H}]\{\hat{\boldsymbol{\tau}}_k\} = \{\boldsymbol{\tau}_k\}, \qquad k = 1, 2, 3 \tag{10}$$

Equations (10) relate the nodal approximations of u and τ to their nodal values.

2.2 Surface Derivatives

Surface derivatives of the potential (or displacement) field u are required for the HBIE. These are computed as follows. With

$$\mathbf{C} = \mathbf{A}^{-1}\mathbf{B}$$

equations (8) and (9) give:

$$u(\mathbf{s}) = \sum_{I=1}^{n}\sum_{j=1}^{m} p_j(\mathbf{s} - \mathbf{s}^E)\mathbf{C}_{jI}(\mathbf{s})\hat{u}_I \tag{11}$$

and the tangential derivatives of u can be written as:

$$\frac{\partial u(\mathbf{s})}{\partial s_k} = \sum_{I=1}^{n}\sum_{j=1}^{m}\left[\frac{\partial p_j}{\partial s_k}(\mathbf{s} - \mathbf{s}^E)\mathbf{C}_{jI}(\mathbf{s}) + p_j(\mathbf{s} - \mathbf{s}^E)\frac{\partial \mathbf{C}_{jI}(\mathbf{s})}{\partial s_k}\right]\hat{u}_I \qquad k = 1, 2 \tag{12}$$

The derivatives of the monomials p_j can be promptly computed. These are:

$$\frac{\partial \mathbf{p}^T}{\partial s_1}(s_1 - s_1^E, s_2 - s_2^E) = [0, 1, 0, 2(s_1 - s_1^E), 0, (s_2 - s_2^E)] \tag{13}$$

$$\frac{\partial \mathbf{p}^T}{\partial s_2}(s_1 - s_1^E, s_2 - s_2^E) = [0, 0, 1, 0, 2(s_2 - s_2^E), (s_1 - s_1^E)] \tag{14}$$

After some simple algebra ([Chat99]), the derivatives of the matrix $\mathbf{C}$ with respect to s_k take the form:

$$\frac{\partial \mathbf{C}(\mathbf{s})}{\partial s_k} = -\mathbf{A}^{-1}(\mathbf{s})\frac{\partial \mathbf{B}(\mathbf{s})}{\partial s_k}\mathbf{P}(\mathbf{s}^I - \mathbf{s}^E)\mathbf{A}^{-1}(\mathbf{s})\mathbf{B}(\mathbf{s}) + \mathbf{A}^{-1}(\mathbf{s})\frac{\partial \mathbf{B}(\mathbf{s})}{\partial s_k}, \qquad k = 1, 2 \tag{15}$$

with

$$\frac{\partial \mathbf{B}(\mathbf{s})}{\partial s_k} = \mathbf{P}^T(\mathbf{s}^I - \mathbf{s}^E)\frac{\partial \mathbf{W}(\mathbf{s}, \mathbf{s}^I)}{\partial s_k} \tag{16}$$

In deriving equation (15), the following identity has been used:

$$\frac{\partial \mathbf{A}^{-1}(\mathbf{s})}{\partial s_k} = -\mathbf{A}^{-1}(\mathbf{s})\frac{\partial \mathbf{A}(\mathbf{s})}{\partial s_k}\mathbf{A}^{-1}(\mathbf{s}) , \qquad k = 1, 2 \tag{17}$$

Tangential derivatives of the weight functions (described in Section 2.3) are easily computed ([Chat99]). The final form of the tangential derivatives of the potential (or displacement) u, at an evaluation point E, takes the form:

$$\begin{aligned}\frac{\partial u}{\partial s_k}(\mathbf{s}^E) &= \sum_{I=1}^{n}\sum_{j=1}^{m}\left[\frac{\partial p_j}{\partial s_k}(0,0)\mathbf{C}_{jI}(\mathbf{s}^E)\right]\hat{u}_I \\ &+ \sum_{I=1}^{n}\sum_{j=1}^{m}\left[p_j(0,0)\left\{\mathbf{A}^{-1}(\mathbf{s}^E)\frac{\partial \mathbf{B}}{\partial s_k}(\mathbf{s}^E)\left(\mathbf{I} - \mathbf{P}(\mathbf{s}^I - \mathbf{s}^E)\mathbf{A}^{-1}(\mathbf{s}^E)\mathbf{B}(\mathbf{s}^E)\right)\right\}\right]\hat{u}_I \end{aligned} \tag{18}$$

with $k = 1, 2$. In the above equation, $\mathbf{I}$ is the identity matrix.

One also needs the spatial gradient of the function u in order to solve the HBIE (see Section 3). For problems in potential theory, this is easily obtained from its tangential and normal derivatives $\partial u/\partial s_k$ and $\partial u/\partial n$, respectively. For elasticity problems, however, one must also use Hooke's law at a point on the surface ∂B. Details of this procedure are given in the Appendix of [ChatMP01a].

Equation (18) can be rewritten in compact form as:

$$\frac{\partial u}{\partial s_k}(\mathbf{s}^E) = \sum_{I=1}^{n}\Psi_I^{(k)}(\mathbf{s}^E)\hat{u}_I \quad ; \quad k = 1, 2 \tag{19}$$

where the approximating functions Ψ_I^k are:

$$\begin{aligned}\Psi_I^{(k)}(\mathbf{s}^E) &= \sum_{j=1}^{m}\left[\frac{\partial p_j}{\partial s_k}(0,0)\mathbf{C}_{jI}(\mathbf{s}^E)\right] \\ &+ \sum_{j=1}^{m}\left[p_j(0,0)\left\{\mathbf{A}^{-1}(\mathbf{s}^E)\frac{\partial \mathbf{B}}{\partial s_k}(\mathbf{s}^E)\left(\mathbf{I} - \mathbf{P}(\mathbf{s}^I - \mathbf{s}^E)\mathbf{A}^{-1}(\mathbf{s}^E)\mathbf{B}(\mathbf{s}^E)\right)\right\}\right] \end{aligned} \tag{20}$$

2.3 Weight Functions

The basic idea behind the choice of a weight function is that its value should decrease with distance from a node and that it should have compact support so that the region of influence of a node is of finite extent (Figure 2(b)). A possible choice is the Gaussian weight function:

$$w_I(d) = \begin{cases} e^{-(d/d_I)^2} & \text{for } d \leq d_I \\ 0 & \text{for } d > d_I \end{cases} \tag{21}$$

Here $d = g(\mathbf{s}, \mathbf{s^I})$ is the *minimum distance*, measured on the surface ∂B, (i.e. the geodesic) between a point $\mathbf{s}$ and the collocation node I. In the research performed to date, the region of influence of a node has been truncated at the edge of a panel (Figure 2(a)) so that geodesics, and their derivatives (for use in equation (21)), need only be computed on piecewise smooth surfaces. Finally, the quantities d_I determine the extent of the region of influence (the compact support) of node I. They can be made globally uniform, or can be adjusted such that approximately the same number of nodes get included in the region of influence of any given node I or in the domain of dependence of a given evaluation point E. Such ideas have been successfully implemented in Chati and Mukherjee [ChatM00] and Chati et al. [ChatMM99].

3 The BNM and the HBNM

Derivations of BNM and HBNM formulations, for 3-D linear elasticity, are presented below.

3.1 The Boundary Node Method

The standard BIE for linear elasticity, and its coupling with MLS interpolants to yield the BNM, is discussed below.

3.1.1 BIE.

For 3-D linear elasticity, the standard boundary integral equation, in regularized form, and in the absence of body forces, can be written as ([Rizz67])

$$0 = \int_{\partial B} [U_{ik}(\mathbf{x}, \mathbf{y})\tau_i(\mathbf{y}) - T_{ik}(\mathbf{x}, \mathbf{y})(u_i(\mathbf{y}) - u_i(\mathbf{x}))] \, dS(\mathbf{y}) \tag{22}$$

where u_k and τ_k are the components of the displacement and traction respectively, and the Kelvin kernels are:

$$U_{ik} = \frac{1}{16\pi(1-\nu)Gr} [(3-4\nu)\delta_{ik} + r_{,i}r_{,k}] \tag{23}$$

$$T_{ik} = \frac{-1}{8\pi(1-\nu)r^2} \left[\{(1-2\nu)\delta_{ik} + 3r_{,i}r_{,k}\} \frac{\partial r}{\partial n} + (1-2\nu)(r_{,i}n_k - r_{,k}n_i) \right] \tag{24}$$

In the above, $\mathbf{x}$ and $\mathbf{y}$ are source and field points, respectively, r is the Euclidean distance between $\mathbf{x}$ and $\mathbf{y}$, n_i are the components of the unit normal at a boundary field point $\mathbf{y}$, G is the shear modulus, ν is the Poisson ratio and δ_{ij} denotes the Kronecker delta. (Source and field points are also referred to as p and q (for internal points) and as P and Q (for boundary points), respectively, in this work). A comma denotes a derivative with respect to a field point, i.e.

$$r_{,i} = \frac{\partial r}{\partial y_i} = \frac{y_i - x_i}{r} \tag{25}$$

Note that the stress kernel $\mathbf{\Sigma}$, defined, for example, in [NagaML96], is related to the traction kernel $\mathbf{T}$ according to:

$$T_{ik} = \Sigma_{ijk} n_j \tag{26}$$

3.1.2 BNM.

The MLS approximants derived in Section 2 are used to approximate u and τ on the boundary ∂B. In order to carry out the integrations, the bounding surface is discretized into N_c cells. A variety of shape functions have been used in this work in order to interpolate the geometry. In particular, the bilinear (Q4) and quadratic (T6) triangle cells have been employed. These "geometric" shape functions can be found in any standard text on the FEM (e.g. [Ziek77]).

The BNM equation for elasticity is obtained by substituting the expressions for u_k and τ_k (equation (8) into equation (22), leading to:

$$0 = \sum_{m=1}^{N_c} \int_{\partial B_m} \left[U_{ik}(\mathbf{x},\mathbf{y}) \sum_{I=1}^{N_Q} \Phi_I(\mathbf{y}) \hat{\tau}_{iI} - T_{ik}(\mathbf{x},\mathbf{y}) \left\{ \sum_{I=1}^{N_Q} \Phi_I(\mathbf{y}) \hat{u}_{iI} - \sum_{I=1}^{N_P} \Phi_I(\mathbf{x}) \hat{u}_{iI} \right\} \right] dS(\mathbf{y}) \tag{27}$$

In the above, N_Q nodes lie in the domain of dependence of a field point $\mathbf{y} \equiv Q$, and similarly for the source point $\mathbf{x} \equiv P$.

3.1.3 Discretization

In order to evaluate the nonsingular integrals in equation (27) over (possibly curved) triangular or rectangular surface cells, 7 point and 3×3 Gauss quadrature are used, respectively. However, as $Q \to P$ the kernel U_{ik} becomes weakly singular and the kernel T_{ik} become strongly singular. As shown in equation (27), the strongly singular integrands are regularized by using rigid body modes and the regularized versions are weakly singular. Finally, special integration techniques are used to evaluate the resulting weakly singular integrals in equation (27) ([NagaM93], [ChatM00]).

The final discretized version of equation (27) has the form:

$$[\mathbf{A}_{(\hat{u})}]\{\hat{\mathbf{u}}\} + [\mathbf{A}_{(\hat{\tau})}]\{\hat{\tau}\} = \{\mathbf{0}\} \tag{28}$$

With respect to elasticity theory, the count for the number of equations and unknowns follows. For N_B nodes on the bounding surface, there are a total of $12N_B$ quantities on the boundary, i.e. $3N_B$ values for each of u_i and its nodal approximation $\hat{u}_i$, and similarly for τ_i. For a well posed problem, values of either u_i or τ_i are known at each node on the boundary, so $3N_B$ nodal values are given. Therefore, $9N_B$ equations are needed to solve for the $9N_B$ remaining unknowns. Equation (28) consists of $3N_B$ equations and equations (10) consist of $3N_B$ equations each. Thus, a well posed boundary value problem can be solved using equation (28), in combination with equations (10).

3.2 The Hypersingular Boundary Node Method

The HBIE for linear elasticity, and its coupling with MLS interpolants to yield the HBNM, is discussed below.

3.2.1 HBIE.

The fully regularized HBIE for linear elasticity can be written as ([CrusR96]):

$$\begin{aligned} 0 \;=\; & \int_{\partial B} D_{ijk}(\mathbf{x},\mathbf{y})(\tau_k(\mathbf{y}) - \tau_k(\mathbf{x}))\, dS(\mathbf{y}) \\ & -\sigma_{km}(\mathbf{x}) \int_{\partial B} D_{ijk}(\mathbf{x},\mathbf{y})(n_m(\mathbf{y}) - n_m(\mathbf{x}))\, dS(\mathbf{y}) \\ & - \int_{\partial B} S_{ijk}(\mathbf{x},\mathbf{y})\left[u_k(\mathbf{y}) - u_k(\mathbf{x}) - u_{k,m}(\mathbf{x})(y_m - x_m)\right] dS(\mathbf{y}) \end{aligned} \tag{29}$$

where the (strongly singular) kernel D_{ijk} and (hypersingular) kernel S_{ijk} are:

$$D_{ijk} = \frac{1}{8\pi(1-\nu)r^2}\left[(1-2\nu)(\delta_{ki}r_{,j} + \delta_{kj}r_{,i} - \delta_{ij}r_{,k}) + 3r_{,i}r_{,j}r_{,k}\right] \tag{30}$$

$$\begin{aligned} S_{ijk} \;=\; & \frac{G}{4\pi(1-\nu)r^3}\left[3\frac{\partial r}{\partial n}\left[(1-2\nu)\delta_{ij}r_{,k} + \nu(\delta_{ik}r_{,j} + \delta_{jk}r_{,i}) - 5r_{,i}r_{,j}r_{,k}\right]\right] \\ & + \frac{G}{4\pi(1-\nu)r^3}[3\nu(n_i r_{,j}r_{,k} + n_j r_{,i}r_{,k}) \\ & + (1-2\nu)(3n_k r_{,i}r_{,j} + n_j\delta_{ik} + n_i\delta_{jk}) - (1-4\nu)n_k\delta_{ij}] \end{aligned} \tag{31}$$

Taking the inner product of equation (29) with the source normal, one gets the equation:

$$\begin{aligned} 0 \;=\; & \int_{\partial B} D_{ijk}(\mathbf{x},\mathbf{y})n_j(\mathbf{x})(\tau_k(\mathbf{y}) - \tau_k(\mathbf{x}))\, dS(\mathbf{y}) \\ & -\sigma_{km}(\mathbf{x}) \int_{\partial B} D_{ijk}(\mathbf{x},\mathbf{y})n_j(\mathbf{x})(n_m(\mathbf{y}) - n_m(\mathbf{x}))\, dS(\mathbf{y}) \\ & - \int_{\partial B} S_{ijk}(\mathbf{x},\mathbf{y})n_j(\mathbf{x})\left[u_k(\mathbf{y}) - u_k(\mathbf{x}) - u_{k,m}(\mathbf{x})(y_m - x_m)\right] dS(\mathbf{y}) \end{aligned} \tag{32}$$

The procedure for obtaining the displacement gradients $u_{k,m}$, that are required in equations (29) and (32), is described in the Appendix of [ChatMP01a]. The stress components σ_{ij} can be easily obtained from the displacement gradients using Hooke's law.

3.2.2 HBNM.

Discretized versions of equations (29) and (32) are readily obtained as:

$$
\begin{aligned}
0 \;=\; & \sum_{l=1}^{N_c} \int_{\partial B_l} D_{ijk}(\mathbf{x},\mathbf{y}) \left[\sum_{I=1}^{N_Q} \Phi_I(\mathbf{y}) \hat{\tau}_{kI} - \sum_{I=1}^{N_P} \Phi_I(\mathbf{x}) \hat{\tau}_{kI} \right] dS(\mathbf{y}) \\
- \; & \sigma_{km}(\mathbf{x}) \int_{\partial B_l} D_{ijk}(\mathbf{x},\mathbf{y}) (n_m(\mathbf{y}) - n_m(\mathbf{x}))\, dS(\mathbf{y}) \\
- & \int_{\partial B_l} S_{ijk}(\mathbf{x},\mathbf{y}) \left[\sum_{I=1}^{N_Q} \Phi_I(\mathbf{y}) \hat{u}_{kI} - \sum_{I=1}^{N_P} \Phi_I(\mathbf{x}) \hat{u}_{kI} - u_{k,m}(\mathbf{x})(y_m - x_m) \right] dS(\mathbf{y}) \qquad (33)
\end{aligned}
$$

and

$$
\begin{aligned}
0 \;=\; & \sum_{l=1}^{N_c} \int_{\partial B_l} D_{ijk}(\mathbf{x},\mathbf{y}) n_j(\mathbf{x}) \left[\sum_{I=1}^{N_Q} \Phi_I(\mathbf{y}) \hat{\tau}_{kI} - \sum_{I=1}^{N_P} \Phi_I(\mathbf{x}) \hat{\tau}_{kI} \right] dS(\mathbf{y}) \\
- \; & \sigma_{km}(\mathbf{x}) \int_{\partial B_l} D_{ijk}(\mathbf{x},\mathbf{y}) n_j(\mathbf{x}) (n_m(\mathbf{y}) - n_m(\mathbf{x}))\, dS(\mathbf{y}) \\
- & \int_{\partial B_l} S_{ijk}(\mathbf{x},\mathbf{y}) n_j(\mathbf{x}) \left[\sum_{I=1}^{N_Q} \Phi_I(\mathbf{y}) \hat{u}_{kI} - \sum_{I=1}^{N_P} \Phi_I(\mathbf{x}) \hat{u}_{kI} - u_{k,m}(\mathbf{x})(y_m - x_m) \right] dS(\mathbf{y}) \qquad (34)
\end{aligned}
$$

3.2.3 Discretization.

Equations (33) and (34) are the HBNM equations for linear elasticity.

The procedure followed for discretization of (34) is quite analogous to the BNM case described before in Section 3.1.3. These equations are fully regularized and contain either nonsingular or weakly singular integrands. Nonsingular integrals are evaluated using the usual Gauss quadrature over surface cells, while the weakly singular integrals are evaluated using the procedure outlined in Nagarajan and Mukherjee [NagaM93] and Chati and Mukherjee [ChatM00]. The discretized version of either equation (34) has the generic form shown in equation (28).

3.3 Numerical Results

Numerical results from the BNM are available in [ChatMM99], while corresponding numerical results from the HBNM are available in [ChatMP01a]. A few selected examples are presented below.

3.3.1 Hollow sphere under internal pressure.

The first example consists of a hollow sphere, with inner and outer radius a and b, respectively, under internal pressure. Figure 3(a) shows a schematic of the problem under consideration. The numerical solution for this problem has been obtained using the material parameters $E = 1.0$, $\nu = 0.25$, and geometric parameters $a = 1.0, b = 4.0$ with internal pressure $p_i = 1$. The

Table 1: Convergence study for $\epsilon(u_r)$, the L_2 error in radial displacement (u_r), on the inner and outer surfaces of the sphere, for the 3-D Lamé problem

L_2 error	144 cells	256 cells	576 cells
Outer surface	0.404 %	0.241 %	0.0496 %
Inner surface	0.460 %	0.214 %	0.0658 %

numerical solution has been obtained by prescribing tractions all over the boundary and then modifying the resulting singular matrix using the ideas presented in the papers [LutzYM98] and [ChatMM99].

The entire surface of the hollow sphere has been modeled here, i.e. symmetry conditions have not been used. The exact solution for the radial displacement, radial and tangential stresses, as functions of the distance r from the center of the sphere, is available in [TimoG70]). The cell/nodal structure used to obtain the numerical results consists of 72 quadratic T6 cells on each surface of the hollow sphere with 1 node per cell.

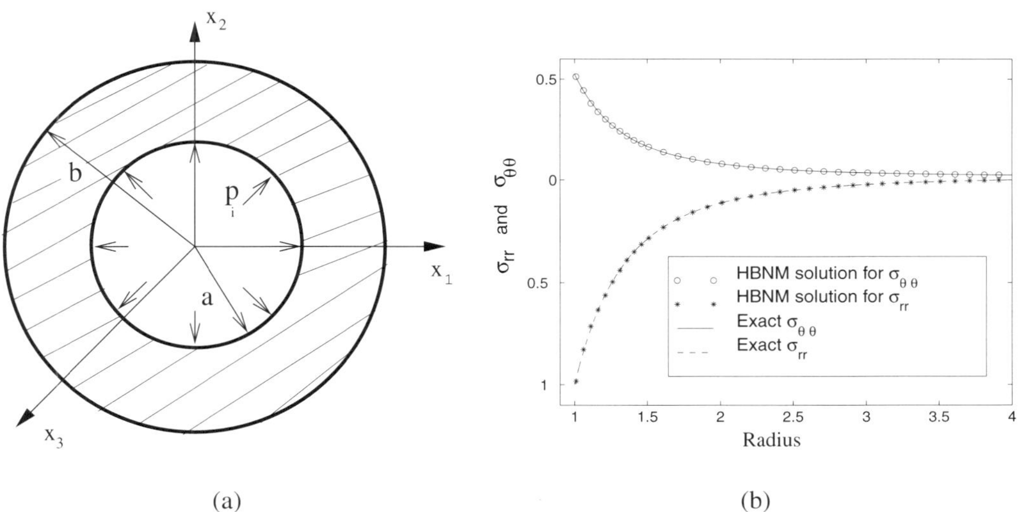

Figure 3: (a) Hollow sphere under internal pressure. (b) Radial and tangential stresses, along the x_1 axis, for the hollow sphere problem : comparison of HBNM and exact solution (from [ChatMP01a])

The radial and tangential stresses, obtained from the HBNM, are compared to the exact solution in Figure 3 (b). It can be clearly seen that the numerical results are in excellent agreement with the analytical solution. Results from the BNM ([ChatMM99]) are essentially identical to those from the HBNM within plotting accuracy.

The algorithm presented in [MukhCS00] is crucial for obtaining accurate solutions for the displacement and stresses at internal points close to the boundary. Also, a convergence study is carried out to demonstrate the robustness of the HBNM. Table 1 shows the L_2 error in the radial displacement (u_r) on the inner and outer surfaces of the sphere. One node per cell is used in all the calculations. As the number of cell increases, the L_2 error decreases on both the outer and inner surfaces of the sphere for the 3-D Lamé problem.

3.3.2 Bimaterial sphere.

The second example, from [ChatMM99], is concerned with a bimaterial sphere. The inner sphere is solid, of radius R_1, with Young's modulus and Poisson's ratio E_1 and ν_1, respectively. The outer sphere is hollow, with inner and outer radii equal to R_1 and R_2, respectively, and with material parameters E_2 and ν_2. Numerical results from the BNM, for this model have been obtained by prescribing displacements on the outer boundary. One could also prescribe tractions over the entire outer surface and then appropriately modify the scheme presented in [LutzYM98] for solving traction prescribed problems. This is planned for the future.

The exact solution for the problem is available in [ChatMM99].

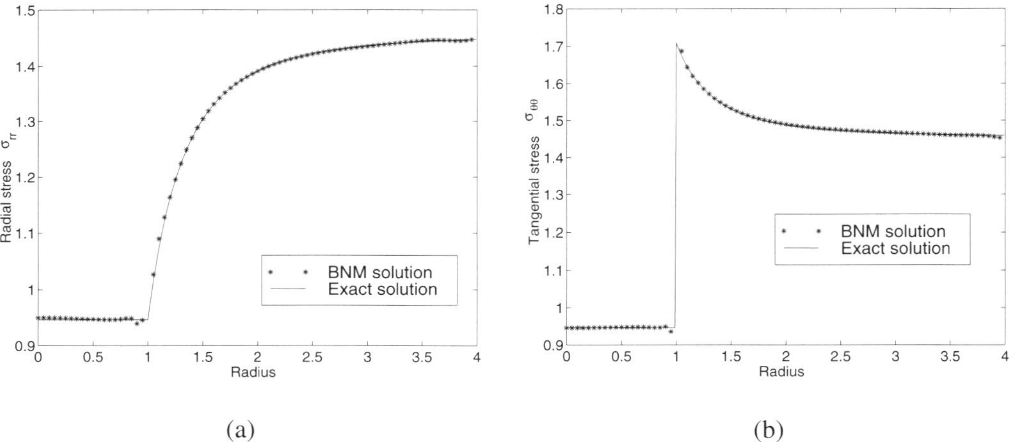

(a) (b)

Figure 4: Bimaterial sphere. (a) Radial stress distribution along the x_1 axis (b) Tangential stress distribution along the x_1 axis (from [ChatMM99])

The material and geometric parameters chosen for the two materials were : for material 1, $E_1 = 1.0, \nu_1 = 0.28, R_1 = 1.0$ and for material 2, $E_2 = 2.0, \nu_2 = 0.33, R_2 = 4.0$. A constant radial displacement is prescribed on the outer boundary of material 2 ($u_0 = 1.0$). Figures 4(a) and (b) show the radial and tangential stresses along the $x_1$axis. It can be seen that the jump in the tangential stress at the bimaterial interface is very nicely captured by the BNM.

3.3.3 3-D Kirsch problem.

The third example (from [ChatMM99] and [ChatMP01a]) concerns a 3-D version of the Kirsch problem. Figure 5(a) shows a cube with a spherical cavity. The cube is loaded by remote uniaxial tension σ_0 in the x_3 direction. Figure 5(b) shows a comparison of the exact solution [TimoG70] for σ_{33}/σ_0 along the x_1 axis, with the BNM and HBNM solutions. The cell structure consists of 96 Q4 cells on the cube surface and 72 T6 cells on the spherical cavity, with one node per cell. The agreement among the three solutions is seen to be excellent.

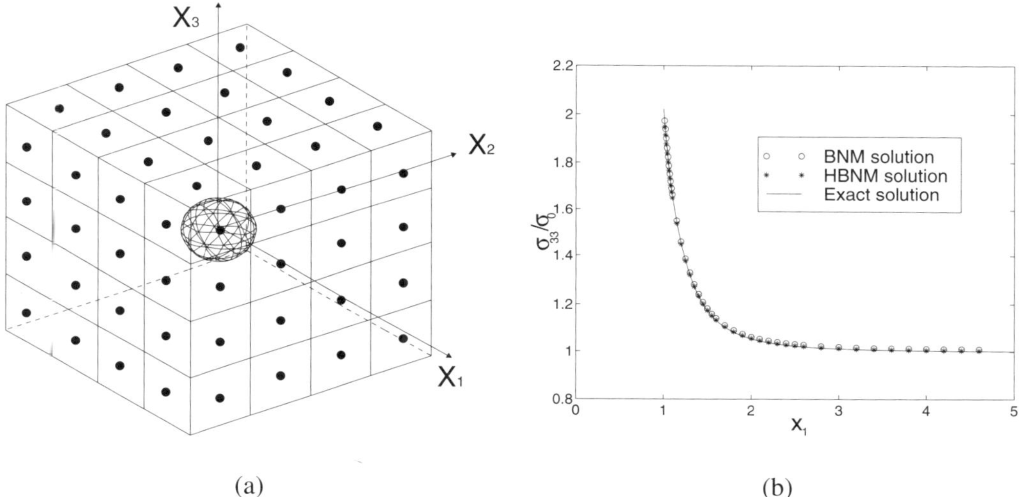

Figure 5: 3-D Kirsch problem. (a) Configuration (b) σ_{33}/σ_0 along the x_1 axis (from [ChatMM99, ChatMP01a])

4 Error Analysis and Adaptivity with the BNM and the HBNM

Error analysis and adaptive meshing for 3-D elasticity problems, with the BNM and the HBNM, is presented in this section. Further details are available in [ChatPM01b].

4.1 Hypersingular Residuals as Local Error Estimators

The idea of using hypersingular residuals, to obtain local error estimates for the BIE, was first proposed by [Paul95] and [PaulGZ96]. This idea has been applied to the collocation BEM ([PaulGZ96], [MenoPM99], [PaulMM01]), to the symmetric Galerkin BEM ([PaulG99]) and to the BNM ([ChatPM01b]). The main idea, applied to the BNM, is as follows.

The usual BNM equation (27) is solved first for the boundary variables - tractions and displacements. Next, these values of the boundary tractions and displacements is input into the right hand side of equation (33) in order to obtain the hypersingular residuals r_{ij} in the stress components σ_{ij}. (Equation (33) yields the hypersingular residuals in the stress components - see the derivation of equation (29) in [CrusR96]).

Finally, a scalar measure r of the residual is defined as:

$$r = r_{ij} r_{ij} \tag{35}$$

A corresponding singular residual in the displacement [ChatPM01b] can be defined by the opposite process, i.e. by first solving (34) for the boundary values of displacements and tractions, and then inputting these values into the right hand side of (27).

It has been proved in [MenoPM99], [PaulMM01] (for the BIE) that, under certain favorable conditions, real positive constants c_1 and c_2 exist such that:

$$c_1 r \leq \epsilon \leq c_2 r \tag{36}$$

where r is some scalar measure of a hypersingular residual and ϵ is a scalar measure of the exact local error. Thus, a hypersingular residual is expected to provide a good estimate of the local error on a boundary element. It should be mentioned here that the definitions of the residuals used in [MenoPM99] and [PaulMM01] are analogous to, but different in detail from, the ones proposed in this work.

In the rest of this chapter, $e = r$, where r, defined in equation (35) (and evaluated at a node - in this work, each cell has a single node located at its centroid), is the hypersingular residual, and e is the local cell error estimator that is used to drive an h-adaptive procedure with the BNM.

4.2 Progressive Adaptive Meshing Strategy

The progressive adaptive meshing strategy adopted in this work is presented below.

4.2.1 Iterative cell design scheme.

There are similarities between adaptive techniques (e.g. h-version) for mesh-based methods (see [PaulMCM99, PaulSMR97]) and meshless methods, however, the latter set of methods provides substantially more flexibility in the (re-)discretization process than the former ones.

The $h-$version iterative self-adaptive procedure employed in this work is presented in the flowchart of Figure 6. The goal is to efficiently develop a final cell configuration which leads to a reliable numerical solution, in as simple a manner as possible.

4.2.2 Cell refinement criterion.

A simple criterion for cell refinement consists of subdividing the cells for which the error indicator is larger than a certain reference value. In this work, the reference quantity is taken as the average value of the error indicator given by:

$$\bar{e} = \frac{1}{N_n} \sum_{i=1}^{N_n} e^{(i)} \tag{37}$$

where N_n is the total number of nodes. If the inequality:

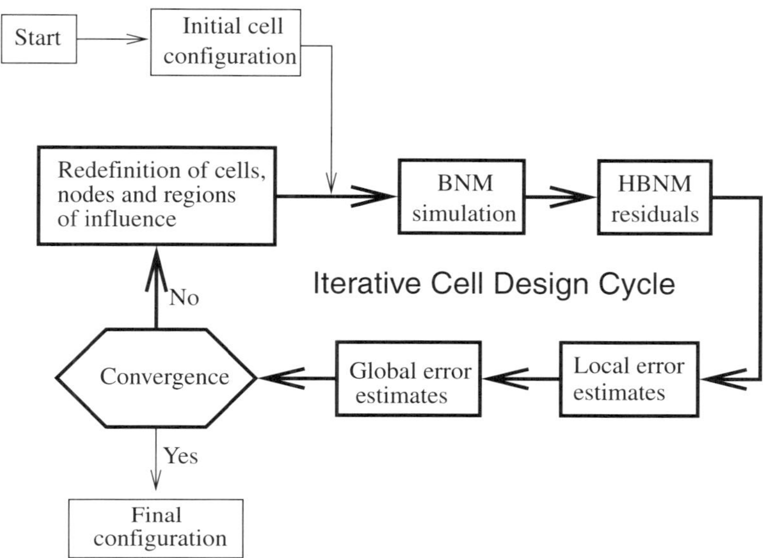

Figure 6: Self-adaptive iterative BNM algorithm (from [ChatPM01b])

$$e > \gamma \, \bar{e} \tag{38}$$

is satisfied, (where e is evaluated at the single (centroidal) node in a cell) then the cell is subdivided into four pieces (see Figure 7). The parameter γ in equation (38) is a weighting coefficient that controls the "cell refinement velocity". The standard procedure consists of using $\gamma = 1.0$. If $\gamma > 1.0$, then the number of cells to be refined is less than with $\gamma = 1.0$. According to Figure 6, the numerical solution of the next iterative step is expected to be more accurate than that of the current step; however, the increase on the total number of cells is comparatively small when $\gamma > 1.0$. If $\gamma < 1.0$, then the number of cells to be refined is larger than that with $\gamma = 1.0$. The advantage in this case is that the refinement rate increases, however, the computational efficiency may decrease owing to likely generation of an excessive number of cells.

4.2.3 Global error estimation and stopping criterion.

Cell refinement (see Figure 7) can be stopped when:

$$\bar{e} \leq e_{global} \tag{39}$$

where e_{global} has a preset value, which depends on the overall level of accuracy desired. The goal of the adaptive procedure is to obtain well-distributed (i.e. near optimal) cell configurations. Ideally, as the iterative cell refinement progresses, the error estimates should decrease both locally and globally.

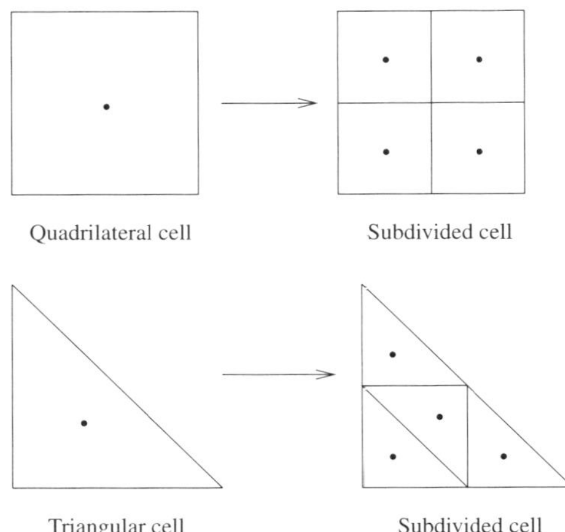

Figure 7: Cell refinement for quadrilateral and triangular cells with one node per cell (from [ChatPM01b])

4.3 ONE-step Adaptive Meshing Strategy

A ONE-step adaptive meshing strategy is presented below.

4.3.1 ONE-step cell design scheme.

Section 4.2 above has dealt with an iterative adaptive technique for cell refinement ($h-$version). Here the interest is on developing a simple ONE-step algorithm for cell refinement in the meshless BNM setting. The flowchart of Figure 8 illustrates this idea which is based on the concept of refinement level (RL) employed by [KrisU93] and [MosaP01].

4.3.2 Cell refinement strategy.

Figures 9(a) and (b) show that different degrees of refinement are carried out for different values of the refinement level. From these figures, the expression relating the final cell size h_f to the refinement level RL is:

$$h_f = \frac{h_i}{2^{RL}} \tag{40}$$

where h_i denotes the initial cell size. *Assuming* that the rate of convergence of the error is $\mathcal{O}(h^p)$, where h is a characteristic cell size in the area covered by the cells, which are of order p, and setting the error estimate $\eta = e/(\gamma\bar{e}) = (h_i/h_f)^p$ (see equations (37) and (38)), one obtains:

$$h_f = \frac{h_i}{\eta^{1/p}} \tag{41}$$

From Equations (40) and (41), the RL is given by:

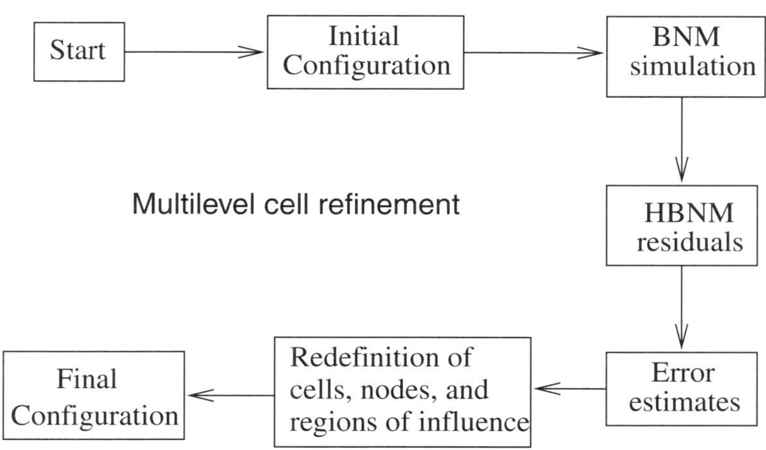

Figure 8: ONE-step adaptive BEM algorithm based on multilevel cell refinement (from [ChatPM01b])

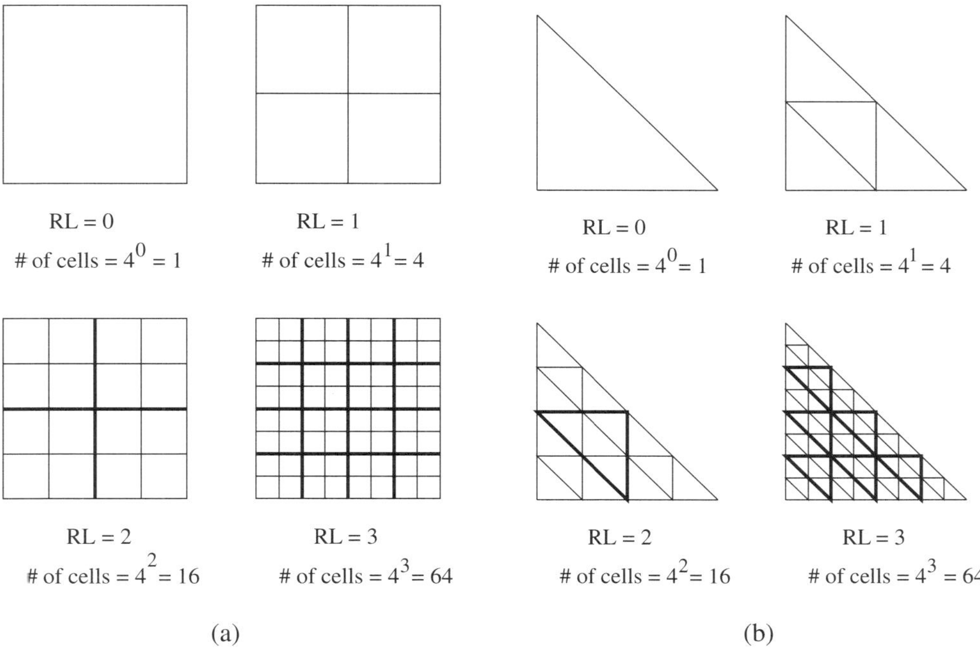

Figure 9: Refinement level RL using (a) rectangular and (b) triangular cells. The bold lines illustrate the idea of cell structure embedding (from [ChatPM01b])

$$RL = \begin{cases} \frac{\log \eta}{p \log 2} & \text{for } \eta \geq 1 \\ 0 & \text{for } \eta < 1 \end{cases} \tag{42}$$

where p is order of the interpolating function. (Of course, RL is an integer, the nearest to the real number obtained from equation (42)). For the interpolation procedure used in this work, $p = m$, the number of monomials used in the basis for u_i or τ_i (see equation (1)). The second condition in equation (42) is enforced because cell structure coarsening is not considered in this work.

4.4 Numerical Results

This example is concerned with a short cylinder which is clamped at the bottom and subjected to unit tensile traction on the top surface (Figure 10(a)). The radius and length of the cylinder are each 2 units, the Young's modulus of the cylinder material is 1.0 and the Poisson's ratio is 0.25 (in consistent units). The initial mesh on the top (loaded) and bottom (clamped) faces of the cylinder are identical and are shown in Figure 10 (b).

It is known ([Crus69], [Pick44]) that, for this problem, the normal stress component σ_{33} varies slowly over much of the clamped face, but exhibits sharp gradients near its boundary. This stress component becomes singular on the boundary of the clamped face. The behavior of the shearing stress component σ_{zr} (here $r, \theta, z \equiv 3$ are the usual polar coordinates) on the clamped face is qualitatively similar to that of σ_{33}. The stresses are uniform on the loaded face.

4.4.1 Progressive refinement results.

It is seen from Figure 10 that this behavior is captured well by the adaptive scheme. The region near the boundary of the clamped face is refined most while the mesh on the loaded face of the cylinder is left unaltered. Also, Figure 17 (c) in [ChatPM01b] shows that most of the mesh refinement takes place on the bottom layer of the curved surface of the cylinder, which is nearest to the clamped face, while the rest of the mesh on it remains substantially unaltered.

In this example, the boundary value problem is solved by using the HBNM (34) and the singular residual is obtained from the standard BNM (27).

4.4.2 ONE-step refinement results.

The same problem, discussed in Section 4.4.1 above, is now solved with the ONE-step algorithm.

Results for the ONE-step multilevel refinement are given in Figure 11. Comparing these results with the ones in Figure 10, one verifies that the overall trends are quite similar in both situations. However, two main differences are noticeable. First, the loaded face is refined (a little) in Figure 11(a), while it is not refined at all in the adapted configuration shown in Figure 10(b). Second, as expected, the multilevel cell refinement does not allow the smooth cell gradation which occurs in progressively adapted cell configurations. Nevertheless, such gradation, which is essential for mesh-based methods, is not required at all in the present meshless methods (BNM and HBNM).

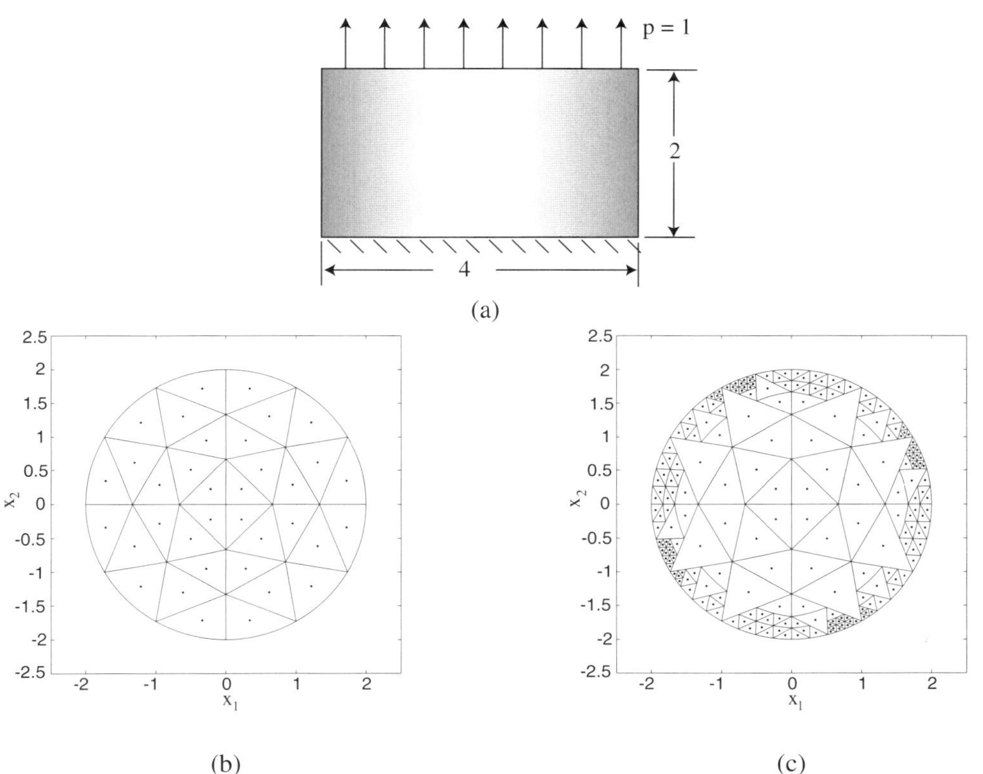

Figure 10: Progressive adaptive meshing of the top and bottom faces of a short clamped cylinder under tension (analysis with the progressive BNM). (a) Geometry and loading (b) Cells and nodes on clamped and loaded faces - initial configuration; as well as on loaded face - final configuration (c) Cells and nodes on clamped face - final configuration. Note that the triangles are cells and NOT elements – standard connectivity information is NOT required in the BNM (from [ChatPM01b])

5 Acknowledgements

Most of this research has been supported by a University Research Programs (URP) grant from Ford Motor Company to Cornell University. Some of the computing has been performed at the Cornell Theory Center. The contributions of Y.X. Mukherjee, M. Chati and G.H. Paulino, to this research, are gratefully acknowledged.

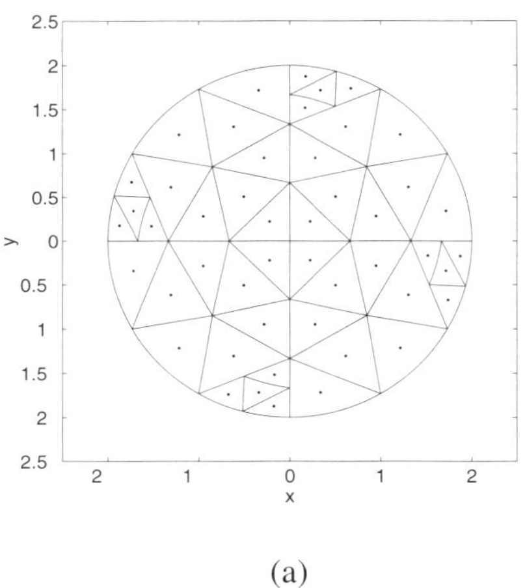

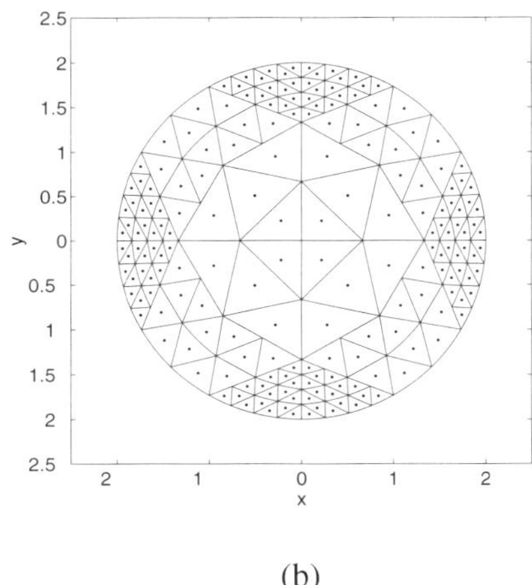

(a) (b)

Figure 11: ONE-step multi-level cell refinement of the top and bottom faces of a short clamped cylinder under tension. Final configurations : (a) loaded face. (b) clamped face. The initial configurations are shown in Figure 10(b) (from [ChatPM01b])

References

[AlurL01] Aluru, N.R., and Li, G. (2001). Finite cloud method: a true meshless technique based on a fixed reproducing kernel approximation. *International Journal for Numerical Methods in Engineering* 50:2373-2410.

[AskePd00] Askes, H., Pamin, J., and de Brost, R. (2000). Dispersion analysis and element-free Galerkin solutions of second- and fourth-order gradient-enhanced damage models. *International Journal for Numerical Methods in Engineering* 49:811-832.

[AtluZ98a] Atluri, S.N., and Zhu, T. (1998). New meshless local Petrov-Galerkin (MPLG) approach in computational mechanics. *Computational Mechanics* 22:117-127.

[AtluZ98b] Atluri, S.N., and Zhu, T. (1998). A new meshless local Petrov-Galerkin (MPLG) approach to nonlinear problems in computer modeling and simulation. *Computer Modelling and Simulation in Engineering* 3:187-196.

[BarrS99] Barry, W., Saigal, S. (1999). A three-dimensional element-free Galerkin elastic and elastoplastic formulation. *International Journal for Numerical Methods in Engineering* 46:671-693.

[BeisB96] Beissel, S., and Belytschko, T. (1996). Nodal integration of the element-free Galerkin method. *Computer Methods in Applied Mechanics and Engineering* 139:49-74.

[BelyLG94] Belytschko, T., Lu, Y.Y., and Gu, L. (1994). Element-free Galerkin methods. *International Journal for Numerical Methods in Engineering* 37:229-256.

[BelyLG95] Belytschko, T., Lu, Y.Y., Gu, L. (1995). Crack propagation by element-free Galerkin methods. *Engineering Fracture Mechanics* 51:295-315.

[BelyLGT95] Belytschko, T., Lu, Y.Y., Gu, L., and Tabbara, M. (1995). Element-free Galerkin methods for static and dynamic fracture. *International Journal of Solids and Structures* 32:2547-2570.

[BelyT96] Belytschko, T., and Tabbara, M. (1996). Dynamic fracture using element-free Galerkin methods. *International Journal for Numerical Methods in Engineering* 39:923-938.

[BelyKOFK96] Belytschko, T., Krongauz, Y., Organ, D., Fleming, M., and Krysl, P. (1996). Meshless methods: An overview and recent developments. *Computer Methods in Applied Mechanics and Engineering* 139:3-47.

[BobaM01a] Bobaru, F., and Mukherjee, S. (2001). Shape sensitivity analysis and shape optimization in planar elasticity using the element-free Galerkin method. *Computer Methods in Applied Mechanics and Engineering*. 190:4319-4337.

[BobaM01b] Bobaru, F., and Mukherjee, S. (2001). Meshless approach to shape optimization of linear thermoelastic solids. *International Journal for Numerical Methods in Engineering*, in press.

[Bonn95] Bonnet, M. (1995). Regularized direct and indirect symmetric variational BIE formulations for three-dimensional elasticity. *Engineering Analysis with Boundary Elements* 15:93-102.

[Chat99] Chati, M.K. (1999). *Meshless standard and hypersingular boundary node method - Applications in three-dimensional potential theory and linear elasticity*. Ph.D. Dissertation, Cornell University, Ithaca, NY.

[ChatM99] Chati, M.K., and Mukherjee, S. (1999). Evaluation of gradients on the boundary using fully regularized hypersingular boundary integral equations. *Acta Mechanica* 135:41-45.

[ChatMM99] Chati, M.K., Mukherjee, S., and Mukherjee, Y.X. (1999). The boundary node method for three-dimensional linear elasticity. *International Journal for Numerical Methods in Engineering* 46:1163-1184.

[ChatM00] Chati, M.K, and Mukherjee, S. (2000). The boundary node method for three-dimensional problems in potential theory. *International Journal for Numerical Methods in Engineering* 47:1523-1547.

[ChatMP01a] Chati, M.K., Mukherjee, S., and Paulino, G.H. (2001). The meshless hypersingular boundary node method for three-dimensional potential theory and linear elasticity problems. *Engineering Analysis with Boundary Elements* 25:639-653.

[ChatPM01b] Chati, M.K., Paulino, G.H., and Mukherjee, S. (2001). The meshless standard and hypersingular boundary node methods - applications to error estimation and adaptivity in three-dimensional problems. *International Journal for Numerical methods in Engineering* 50:2233-2269.

[ChenH99] Chen, J.T., and Hong, H-K. (1999). Review of dual boundary element methods with emphasis on hypersingular integrals and divergent series. *ASME Applied Mechanics Reviews* 52:17-33.

[ChenPWL96] Chen, J-S., Pan, C., Wu, C.T., and Liu, W.K. (1996). Reproducing kernel particle methods for large deformation analysis of non-linear structures. *Computer Methods in Applied Mechanics and Engineering* 139:195-227.

[ChenWYY01] Chen, J-S, Wu, C-T, Yoon, S., and You, Y. (2001). A stabilized conforming nodal integration for Galerkin meshfree methods. *International Journal for Numerical Methods in Engineering* 50:435-466.

[Crus69] Cruse, T.A. (1969). Numerical solutions in three-dimensional elastostatics. *International Journal of Solids and Structures* 5:1259-1274.

[CrusR96] Cruse, T.A., and Richardson, J.D. (1996). Non-singular Somigliana stress identities in elasticity. *International Journal for Numerical methods in Engineering* 39:3273-3304.

[DolbMB00] Dolbow, J., Möes, N., and Belytschko, T. (2000). Modeling fracture in Mindlin-Reissner plates with the extended finite element method. *International Journal of Solids and Structures* 37:7161-7183.

[DuarO96a] Duarte, C.A.M., and Oden, J.T. (1996). H-p clouds – An *h-p* meshless method. *Numerical Methods for Partial Differential Equations* 12:673-705.

[DuarO96b] Duarte, C.A.M., and Oden, J.T. (1996). An h-p adaptive method using clouds. *Computer Methods in Applied Mechanics and Engineering* 139:237-262.

[GowrM01] Gowrishankar, R., and Mukherjee, S. (2001). A "pure" boundary node method for potential theory. *Communications in Numerical Methods in Engineering*, submitted.

[GrayMI90] Gray, L.J., Martha, L.F., and Ingraffea A.R. (1990). Hypersingular integrals in boundary element fracture analysis. *International Journal for Numerical methods in Engineering* 29:1135-1158.

[GrayBK95] Gray, L.J., Balakrishna, C., and Kane, J.H. (1995). Symmetric Galerkin fracture analysis. *Engineering Analysis with Boundary Elements* 15:103-109.

[GrayP97a] Gray, L.J., and Paulino, G.H. (1997). Symmetric Galerkin boundary integral formulation for interface and multizone problems. *International Journal for Numerical methods in Engineering* 40:3085-3101.

[GrayP97b] Gray, L.J., and Paulino, G.H. (1997). Symmetric Galerkin boundary integral fracture analysis for plane orthotropic elasticity. *Computational Mechanics* 20:26-33.

[GrayP98] Gray, L.J., and Paulino, G.H. (1998). Crack tip interpolation revisited. *SIAM Journal of Applied Mathematics* 58:428-455.

[Guig94] Guiggiani, M. (1994). Hypersingular formulation for boundary stress evaluation. *Engineering Analysis with Boundary Elements* 13:169-179.

[KaljS97] Kaljevic, I., and Saigal, S. (1997). An improved element free Galerkin formulation. *International Journal for Numerical Methods in Engineering* 40:2953-2974.

[KothMM99] Kothnur, V.S., Mukherjee, S., and Mukherjee, Y.X. (1999). Two dimensional linear elasticity by the boundary node method. *International Journal of Solids and Structures* 36:1129-1147.

[KrisSRR90] Krishnasamy, G., Schmerr, L.W., Rudolphi, T.J., and Rizzo, F.J. (1990). Hypersingular boundary integral equations : some applications in acoustic and elastic wave scattering. *ASME Journal of Applied Mechanics* 57:404-414.

[KrisRR92] Krishnasamy, G., Rizzo, F.J. and Rudolphi, T.J. (1992). Hypersingular boundary integral equations: their occurrence, interpretation, regularization and computation. In Banerjee, P.K., and Kobayashi, S., eds., *Developments in Boundary Element Methods-7*, Elsevier Applied Science: London. 207-252.

[KrisU93] Krishnamoorthy, C.S., and Umesh, K.R. (1993). Adaptive mesh refinement for two-dimensional finite element stress analysis. *Computers and Structures* 48:121-133.

[KrysB95] Krysl, P., and Belytschko, T. (1995). Analysis of thin plates by the element-free Galerkin method. *Computational Mechanics* 17:26-35.

[KrysB96] Krysl, P., and Belytschko, T. (1996). Analysis of thin shells by the element-free Galerkin method. *International Journal of Solids and Structures* 33:3057-3080.

[KrysB99] Krysl, P., and Belytschko, T. (1999). The element free Galerkin method for dynamic propagation of arbitrary 3-D cracks. *International Journal for Numerical Methods in Engineering* 44:767-800.

[LancS90] Lancaster, P., and Salkauskas, K. (1990). *Curve and Surface Fitting - An Introduction.* London:Academic Press.

[LiA01] Li, G., and Aluru, N.R. (2001). Boundary cloud method: a combined scattered point/boundary integral approach for boundary-only analysis. *Computer Methods in Applied Mechanics and Engineering*, submitted.

[LiuJZ95] Liu, W.K., Jun, S., and Zhang, Y. (1995). Reproducing kernel particle methods. *International Journal for Numerical Methods in Fluids* 20:1081-1106.

[LiuCJBPUC96] Liu, W.K., Chen, Y., Jun, S., Belytschko, T., Pan, C., Uras, R.A. and Chang, C.T. (1996). Overview and applications of the reproducing kernel particle methods. *Archives of Computational Methods in Engineering* 3:3-80.

[LutzIG92] Lutz, E.D., Ingraffea, A.R., and Gray, L.J. (1992). Use of 'simple solutions' for boundary integral methods in elasticity and fracture analysis. *International Journal for Numerical methods in Engineering* 35:1737-1751.

[LutzYM98] Lutz, E.D., Ye, W., and Mukherjee, S. (1998). Elimination of rigid body modes from discretized boundary integral equations. *International Journal of Solids and Structures* 35:4427-4436.

[Mack93] Mackerle, J. (1993). Mesh generation and refinement for FEM and BEM - A bibliography (1990 - 1993). *Finite Elements in Analysis and Design* 15:177-188.

[MartR96] Martin, P.A., and Rizzo, F.J. (1996). Hypersingular integrals : how smooth must the density be ? *International Journal for Numerical methods in Engineering* 39:687-704.

[MartRC98] Martin, P.A., Rizzo, F.J., and Cruse, T.A. (1998). Smoothness-relaxation strategies for singular and hypersingular integral equations. *International Journal for Numerical methods in Engineering* 42:885-906.

[Meno96] Menon, G. (1996). *Hypersingular error estimates in boundary element methods.* M.S. Thesis, Cornell University, Ithaca, NY

[MenoPM99] Menon, G., Paulino, G.H., and Mukherjee, S. (1999). Analysis of hypersingular residual error estimates in boundary element methods for potential problems. *Computer Methods in Applied Mechanics and Engineering* 173:449-473.

[MoesDB99] Moës N., Dolbow, J., and Belytschko, T. (1999). A finite element method for crack growth without remeshing. *International Journal for Numerical Methods in Engineering* 46:131-150.

[MosaP01] Mosalam, K.M., and Paulino, G.H. (2001). Towards adaptive nonlinear finite element analysis of reinforced concrete structures considering smeared cracking effects. Submitted.

[Mukh00a] Mukherjee, S. (2000). CPV and HFP integrals and their applications in the boundary element method. *International Journal of Solids and Structures* 37:6623-6634.

[Mukh00b] Mukherjee, S. (2000). Finite parts of singular and hypersingular integrals with irregular boundary source points. *Engineering Analysis with Boundary Elements* 24:767-776.

[Mukh01] Mukherjee, S. (2001). On boundary integral equations for cracked and for thin bodies. *Mathematics and Mechanics of Solids* 6:47-64.

[MukhM97a] Mukherjee, Y.X., and Mukherjee, S. (1997) The boundary node method for potential problems. *International Journal for Numerical Methods in Engineering* 40:797-815.

[MukhM97b] Mukherjee, Y.X., and Mukherjee, S. (1997). On boundary conditions in the element-free Galerkin method. *Computational Mechanics* 19:264-270.

[MukhM98] Mukherjee, S., and Mukherjee, Y. X. (1998). The hypersingular boundary contour method for three-dimensional linear elasticity. *ASME Journal of Applied Mechanics* 65:300-309.

[MukhCS00] Mukherjee, S., Chati, M.K., and Shi, X. (2000). Evaluation of nearly singular integrals in boundary element, contour and node methods for three-dimensional linear elasticity. *International Journal of Solids and Structures* 37:7633-7654.

[MukhM01] Mukherjee, Y.X., and Mukherjee, S. (2001). Error analysis and adaptivity in three-dimensional linear elasticity by the usual and hypersingular boundary contour method. *International Journal of Solids and Structures* 38:161-178.

[NagaM93] Nagarajan, A., and Mukherjee, S. (1993). A mapping method for numerical evaluation of two-dimensional integrals with 1/r singularity. *Computational Mechanics* 12:19-26.

[NagaML96] Nagarajan, A., Mukherjee, S., and Lutz, E.D. (1996). The boundary contour method for three-dimensional linear elasticity. *ASME Journal of Applied Mechanics* 63:278-286.

[NayrTV92] Nayroles, B., Touzot, G., and Villon, P. (1992). Generalizing the finite element method : diffuse approximation and diffuse elements. *Computational Mechanics* 10:307-318.

[OdenDZ98] Oden, J.T., Duarte, C.A.M., and Zienkiewicz, O.C. (1998). A new cloud-based *hp* finite element method. *Computer Methods in Applied Mechanics and Engineering* 153:117-126.

[OnatIZT96] Oñate, E., Idelsohn, S., Zienkiewicz, O.C., and Taylor, R.L. (1996). A finite point method in computational mechanics. Applications to convective transport and fluid flow. *International Journal for Numerical Methods in Engineering* 39:3839-3866.

[OnatIZTS96] Oñate, E., Idelsohn, S., Zienkiewicz, O.C., Taylor, R.L., and Sacco, C. (1996). A stabilized finite point method for analysis of fluid mechanics problems. *Computer Methods in Applied Mechanics and Engineering* 139:315-346.

[Paul95] Paulino, G.H. (1995). *Novel formulations of the boundary element method for fracture mechanics and error estimation.* Ph.D. dissertation, Cornell University, Ithaca, NY

[PaulGZ96] Paulino, G.H., Gray, L.J., and Zarkian, V. (1996). Hypersingular residuals - a new approach for error estimation in the boundary element method. *International Journal for Numerical methods in Engineering* 39:2005-2029.

[PaulSMR97] Paulino, G.H., Shi, F., Mukherjee, S., and Ramesh P. (1997). Nodal sensitivities as error estimates in computational mechanics. *Acta Mechanica* 121:191-213.

[PaulMCM99] Paulino, G.H., Menezes, I.F.M., Cavalcante Neto, J.B., and Martha L.F. (1999). A methodology for adaptive finite element analysis: towards an integrated computational environment. *Computational Mechanics* 23:361-388.

[PaulG99] Paulino, G.H., and Gray, L.J. (1999). Galerkin residuals for adaptive symmetric-Galerkin boundary element methods. *ASCE Journal of Engineering Mechanics* 125:575-585.

[PaulMM01] Paulino, G.H., Menon, G., and Mukherjee S. (2001). Error estimation using hypersingular integrals in boundary element methods for linear elasticity. *Engineering Analysis with Boundary Elements* Manuscript in print.

[PhanMM98b] Phan, A-V., Mukherjee, S., and Mayer J.R.R. (1998). The hypersingular boundary contour method for two-dimensional linear elasticity. *Acta Mechanica* 130:209-225.

[Pick44] Pickett, G. (1944). Application of the Fourier method to the solution of certain boundary value problems in the theory of elasticity. *ASME Journal of Applied Mechanics* 66:A176-A182.

[Rizz67] Rizzo, F.J. (1967). An integral equation approach to boundary value problems of classical elastostatics. *Quarterly of Applied Mathematics* 25:83-95.

[SladSA00] Sladek, J., Sladek, V., and Atluri, S.N. (2000). Local boundary integral equation (LBIE) method for solving problems of elasticity with nonhomogeneous material properties. *Computational Mechanics* 24:456-462.

[StroBC00] Strouboulis, T., Babuška, I., and Copps, K. (2000). The design and analysis of the Generalized Finite Element Method. *Computer Methods in Applied Mechanics and Engineering* 181:43-69.

[SukuMBB97] Sukumar, N., Moran, B., Black, T., and Belytschko, T. (1997). An element-free Galerkin method for three-dimensional fracture mechanics. *Computational Mechanics* 20:170-175.

[SukuMB98] Sukumar, N., Moran, B., and Belytschko, T. (1998). The natural element method. *International Journal for Numerical Methods in Engineering* 43:839-887.

[SukuMMB00] Sukumar N., Möes, N., Moran B., and Belytschko T. (2000). Extended finite element method for three-dimensional crack modeling. *International Journal for Numerical Methods in Engineering* 48:1549-1570.

[SukuMSB01] Sukumar N., Moran B., Semenov, A.Y., and Belytschko T. (2001). Natural neighbour Galerkin methods. *International Journal for Numerical Methods in Engineering* 50:1-27.

[TanaSS94] Tanaka, M., Sladek, V., and Sladek, J. (1994). Regularization techniques applied to boundary element methods. *ASME Applied Mechanics Reviews* 47:457-499.

[TimoG70] Timoshenko, S.P., Goodier, J.N. (1976). *Theory of Elasticity*, 3rd. ed. New York:McGraw Hill.

[TohM94] Toh, K-C., and Mukherjee, S. (1994). Hypersingular and finite part integrals in the boundary element method. *International Journal of Solids and Structures* 31:2299-2312.

[WildA98] Wilde, A.J., and Aliabadi, M.H. (1998). Direct evaluation of boundary stresses in the 3D BEM of elastostatics. *Communications in Numerical Methods in Engineering* 14:505-517.

[XuS98a] Xu, Y., and Saigal, S. (1998). An element-free Galerkin formulation for stable crack growth in elastic solids. *Computer Methods in Applied Mechanics and Engineering* 154:331-343.

[XuS98b] Xu, Y., and Saigal, S. (1998). Element-free Galerkin study of steady quasi-static crack growth in plane strain tension in elastic-plastic materials. *Computational Mechanics* 21:276-282.

[XuS99] Xu, Y., and Saigal, S. (1999). An element-free Galerkin analysis of steady dynamic growth of a mode I crack in elastic-plastic materials. *International Journal of Solids and Structures* 36:1045-1079.

[Ziek77] Zienkiewicz, O.C. (1977). *The Finite Element Method*, 3rd edn, New York:McGraw Hill.

[ZhaoL99] Zhao, Z.Y., and Lan, S.R. (1999). Boundary stress calculation - a comparison study. *Computers and Structures* 71:77-85.

[ZhuZA98] Zhu, T., Zhang, J-D., and Atluri, S.N. (1998). A local boundary integral equation (LBIE) method in computational mechanics, and a meshless discretization approach. *Computational Mechanics* 21:223-235.

[Zhu99] Zhu, T. (1999). A new meshless regular local boundary integral equation (MRLBIE) method. *International Journal for Numerical Methods in Engineering* 46:1237-1252.

CHAPTER 13

Symmetric Galerkin BEM in 3D Elasticity: Computational Aspects and Applications to Fracture Mechanics

Novati Giorgio[1] and Frangi Attilio[2]

[1] Department of Mechanical and Structural Engineering, University of Trento, Italy
[2] Department of Structural Engineering, Politecnico of Milan, Italy

Abstract. The formulation of the symmetric Galerkin BEM for 3D elastic fracture mechanics problems and some relevant computational aspects are presented in this paper; the method is employed for the evaluation of stress intensity factors and for the modeling of fatigue crack growth. In the latter context a propagation algorithm has been developed and implemented into a fully automated numerical code which is used to analyze two example problems concerning the fatigue growth of surface breaking cracks.

1 Introduction

The evaluation of stress intensity factors for arbitrary three-dimensional surface or internal cracks often represents a central issue in the integrity assessment of structural components. Well established finite-element procedures can be used to obtain the stress intensity factors (SIFs) with good accuracy but the human labor cost associated to the generation of appropriate meshes for arbitrary non-planar cracks can be very high. With boundary element approaches the stress analysis and the SIF evaluation for a cracked body can be carried out by discretizing appropriate unknown fields pertaining to the boundary of the structure and to the crack surface only; hence the generation of the corresponding surface-mesh is accomplished at a comparatively much lower cost.

For fracture propagation problems, where remeshing is need as the crack evolves, the advantage of integral equation approaches over domain methods become even more evident.

A few contributions have appeared recently concerning crack-growth simulations using boundary elements (see e.g. Li and Keer, 1992, Mi and Aliabadi, 1994, Mi, 1996, Wawrzynek *et al.*, 1988, Xu and Ortiz, 1993). In some cases pure Mode I behaviour is assumed. In other more advanced applications (Mi and Aliabadi, 1994, Mi, 1996) the dual approach is adopted: essentially this formulation enforces the traction equation (in addition to the traditional Somigliana displacement identity) at points of the fracture surface.

The symmetric Galerkin boundary element method (SGBEM), based on a weak-form enforcement of the displacement and traction integral equations and thus entailing double integrations, is well suited for the analysis of fracture mechanics problems and is characterized by the following attractive features (not shared by other versions of the BEM): i) it leads to a final equation system with a symmetric coefficient matrix; ii) after carrying out a regularization process, all the kernels involved in the formulation turn out to be weakly singular; iii) only C^0 continuity of the kinematic fields (displacements on the boundary and displacement discontinuities on the crack) is required. This contrasts with usual hypersingular collocation BE approaches where C^1 continuity is mandatory, at least from a purely theoretical point of view.

With reference to 3D fracture mechanics problems and the SIF evaluation, the symmetric Galerkin approach has been recently implemented and numerically tested by Li *et al.* (1998) and Frangi *et al.* (2001). The two contributions, which differ in various computational aspects, show the potential of the method and demonstrate that an accurate evaluation of the SIFs, through extrapolation from displacement discontinuities in the vicinity of the crack front, is obtained even with relatively coarse meshes.

Here, starting from these experiences, the SGBEM is employed to simulate fracture propagation, with reference to fatigue crack growth at high number of cycles governed by Paris law and a simple propagation criterion.

A detailed explanation of the theoretical background for the SGBEM can be found e.g. in Bonnet *et al.* (1998), Frangi (1998), Li *et al.* (1998) and is only briefly recalled in the sequel.

Let Ω denote a generic body of boundary S in the orthonormal cartesian reference system (x_1, x_2, x_3), subject to tractions $p_i(\mathbf{x}) = \bar{p}_i(\mathbf{x})$ on S_p, with displacements $u_i(\mathbf{x}) = \bar{u}_i(\mathbf{x})$ enforced on $S_u = S \backslash S_p$. Let surface S_c denote a crack inside Ω, conceived as a locus of displacement discontinuity $w(\mathbf{x}) = u(\mathbf{x}^+) - u(\mathbf{x}^-)$, with $\mathbf{x}^+ \in S_c^+$ and $\mathbf{x}^- \in S_c^-$, S_c^+ and S_c^- being the upper and lower faces of the crack. The positive orientation of S_c is associated with the normal unit vector to S_c pointing from S_c^+ to S_c^-. Equal and opposite tractions can be applied to the crack surfaces: $p_i(\mathbf{x}) \equiv p_i(\mathbf{x}^-) = -p_i(\mathbf{x}^+)$ on S_c.

Two different integral equations serve as starting point for the formulation of the SGBEM: the classical Somigliana equation

$$u_i(\tilde{\mathbf{x}}) = \int_S \left[G_{ik}^{uu}(\mathbf{x} - \tilde{\mathbf{x}}) p_k(\mathbf{x}) - G_{ikj}^{u\sigma}(\mathbf{x} - \tilde{\mathbf{x}}) n_j u_k(\mathbf{x}) \right] \mathrm{d}S_x + \int_{S_c} G_{ikj}^{u\sigma}(\mathbf{x} - \tilde{\mathbf{x}}) n_j w_k(\mathbf{x}) \, \mathrm{d}S_x \tag{1}$$

and the "traction" equation

$$p_i(\tilde{\mathbf{x}}) = \int_S \left[G_{ijk}^{\sigma u}(\mathbf{x} - \tilde{\mathbf{x}}) \tilde{n}_j(\tilde{\mathbf{x}}) p_k(\mathbf{x}) - G_{ijkp}^{\sigma\sigma}(\mathbf{x} - \tilde{\mathbf{x}}) \tilde{n}_j(\tilde{\mathbf{x}}) n_p(\mathbf{x}) u_k(\mathbf{x}) \right] \mathrm{d}S_x + \int_{S_c} G_{ijkp}^{\sigma\sigma}(\mathbf{x} - \tilde{\mathbf{x}}) \tilde{n}_j(\tilde{\mathbf{x}}) n_p(\mathbf{x}) w_k(\mathbf{x}) \, \mathrm{d}S_x \tag{2}$$

Indeed, eqns. (1) and (2) hold for a generic point $\tilde{\mathbf{x}} \in \Omega$; the two-point Kelvin kernel $G_{ik}^{uu}(\mathbf{x} - \tilde{\mathbf{x}})$ expresses the displacement at $\mathbf{x}$ in the kth direction due to a concentrated force acting at $\tilde{\mathbf{x}}$ in the ith direction:

$$G_{ik}^{uu}(\mathbf{x} - \tilde{\mathbf{x}}) = \frac{1}{16\pi\mu(1-\nu)} \frac{1}{r} \left[(3 - 4\nu)\delta_{ik} + r_{,i} r_{,k} \right] \tag{3}$$

Kernel $G_{ikj}^{u\sigma}(\mathbf{x} - \tilde{\mathbf{x}})$ denotes the kj component of stresses at $\mathbf{x}$ due to the same source, obtained by differentiation of G_{ik}^{uu} and application of Hooke's law, C_{kjmn} being the elastic stiffness tensor:

$$G_{ikj}^{u\sigma}(\mathbf{x} - \tilde{\mathbf{x}}) = C_{kjmn} G_{im,n}^{uu}(\mathbf{x} - \tilde{\mathbf{x}}). \tag{4}$$

Moreover

$$G^{\sigma u}_{ijk} = C_{ijab} G^{uu}_{ak,\tilde{b}} \qquad G^{\sigma\sigma}_{ijkp} = C_{ijab} C_{kpcd} G^{uu}_{ac,b\tilde{d}} \tag{5}$$

are the so-called double layer kernels. Unit vector $\mathbf{n}$ defines the outward normal to S at point $\mathbf{x}$ while $\tilde{\mathbf{n}}$ defines a reference normal associated to $\tilde{\mathbf{x}}$; symbol $(\cdot)_{,\tilde{i}}$ denotes differentiation with respect to $\tilde{x}_i$.

In customary collocation approaches eqns. (1) and (2) are enforced pointwise at given points on the boundary with $\tilde{\mathbf{n}}$ in eqn. (2) being set equal to the actual outward normal at the collocation boundary point. In view of the singular nature of the kernels for $\tilde{\mathbf{x}} \to \mathbf{x}$, an infinitesimal domain Ω_ε (of linear dimension ε) in the neighborhood of the boundary point $\tilde{\mathbf{x}}$ is often excluded from Ω and the limit form, for $\varepsilon \to 0$, of the above integral equation are actually utilized in the computations.

A totally different approach is followed in the symmetric Galerkin method. Let us introduce a surface $\tilde{S}$ representing a fictitious contour *internal* to Ω. We assume the existence of a one-to-one correspondence between points $\mathbf{x} \in S$ and $\tilde{\mathbf{x}} \in \tilde{S}$: $\tilde{\mathbf{x}} = \boldsymbol{\mathcal{X}}(\mathbf{x}, h)$, where h is a parameter such that $\mathbf{x} = \boldsymbol{\mathcal{X}}(\mathbf{x}, 0)$. Hence the two surfaces coincide for $h = 0$. In particular $\tilde{S}$ will consist of portions $\tilde{S}_u$, $\tilde{S}_p$ and, in the presence of an internal crack, also of $\tilde{S}^+_c$, $\tilde{S}^-_c$ mapped by one-to-one correspondences onto the respective portions of S and of S_c.

The procedure basically consists of two distinct steps: first eqns. (1) and (2) are enforced in a weak sense on the auxiliary contour $\tilde{S}$ *distinct* from S (i.e. with $h \neq 0$) and an analytical regularization procedure is performed via integration by parts removing all higher-order singularities. Secondly the limit $\tilde{S} \to S$ ($h \to 0$) is performed and the discretization procedure is initiated. The definition of an auxiliary surface $\tilde{S}$ separated from S is hence only an artifice which proves useful in the rather involved issue of guaranteeing a firm mathematical and computational basis for the evaluation of the singular double integrals involved; however $\tilde{S}$ does not play any role in the final implementation of the method, since for $h \to 0$, $\tilde{S} \equiv S$.

More specifically eqn. (1) is enforced on $\tilde{S}_u$ using as test function the static field $\tilde{p}_i(\tilde{\mathbf{x}})$ while eqn. (2) is enforced both on $\tilde{S}_p$ and on $\tilde{S}_c$ using as test function the kinematic field $\tilde{u}_i$. By applying the regularization procedure detailed by Frangi (1998) and Bonnet *et al.* (1998), and taking the limit $\tilde{S} \to S$ (hence $h \to 0$), the variational equations listed in the following paragraphs are obtained.

Regularized weak form of the displacement integral equation on S_u.

$$\begin{aligned}\frac{1}{2}\int_{S_u} u_i(\mathbf{x})\tilde{p}_i(\mathbf{x})\,\mathrm{d}S_x &= \int_{S_u}\int_S \tilde{p}_i(\tilde{\mathbf{x}}) G^{uu}_{ik}(\mathbf{x}-\tilde{\mathbf{x}}) p_k(\mathbf{x})\,\mathrm{d}S_x\,\mathrm{d}S_{\tilde{x}} \\ &+ \int_{S_u}\int_S \tilde{p}_i(\tilde{\mathbf{x}})\Big[\frac{1}{4\pi}\frac{1}{r^2} r_{,i}\, n_i(\mathbf{x}) u_i(\mathbf{x}) + G^{u\varphi}_{ikj}(\mathbf{x}-\tilde{\mathbf{x}}) R_j[u_k](\mathbf{x})\Big]\,\mathrm{d}S_x\,\mathrm{d}S_{\tilde{x}} \\ &- \int_{S_u}\int_{S_c} \tilde{p}_i(\tilde{\mathbf{x}})\Big[\frac{1}{4\pi}\frac{1}{r^2} r_{,i}\, n_i(\mathbf{x}) w_i(\mathbf{x}) + G^{u\varphi}_{ikj}(\mathbf{x}-\tilde{\mathbf{x}}) R_j[w_k](\mathbf{x})\Big]\,\mathrm{d}S_x\,\mathrm{d}S_{\tilde{x}}\end{aligned} \tag{6}$$

where

$$G^{u\varphi}_{ikj}(\mathbf{x}-\tilde{\mathbf{x}}) = \frac{1}{8\pi(1-\nu)r}\left[(1-2\nu)\, e_{ikj} - e_{ipj} r_{,p} r_{,k}\right], \tag{7}$$

and R_j denotes the surface rotor operator (see Bonnet *et al.*, 1998):

$$R_j[u_k](\mathbf{x}) = e_{bcj} n_b(\mathbf{x}) \frac{\partial u_k}{\partial x_c}(\mathbf{x}) \tag{8}$$

It should be stressed that both G^{uu} and the auxiliary kernel $G^{u\varphi}$ are weakly singular; the additional terms in eqn. (6) (which coincide with the double layer kernel for potential problems) are, in actual fact, weakly singular as well since

$$\frac{1}{r^2} r_{,i}\, n_i \, \mathrm{d}S_x = \mathrm{d}\Theta$$

represents the differential solid angle under which $\mathrm{d}S_x$ is seen from $\tilde{\mathbf{x}}$.

Regularized weak form of the traction integral equation on S_p.

$$\begin{aligned} \frac{1}{2} \int_{S_p} p_i(\mathbf{x}) \tilde{u}_i(\mathbf{x}) \, \mathrm{d}S_x = & \int_{S_p} \int_S \Bigg[-\frac{1}{4\pi} \tilde{u}_i(\tilde{\mathbf{x}}) \frac{1}{r^2} r_{,\tilde{p}}\, \tilde{n}_p(\tilde{\mathbf{x}}) p_i(\mathbf{x}) \\ & - \tilde{R}_j[\tilde{u}_i](\tilde{\mathbf{x}}) G^{\varphi u}_{ijk}(\mathbf{x} - \tilde{\mathbf{x}}) p_k(\mathbf{x}) - \tilde{R}_j[\tilde{u}_i](\tilde{\mathbf{x}}) G^{\varphi\varphi}_{ijkq}(\mathbf{x} - \tilde{\mathbf{x}}) R_q[u_k](\mathbf{x}) \Bigg] \mathrm{d}S_x \, \mathrm{d}S_{\tilde{x}} \\ & + \int_{S_p} \int_{S_c} \tilde{R}_j[\tilde{u}_i](\tilde{\mathbf{x}}) G^{\varphi\varphi}_{ijkq}(\mathbf{x} - \tilde{\mathbf{x}}) R_q[w_k](\mathbf{x}) \, \mathrm{d}S_x \, \mathrm{d}S_{\tilde{x}} \end{aligned} \tag{9}$$

where:

$$\begin{aligned} G^{\varphi u}_{ijk} &= \frac{1}{8\pi(1-\nu) r} \left[(1-2\nu)\, e_{kij} - e_{kpj} r_{,p} r_{,i} \right] \\ G^{\varphi\varphi}_{ijkq} &= -\frac{\mu}{8\pi} \frac{1}{r} \left(\delta_{eg} - r_{,e}\, r_{,g} \right) e_{iep} e_{jgr} \left[\frac{2\nu}{1-\nu} \delta_{pk} \delta_{rq} + \delta_{pr} \delta_{kq} + \delta_{pq} \delta_{kr} \right] \end{aligned} \tag{10}$$

and:

$$\begin{aligned} \tilde{R}_j[\tilde{u}_i](\tilde{\mathbf{x}}) &= e_{bcj} \tilde{n}_b(\tilde{\mathbf{x}}) \frac{\partial \tilde{u}_i}{\partial \tilde{x}_c}(\tilde{\mathbf{x}}) \\ \frac{1}{r^2} r_{,\tilde{i}}\, \tilde{n}_i \, \mathrm{d}S_{\tilde{x}} &= \mathrm{d}\tilde{\Theta} \end{aligned}$$

$\mathrm{d}\tilde{\Theta}$ represents the differential solid angle under which $\mathrm{d}S_{\tilde{x}}$ is seen from $\mathbf{x}$. Hence, also all the kernels in eqn. (9) are weakly singular.

Regularized weak form for the traction integral equation on S_c. If we define the auxiliary displacement discontinuity field $\tilde{w}_i = \tilde{u}_i^+ - \tilde{u}_i^-$ (naturally flowing from the $\tilde{u}_i$ field after the limit process $h \rightarrow 0$ forces $\tilde{S}^+$ to coincide with $\tilde{S}^-$) the variational traction equation on S_c can be written as:

$$\int_{S_c} p_i(\mathbf{x})\tilde{w}_i(\mathbf{x})\,\mathrm{d}S_x = \int_{S_c}\int_S \Big[-\frac{1}{4\pi}\tilde{w}_i(\tilde{\mathbf{x}})\frac{1}{r^2} r_{,\tilde{p}}\,\tilde{n}_p(\tilde{\mathbf{x}})p_i(\mathbf{x})$$

$$- \tilde{R}_j[\tilde{w}_i](\tilde{\mathbf{x}})G^{\varphi u}_{ijk}(\mathbf{x}-\tilde{\mathbf{x}})p_k(\mathbf{x}) - \tilde{R}_j[\tilde{w}_i](\tilde{\mathbf{x}})G^{\varphi\varphi}_{ijkq}(\mathbf{x}-\tilde{\mathbf{x}})R_q[u_k](\mathbf{x})\Big]\,\mathrm{d}S_x\,\mathrm{d}S_{\tilde{x}}$$

$$+ \int_{S_c}\int_{S_c} \tilde{R}_j[\tilde{w}_i](\tilde{\mathbf{x}})G^{\varphi\varphi}_{ijkq}(\mathbf{x}-\tilde{\mathbf{x}})R_q[w_k](\mathbf{x})\,\mathrm{d}S_x\,\mathrm{d}S_{\tilde{x}} \tag{11}$$

Continuity requirements. Weak continuity requirements must be enforced on $u_k, w_k,\ \tilde{u}_k, \tilde{w}_k$ to guarantee the validity of equations (6)-(11), since they must belong to the class of $C^0_0(S)$ continuous functions. The class $C^0_0(S)$ is defined as follows: $C^0_0(S) \equiv C^0(S)$ if S is closed, and $C^0_0(S) = \{f(\mathbf{x}) \in C^0(S) : f|_{\partial S} = 0\}$ if S is open (e.g. admissible kinematic fields for cracks inside bodies are assumed to be in $C^0_0(S)$). On the contrary no special constraints are set on the static fields p_k and $\tilde{p}_k$.

The set of the three previous integral equations can be represented in the compact operatorial form:

$$\mathcal{A}(\tilde{\mathbf{u}},\tilde{\mathbf{p}},\tilde{\mathbf{w}},\mathbf{u},\mathbf{p},\mathbf{w}) = f(\tilde{\mathbf{u}},\tilde{\mathbf{p}},\tilde{\mathbf{w}},\bar{\mathbf{u}},\bar{\mathbf{p}}) \qquad \forall\ \tilde{\mathbf{u}},\tilde{\mathbf{p}},\tilde{\mathbf{w}} \tag{12}$$
$$\text{with}\qquad \mathcal{A}(\tilde{\mathbf{u}},\tilde{\mathbf{p}},\tilde{\mathbf{w}},\mathbf{u},\mathbf{p},\mathbf{w}) = \mathcal{A}(\mathbf{u},\mathbf{p},\mathbf{w},\tilde{\mathbf{u}},\tilde{\mathbf{p}},\tilde{\mathbf{w}})$$

where $\mathcal{A}$ is a self-adjoint bilinear form.

Through the adoption of a Galerkin discretization scheme, eqn. (12) leads to a symmetric linear equation system.

The evaluation of the double surface integrals represents probably the main obstacle which has hampered the application of the method in the 3D context. However, recent results obtained by applied mathematicians (Erichsen and Sauter, 1998, Sauter and Schwab, 1997) have led to innovative algorithms which are now being adopted by the engineering BE community and have served as a basis for the fracture-oriented implementation of the SGBEM in 3D recently presented by Frangi *et al.* (2001).

Discretization. At this stage the boundary surface S and the crack S_c are discretized into boundary elements and symmetry is preserved also in the discrete formulation if the auxiliary and real fields are interpolated over the BEs according to a Galerkin scheme. Further details relevant to the discretization phase and proofs of symmetry properties can be found in Bonnet *et al.* (1998).

2 Numerical Evaluation of Weakly Singular Integrals

Let us assume that the surface S has been partitioned into 9-noded quadrilateral and/or 6-noded triangular isoparametric elements; f and $\tilde{f}$ are given functions of $\mathbf{x}$ and $\tilde{\mathbf{x}}$, respectively, and $B(\mathbf{x}-\tilde{\mathbf{x}})$ is a generic weakly singular kernel. Our aim is to compute

$$\mathcal{I}_{mn} = \int_{S_m}\int_{S_n} \tilde{f}(\tilde{\mathbf{x}})B(\mathbf{x}-\tilde{\mathbf{x}})f(\mathbf{x})\,\mathrm{d}S_x\,\mathrm{d}S_{\tilde{x}} = \int_{S_m}\int_{S_n} \mathcal{B}(f,\tilde{f},\mathbf{x},\tilde{\mathbf{x}})\,\mathrm{d}S_x\,\mathrm{d}S_{\tilde{x}} \tag{13}$$

where S_m ($\tilde{\mathbf{x}} \in S_m$) and S_n ($\mathbf{x} \in S_n$) represent a generic element pair. Intrinsic parameters η_1, η_2 and $\tilde{\eta}_1, \tilde{\eta}_2$ are introduced on the parent (master) elements, such that on the physical element S_m, $\tilde{\mathbf{x}} = \tilde{\boldsymbol{\mathcal{X}}}(\tilde{\eta}_1, \tilde{\eta}_2)$ and on S_n, $\mathbf{x} = \boldsymbol{\mathcal{X}}(\eta_1, \eta_2)$.
E.g. for S_m and S_n, quadrilateral source and field elements, eqn. (13) becomes:

$$\mathcal{I}_{mn} = \int_0^1 \int_0^1 \int_0^1 \int_0^1 \tilde{\mathcal{B}}\big[\tilde{f}(\tilde{\boldsymbol{\eta}}), f(\boldsymbol{\eta}), \tilde{\mathbf{x}}(\tilde{\boldsymbol{\eta}}), \mathbf{x}(\boldsymbol{\eta})\big] \, \mathrm{d}\eta_1 \, \mathrm{d}\eta_2 \, \mathrm{d}\tilde{\eta}_1 \, \mathrm{d}\tilde{\eta}_2 \tag{14}$$

where $\tilde{\mathcal{B}}$ also includes the jacobians of the transformations. The evaluation of such double surface integrals represents a crucial aspect of the method and is computationally expensive. In this paper the approach described by Erichsen and Sauter (1998) and Sauter and Schwab (1997) is adopted. Four different situations must be accounted for, in general, according to whether S_m and S_n are: (i) coincident elements, (ii) adjacent elements sharing one edge, (iii) adjacent elements sharing one vertex; (iv) distinct elements (see Figure 1 for the case of quadrilateral-quadrilateral elements).

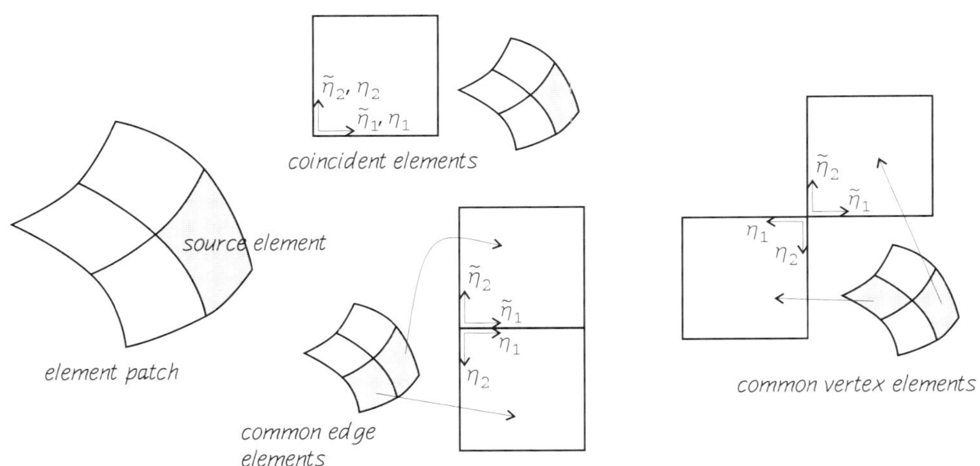

Figure 1. Singular cases for quadrilateral element pairs

In the case of distinct elements, standard product Gauss formulae are employed choosing an appropriate number of Gauss points. For the first three cases the procedure can be outlined as follows. (i) The domain of integration is expressed, via suitable coordinates transformations, as the sum of "pyramidal" shaped subdomains in which the singularity is concentrated at one vertex. (ii) For each subdomain a regularizing variable transformation (involving Duffy generalized coordinates) renders the integrand analytic, introducing a jacobian which cancels the singularity in the kernel. With reference to the case of quadrilateral elements the following sections give the final ready-to-implement expressions of the regular integrals to which original integrals of type (13) turn out to reduce.

2.1 Coincident Elements

A somewhat unorthodox notation will be employed for parameter domains (treated as functions of suitable cartesian coordinates).
The symbol $\mathcal{D}(\tilde{\eta}_1, \tilde{\eta}_2, \eta_1, \eta_2)$ denotes a four-dimensional polyhedron collecting all the points $\{\tilde{\eta}_1, \tilde{\eta}_2, \eta_1, \eta_2\}$ spanned by the multivariate integral in eqn. (13).

Let us introduce the relative variables $u_1 \equiv \eta_1 - \tilde{\eta}_1$, $u_2 \equiv \eta_2 - \tilde{\eta}_2$. The singularity in eqn. (13) is activated whenever $u_1 = u_2 = 0$, being the integrand regular with respect to $\tilde{\eta}_1, \tilde{\eta}_2$.

The procedure can be outlined as follows. (i) The domain $\mathcal{D}$ is expressed in terms of $\tilde{\eta}_1, \tilde{\eta}_2, u_1, u_2$ and algebraic manipulations are performed in order to make the inequalities concerning u_1, u_2 the outermost ones (so exchanging the integration order) and to provide a partition of $\mathcal{D}$ into subdomains sharing the point $u_1 = u_2 = 0$ as common vertex. (ii) For each subdomain a regularizing variable transformation (involving Duffy generalized coordinates) renders the integrand analytic, introducing a jacobian which cancels the singularity in the kernel.

Domain partition.

$$\mathcal{D}(\tilde{\eta}_1, \tilde{\eta}_2, \eta_1, \eta_2) \equiv \begin{cases} 0 \le \tilde{\eta}_1 \le 1 \\ 0 \le \tilde{\eta}_2 \le 1 \\ 0 \le \eta_1 \le 1 \\ 0 \le \eta_2 \le 1 \end{cases} = \begin{cases} 0 \le \tilde{\eta}_1 \le 1 \\ -\tilde{\eta}_1 \le u_1 \le 1 - \tilde{\eta}_1 \\ 0 \le \tilde{\eta}_2 \le 1 \\ 0 \le \eta_2 \le 1 \end{cases}$$

$$\equiv \begin{cases} 0 \le u_1 \le 1 \\ 0 \le \tilde{\eta}_1 \le 1 - u_1 \\ 0 \le \tilde{\eta}_2 \le 1 \\ 0 \le \eta_2 \le 1 \end{cases} \cup \begin{cases} 0 \le -u_1 \le 1 \\ -u_1 \le \tilde{\eta}_1 \le 1 \\ 0 \le \tilde{\eta}_2 \le 1 \\ 0 \le \eta_2 \le 1 \end{cases} \equiv \mathcal{D}_1^c \cup \mathcal{D}_2^c \tag{15}$$

where Transformation C of Appendix A has been exploited setting $s = \tilde{\eta}_1, t = u_1$. In Appendix B it is shown that $\mathcal{D}_2^c(\tilde{\eta}_1, \tilde{\eta}_2, \eta_1, \eta_2) = \mathcal{D}_1^c(\eta_1, \tilde{\eta}_2, \tilde{\eta}_1, \eta_2)$, i.e. $\mathcal{D}_2$ coincides with $\mathcal{D}_1$ provided that $\tilde{\eta}_1$ is exchanged with η_1 (and, hence, u_1 with $-u_1$). The above statement of equivalence must be understood as follows: let us imagine two reference systems (with superscripts a and b respectively) in the four-dimensional space sharing the same origin and oriented such that the $\tilde{\eta}_1^a$ axis coincides with the η_1^b one, η_1^a with $\tilde{\eta}_1^b$, $\tilde{\eta}_2^a$ with $\tilde{\eta}_2^b$, η_2^a with η_2^b and so on. Hence, domain $\mathcal{D}_1^c$ (imagined plotted in the reference system a) coincides with $\mathcal{D}_2^c$ (imagined plotted in the reference system b). Only $\mathcal{D}_1^c$ will be considered hereafter and the conclusions extended immediately to $\mathcal{D}_2^c$.

Once expressed in terms of $u_2, \tilde{\eta}_2$, the domain $\mathcal{D}_1^c$ is further partitioned into $\mathcal{D}_{11}^c$ and $\mathcal{D}_{12}^c$ exploiting once again Transformation C in Appendix A:

$$\mathcal{D}_1^c = \begin{cases} 0 \le u_1 \le 1 \\ 0 \le \tilde{\eta}_1 \le 1 - u_1 \\ 0 \le \tilde{\eta}_2 \le 1 \\ -\tilde{\eta}_2 \le u_2 \le 1 - \tilde{\eta}_2 \end{cases} \tag{16}$$

$$= \begin{cases} 0 \le u_1 \le 1 \\ 0 \le \tilde{\eta}_1 \le 1 - u_1 \\ 0 \le u_2 \le 1 \\ 0 \le \tilde{\eta}_2 \le 1 - u_2 \end{cases} \cup \begin{cases} 0 \le u_1 \le 1 \\ 0 \le \tilde{\eta}_1 \le 1 - u_1 \\ 0 \le -u_2 \le 1 \\ -u_2 \le \tilde{\eta}_2 \le 1 \end{cases} = \mathcal{D}_{11}^c \cup \mathcal{D}_{12}^c$$

Also in this case it can be verified that $\mathcal{D}^c_{12}(\tilde\eta_1, \tilde\eta_2, \eta_1, \eta_2)= \mathcal{D}^c_{11}(\tilde\eta_1, \eta_2, \eta_1, \tilde\eta_2)$, i.e. the two domains formally coincide if $\tilde\eta_2$ and η_2 are exchanged.

Let us focus on $\mathcal{D}^c_{11}(\tilde\eta_1, \tilde\eta_2, \eta_1, \eta_2)$. The singular point $u_1 = u_2 = 0$ is a vertex for the square $0 \le u_1, u_2 \le 1$.

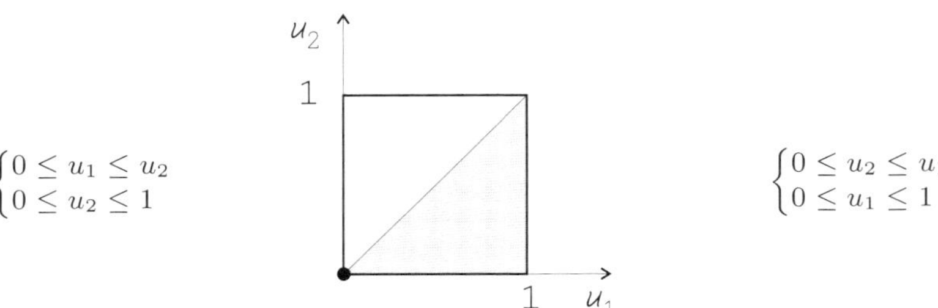

Figure 2. Partition of the square

The square is partitioned into two triangles separated by the diagonal $u_1 = u_2$ as shown in Figure 2:

$$\mathcal{D}^c_{11} = \begin{cases} 0 \le u_1 \le 1 \\ 0 \le u_2 \le u_1 \\ 0 \le \tilde\eta_1 \le 1 - u_1 \\ 0 \le \tilde\eta_2 \le 1 - u_2 \end{cases} \cup \begin{cases} 0 \le u_1 \le 1 \\ u_1 \le u_2 \le 1 \\ 0 \le \tilde\eta_1 \le 1 - u_1 \\ 0 \le \tilde\eta_2 \le 1 - u_2 \end{cases} \equiv \mathcal{D}^c_{111} \cup \mathcal{D}^c_{112} \tag{17}$$

Subdomain $\mathcal{D}^c_{112}$ coincides with $\mathcal{D}^c_{111}$ when exchanging $\eta_1, \tilde\eta_1$ with $\eta_2, \tilde\eta_2$. As a conclusion the original domain $\mathcal{D}$ is obtained as the union of 8 "rotated" subdomains:

$$\begin{aligned}\mathcal{D} \equiv & \mathcal{D}^c_{111}(\tilde\eta_1, \tilde\eta_2, \eta_1, \eta_2) \cup \mathcal{D}^c_{111}(\tilde\eta_1, \eta_2, \eta_1, \tilde\eta_2) \cup \mathcal{D}^c_{111}(\eta_1, \eta_2, \tilde\eta_1, \tilde\eta_2) \cup \\ & \mathcal{D}^c_{111}(\eta_1, \tilde\eta_2, \tilde\eta_1, \eta_2) \cup \mathcal{D}^c_{111}(\tilde\eta_2, \tilde\eta_1, \eta_2, \eta_1) \cup \mathcal{D}^c_{111}(\eta_2, \tilde\eta_1, \tilde\eta_2, \eta_1) \cup \\ & \mathcal{D}^c_{111}(\eta_2, \eta_1, \tilde\eta_2, \tilde\eta_1) \cup \mathcal{D}^c_{111}(\tilde\eta_2, \eta_1, \eta_2, \tilde\eta_1)\end{aligned} \tag{18}$$

Regularizing coordinates and final formula. We seek here a transformation of variables producing a jacobian which might cancel the weak singularity for $u_1 = u_2 = 0$. At this stage, Duffy coordinates make the integrand regular. Let us define the following variables:

$$\begin{cases} v_1 = \omega \\ v_2 = \xi_1 \omega \\ v_3 = \xi_2 (1 - \omega) \\ v_4 = \xi_3 (1 - \xi_1 \omega) \end{cases} \quad \text{with} \quad \begin{cases} 0 \le \omega \le 1 \\ 0 \le \xi_1 \le 1 \\ 0 \le \xi_2 \le 1 \\ 0 \le \xi_3 \le 1 \end{cases} \tag{19}$$

For the 8 subdomains in eqn. (18) the intrinsic variables are expressed as functions of the v_i variables as collected in Table 1.

Moreover, the jacobian of the transformation, $J = \omega(1-\omega)(1-\xi_1\omega)$, cancels the singularity,

case	1	2	3	4
$\tilde{\eta}_1^i=$	v_3	v_3	v_1+v_3	v_1+v_3
$\tilde{\eta}_2^i=$	v_4	v_2+v_4	v_2+v_4	v_4
$\eta_1^i=$	v_1+v_3	v_1+v_3	v_3	v_3
$\eta_2^i=$	v_2+v_4	v_4	v_4	v_2+v_4

case	5	6	7	8
$\tilde{\eta}_1^i=$	v_4	v_2+v_4	v_2+v_4	v_4
$\tilde{\eta}_2^i=$	v_3	v_3	v_1+v_3	v_1+v_3
$\eta_1^i=$	v_2+v_4	v_4	v_4	v_2+v_4
$\eta_2^i=$	v_1+v_3	v_1+v_3	v_3	v_3

Table 1. Definition of intrinsic variables: coincident elements

as required. The rather lengthy procedure outlined leads however to a straightforward implementation formula:

$$\mathcal{I}_{mm} = \int_0^1 \int_0^1 \int_0^1 \int_0^1 \sum_{i=1}^{8} \tilde{\mathcal{B}}\big[\tilde{f}(\tilde{\boldsymbol{\eta}}^i), f(\boldsymbol{\eta}^i), \tilde{\mathbf{x}}(\tilde{\boldsymbol{\eta}}^i), \mathbf{x}(\boldsymbol{\eta}^i)\big] \cdot \\ \cdot\, \omega(1-\omega)(1-\xi_1\omega)\, \mathrm{d}\omega\, \mathrm{d}\xi_1\, \mathrm{d}\xi_2\, \mathrm{d}\xi_3 \tag{20}$$

At this stage eqn. (20) can be evaluated with standard Gaussian numerical schemes, providing a high accuracy even for very low numbers of integration points.

2.2 Common Edge

In this case the singularity is activated whenever the source and field point both lie on the common edge, i.e. when $u_1 = \eta_2 = \tilde{\eta}_2 = 0$.

A procedure similar to the one expounded in the previous section is devised.

Domain partition. Exploiting once more Transformation C in Appendix A, the original intrinsic domain is partitioned as follows:

$$\mathcal{D} \equiv \begin{cases} 0 \le \tilde{\eta}_1 \le 1 \\ 0 \le \eta_1 \le 1 \\ 0 \le \tilde{\eta}_2 \le 1 \\ 0 \le \eta_2 \le 1 \end{cases} = \begin{cases} 0 \le \tilde{\eta}_1 \le 1 \\ -\tilde{\eta}_1 \le u_1 \le 1-\tilde{\eta}_1 \\ 0 \le \tilde{\eta}_2 \le 1 \\ 0 \le \eta_2 \le 1 \end{cases} \tag{21}$$

$$= \begin{cases} 0 \le u_1 \le 1 \\ 0 \le \tilde{\eta}_1 \le 1-u_1 \\ 0 \le \tilde{\eta}_2 \le 1 \\ 0 \le \eta_2 \le 1 \end{cases} \cup \begin{cases} 0 \le -u_1 \le 1 \\ -\tilde{\eta}_1 \le u_1 \le 0 \\ 0 \le \tilde{\eta}_2 \le 1 \\ 0 \le \eta_2 \le 1 \end{cases} \equiv \mathcal{D}_1^{ed} \cup \mathcal{D}_2^{ed} \tag{22}$$

As in the coincident element case $\mathcal{D}_2^{ed}(\tilde{\eta}_1, \tilde{\eta}_2, \eta_1, \eta_2) = \mathcal{D}_1^{ed}(\eta_1, \tilde{\eta}_2, \tilde{\eta}_1, \eta_2)$, i.e. $\mathcal{D}_2^{ed}$ coincides with $\mathcal{D}_1^{ed}$ provided that η_1 is exchanged with $\tilde{\eta}_1$. Hence only $\mathcal{D}_1^{ed}$ will be considered hereafter.

The inequalities concerning $u_1, \tilde{\eta}_2, \eta_2$ define a cube in the $u_1, \tilde{\eta}_2, \eta_2$ space, which is further partitioned into three different pyramidal subdomains having their vertex in $u_1 = \tilde{\eta}_2 = \eta_2 = 0$, as shown in Figure 3.

$$\begin{aligned}\mathcal{D}_1^{ed} = &\begin{cases} 0 \le u_1 \le 1 \\ 0 \le \tilde{\eta}_2 \le u_1 \\ 0 \le \eta_2 \le u_1 \\ 0 \le \tilde{\eta}_1 \le 1 - u_1 \end{cases} \cup \begin{cases} 0 \le \tilde{\eta}_2 \le 1 \\ 0 \le u_1 \le \tilde{\eta}_2 \\ 0 \le \eta_2 \le \tilde{\eta}_2 \\ 0 \le \tilde{\eta}_1 \le 1 - u_1 \end{cases} \cup \begin{cases} 0 \le \eta_2 \le 1 \\ 0 \le u_1 \le \eta_2 \\ 0 \le \tilde{\eta}_2 \le \eta_2 \\ 0 \le \tilde{\eta}_1 \le 1 - u_1 \end{cases} \\ =&\mathcal{D}_{11}^{ed} \cup \mathcal{D}_{12}^{ed} \cup \mathcal{D}_{13}^{ed} \end{aligned} \quad (23)$$

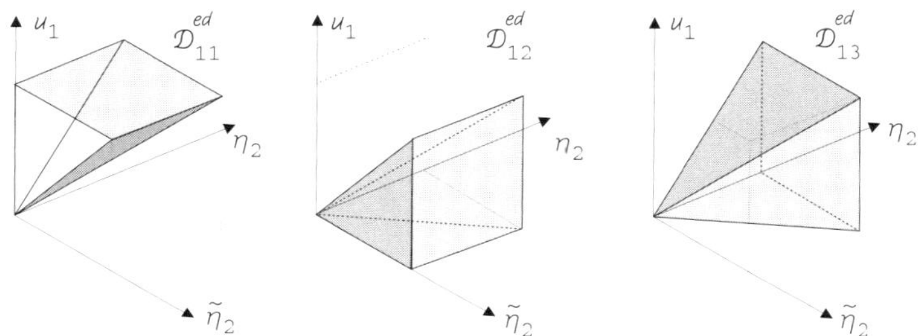

Figure 3. Partition of the unit cube

Hence the original domain $\mathcal{D}$ can be expressed as the union of six subdomains defined by eqns. (21) and (23):

$$\begin{aligned}\mathcal{D} \equiv &\mathcal{D}_{11}^{ed}(\tilde{\eta}_1, \tilde{\eta}_2, \eta_1, \eta_2) \cup \mathcal{D}_{12}^{ed}(\tilde{\eta}_1, \tilde{\eta}_2, \eta_1, \eta_2) \cup \mathcal{D}_{13}^{ed}(\tilde{\eta}_1, \tilde{\eta}_2, \eta_1, \eta_2) \cup \\ &\mathcal{D}_{11}^{ed}(\eta_1, \tilde{\eta}_2, \tilde{\eta}_1, \eta_2) \cup \mathcal{D}_{12}^{ed}(\eta_1, \tilde{\eta}_2, \tilde{\eta}_1, \eta_2) \cup \mathcal{D}_{13}^{ed}(\eta_1, \tilde{\eta}_2, \tilde{\eta}_1, \eta_2)\end{aligned} \quad (24)$$

Regularizing coordinates and final formula.

Let us define variables $v_1, v_2, v_3, v_4, v_5, \mathcal{J}_1$ and $\mathcal{J}_2$ as follows:

$$\begin{cases} v_1 = \omega \\ v_2 = \xi_1 \omega \\ v_3 = \xi_2 \omega \\ v_4 = \xi_3(1 - \omega) \\ v_5 = \xi_3(1 - \xi_1 \omega) \end{cases} \quad \text{with} \quad \begin{cases} 0 \le \omega_1 \le 1 \\ 0 \le \xi_1 \le 1 \\ 0 \le \xi_2 \le 1 \\ 0 \le \xi_3 \le 1 \end{cases} \quad (25)$$

and

$$\mathcal{J}_1 = \omega^2(1 - \omega) \qquad \mathcal{J}_2 = \omega^2(1 - \xi_1 \omega)$$

For each of the 6 cases corresponding to subdomains in eqn. (24), Table 2 collects the definition of intrinsic variables and jacobians.

case	1	2	3	4	5	6
$\tilde{\eta}_1^i=$	v_4	v_5	v_5	v_1+v_4	v_5+v_2	v_5+v_2
$\tilde{\eta}_2^i=$	v_2	v_1	v_3	v_2	v_1	v_3
$\eta_1^i=$	v_1+v_4	v_5+v_2	v_5+v_2	v_4	v_5	v_5
$\eta_2^i=$	v_3	v_3	v_1	v_3	v_3	v_1
$J_i=$	$\mathcal{J}_1$	$\mathcal{J}_2$	$\mathcal{J}_2$	$\mathcal{J}_1$	$\mathcal{J}_2$	$\mathcal{J}_2$

Table 2. Definition of intrinsic variables and jacobians: common edge case

Finally

$$\mathcal{I}_{mm} = \int_0^1\int_0^1\int_0^1\int_0^1 \sum_{i=1}^{6} \tilde{\mathcal{B}}\big[\tilde{f}(\tilde{\boldsymbol{\eta}}^i), f(\boldsymbol{\eta}^i), \tilde{\mathbf{x}}(\tilde{\boldsymbol{\eta}}^i), \mathbf{x}(\boldsymbol{\eta}^i)\big] J_i \, \mathrm{d}\omega_1 \, \mathrm{d}\omega_2 \, \mathrm{d}\xi_1 \, \mathrm{d}\xi_2 \tag{26}$$

and the jacobian of the transformation cancels all singularities.

2.3 Common Vertex

Let us now consider the third case in Figure 1, where the two elements share one vertex. The integrand is singular only if $\tilde{\eta}_1 = \tilde{\eta}_2 = \eta_1 = \eta_2 = 0$.

Domain partition. The domain, a cube in a four-dimensional space, is decomposed in four subdomains on the basis of the partition already devised for the square in $\Re^2$ (Figure 2) and the cube in $\Re^3$ (Figure 3):

$$\mathcal{D} = \begin{cases} 0 \le \tilde{\eta}_1 \le 1 \\ 0 \le \tilde{\eta}_2 \le 1 \\ 0 \le \eta_1 \le 1 \\ 0 \le \eta_2 \le 1 \end{cases} \tag{27}$$

$$= \begin{cases} 0 \le \tilde{\eta}_1 \le 1 \\ 0 \le \tilde{\eta}_2 \le \tilde{\eta}_1 \\ 0 \le \eta_1 \le \tilde{\eta}_1 \\ 0 \le \eta_2 \le \tilde{\eta}_1 \end{cases} \cup \begin{cases} 0 \le \tilde{\eta}_2 \le 1 \\ 0 \le \tilde{\eta}_1 \le \tilde{\eta}_2 \\ 0 \le \eta_1 \le \tilde{\eta}_2 \\ 0 \le \eta_2 \le \tilde{\eta}_2 \end{cases} \cup \begin{cases} 0 \le \eta_1 \le 1 \\ 0 \le \tilde{\eta}_1 \le \eta_1 \\ 0 \le \tilde{\eta}_2 \le \eta_1 \\ 0 \le \eta_2 \le \eta_1 \end{cases} \cup \begin{cases} 0 \le \eta_2 \le 1 \\ 0 \le \tilde{\eta}_1 \le \eta_2 \\ 0 \le \tilde{\eta}_2 \le \eta_2 \\ 0 \le \eta_1 \le \eta_2 \end{cases}$$

$$= \mathcal{D}_1^v \cup \mathcal{D}_2^v \cup \mathcal{D}_3^v \cup \mathcal{D}_4^v$$

Regularizing coordinate transformation and final formula. Duffy coordinates are now introduced on each subdomain; let us define the following functions

$$\begin{cases} v_1 = \omega \\ v_2 = \xi_1 \omega \\ v_3 = \xi_2 \omega \\ v_4 = \xi_3 \omega \end{cases} \quad \text{with} \quad \begin{cases} 0 \le \omega \le 1 \\ 0 \le \xi_1 \le 1 \\ 0 \le \xi_2 \le 1 \\ 0 \le \xi_3 \le 1 \end{cases} \tag{28}$$

case	1	2	3	4
$\tilde{\eta}_1^i=$	v_1	v_2	v_2	v_2
$\tilde{\eta}_2^i=$	v_2	v_1	v_3	v_3
$\eta_1^i=$	v_3	v_3	v_1	v_4
$\eta_2^i=$	v_4	v_4	v_4	v_1

Table 3. Definition of variables: common vertex

According to eqn. (27), the intrinsic variables for the four subdmains are chosen as in Table 3. Finally

$$\mathcal{I}_{mn} = \int_0^1 \int_0^1 \int_0^1 \int_0^1 \sum_{i=1}^{4} \tilde{\mathcal{B}}[\tilde{f}(\tilde{\boldsymbol{\eta}}^i), f(\boldsymbol{\eta}^i), \tilde{\mathbf{x}}(\tilde{\boldsymbol{\eta}}^i), \mathbf{x}(\boldsymbol{\eta}^i)]\, \omega^3 \,\mathrm{d}\omega \,\mathrm{d}\xi_1 \,\mathrm{d}\xi_2 \,\mathrm{d}\xi_3 \tag{29}$$

3 Numerical Example Concerning SIFs Evaluation

The example concerns a circular surface breaking crack in a finite body and demonstrates the accuracy of the numerical procedure in evaluating the SIFs from displacement discontinuities near the crack front. The geometry of the problem is shown in Figure 4; uniform tensile stresses are applied at two opposite faces of the bar (plate) in the direction perpendicular to the crack; a value of Poisson's ratio $\nu = .3$ is adopted.

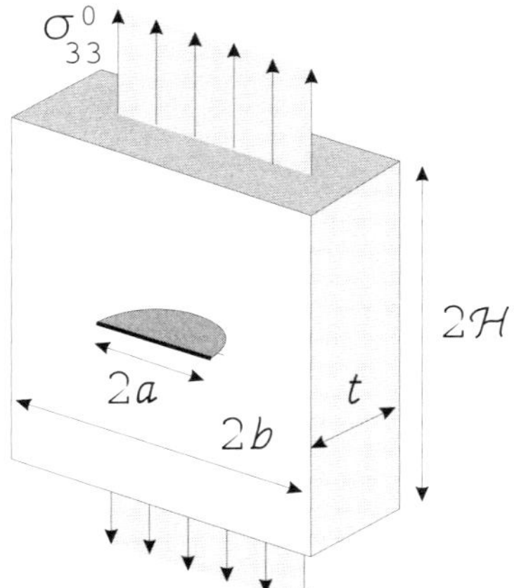

Figure 4. Plate containing a semi-circular surface crack

The configuration considered is characterized by the geometric ratios $H/a = 5$, $b/a = 5$ and $a/t = .4$. The values adopted for H/a and b/a are large enough to effectively represent an edge

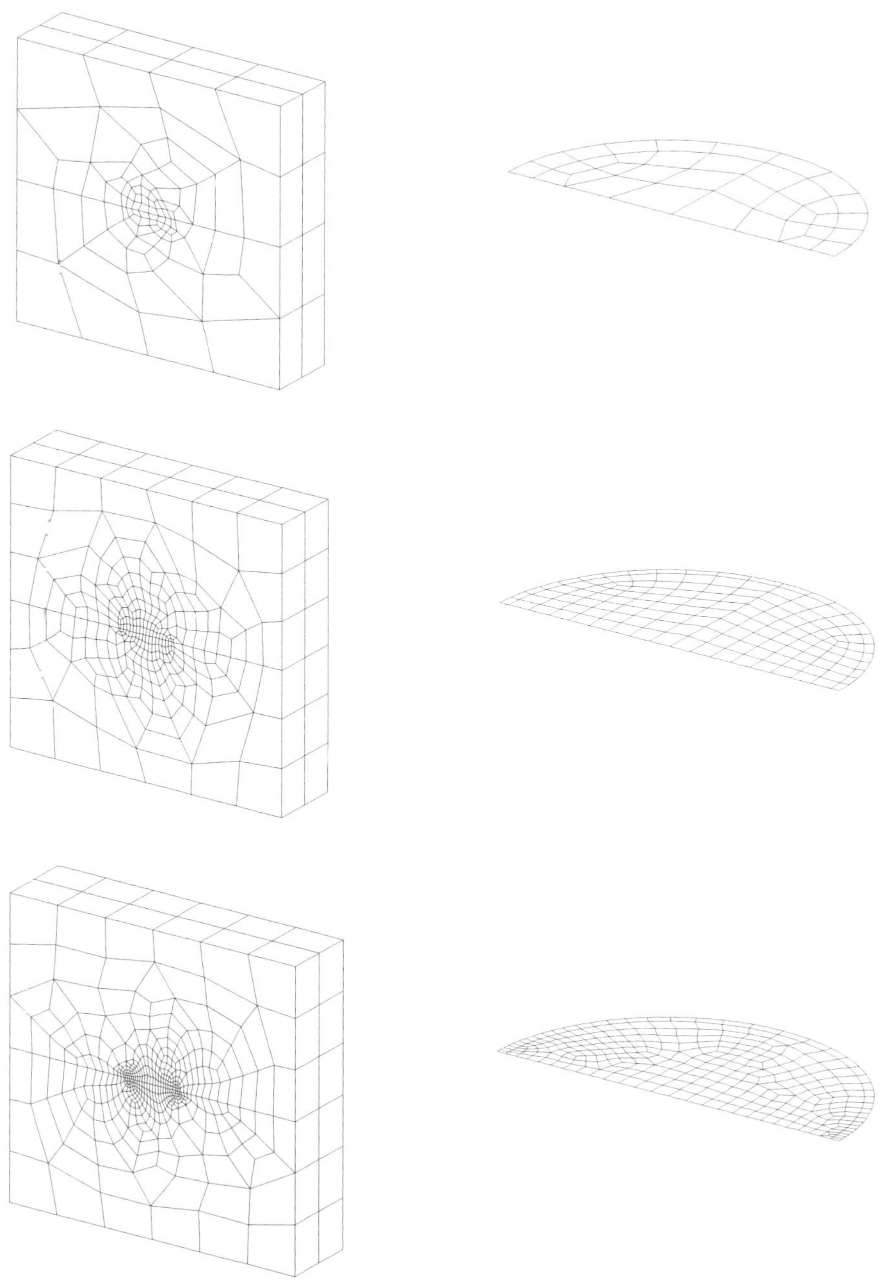

Figure 5. Meshes A, B and C

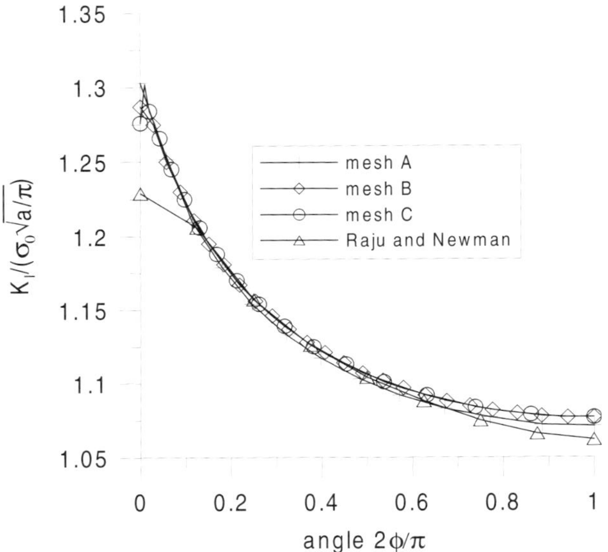

Figure 6. Normalized stress intensity factor for semi-circular surface crack in a plate as function of angular position along crack front; results for the meshes A, B and C and earlier FE results of Raju and Newman (1979).

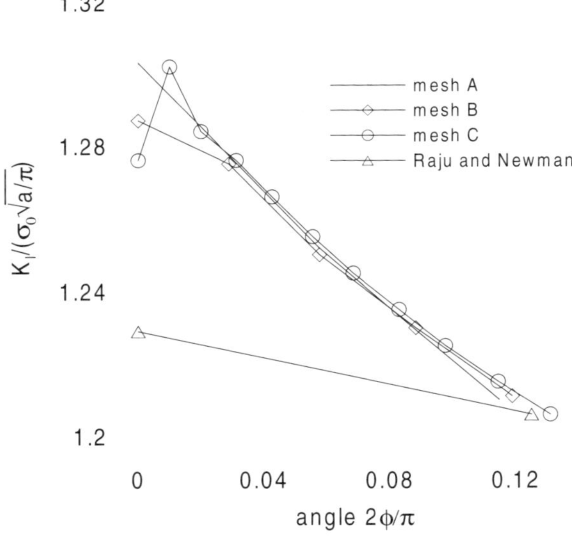

Figure 7. Normalized stress intensity factor for semi-circular surface crack in a plate: "boundary layer" effect for meshes A, B and C and earlier FE results of Raju and Newman (1979).

crack in an infinite plate. The adopted BEs are isoparametric 9-noded quadrilaterals. No quarter point scheme is adopted in this example for the elements at the crack front.

The SIFs extraction at a point Q on the front is carried out, on the basis of the square root asymptotic behaviour in terms of distance from the front, using the displacement discontinuity values at the two nodes behind the front lying in the plane through Q normal to the crack front line: from the two estimates associated to the two nodes, the final estimate for the SIF at Q is obtained by linear extrapolation.

The problem is analyzed using the three meshes, Mesh A, Mesh B and Mesh C, depicted in Figure 5, having 12, 24 and 40 elements along the circular crack front, respectively. Note that the meshes are generated using a commercial finite element pre-processor which has been interfaced to the SGBEM code.

The results obtained in term of mode-I SIF are plotted in Figure 6 as a function of the angular parameter $2\phi/\pi$ (with $\phi = 0$ at the free surface) where they are compared to the finite element results of Raju and Newman (1979). It is well known that, if a surface breaking crack intersects the surface itself at a right angle, the SIFs, as defined on the base of the classical Williams-Westergaard asymptotic formulae, tend to zero as the free surface is approached; this non-standard behaviour shows up in a boundary layer whose thickness depends on the problems geometry and material properties.

From Figure 7 it turns out that the analyses carried out by the SGBEM (in particular with the refined Mesh C) predict a much thinner boundary layer than in Raju and Newman (1979). However SIF values far from the external surface are not significantly affected by the accuracy with which the boundary layer effect is accounted for and can be accurately predicted using even rather coarse meshes.

4 Simulation of Fatigue Crack Growth

The initial configuration of the crack is assigned and the symmetric Galerkin BEM is employed to carry out a stress analysis. Stress intensity factors are extrapolated from the displacement discontinuity field (possibly utilizing quarter-point elements along the crack front). The incremental growth is determined on the basis of the criterion described in the following sections; accordingly a row of boundary elements is then added ahead of the crack front and another analysis cycle is carried out for the new configuration after performing the necessary modifications of the surface mesh as detailed in the sequel. This incremental analysis is repeated until a predefined crack length is reached or the SIFs exceed the fracture toughness of the material, i.e. when unstable propagation of the fracture occurs.

4.1 Propagation Criterion

The formulation of a suitable 3D crack growth criterion, even in the classical context of linear elastic fracture mechanics, still represents an open issue. In this contribution we have chosen to explore potentialities of the SGBEM by simulating a fatigue crack growth governed by Paris law and by a local stress criterion for the selection of the propagation direction.

The procedure can be summarized as follows (for further details and alternative choices see e.g. Mi, 1996): (i) each geometrical point on the crack front moves in the plane perpendicular

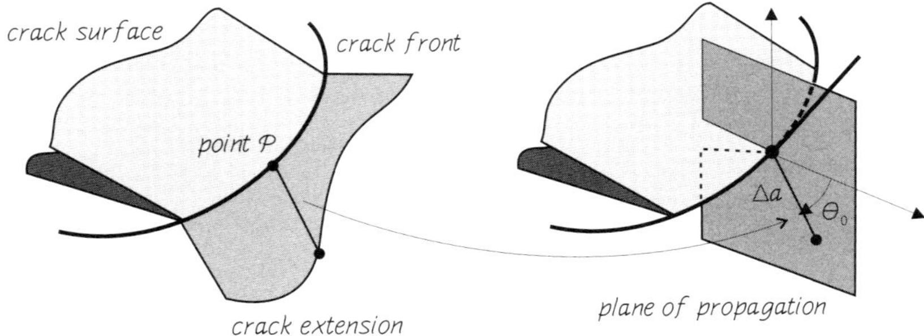

Figure 8. Choice of local crack propagation direction

to the crack front itself (see Figure 8_b); (ii) the propagation angle in this plane is determined through an extension to three dimensions of the classical criterion of the maximum hoop stress for 2D mixed-mode problems:

$$\tan\frac{\theta_0}{2} = \frac{1}{4}\left[\frac{K_{Ieff}}{K_{II}} \pm \sqrt{\left(\frac{K_{Ieff}}{K_{II}}\right)^2 + 8}\right] \tag{30}$$

where $K_{Ieff} = K_I + B|K_{III}|$ is an "effective" or "equivalent" mode I SIF also (partially) accounting for the presence of non-zero K_{III}, B being a material parameter; (iii) the advancement length Δa flows from an application of the Paris law:

$$\frac{\Delta a}{\Delta n} = C\left(\Delta K_{eff}\right)^m \tag{31}$$

where n denotes the number of load-cycles, C and m are material parameters (see Section 5 for comments),

$$\Delta K_{eff} = K_{eff}^{max} - K_{eff}^{min} \qquad K_{eff}^2 = K_{Ieff}^2 + 2K_{II}^2 \tag{32}$$

where K_{eff}^{max} and K_{eff}^{min} are the maximum and minimum values of the effective SIF in the load cycle.

At each step the crack extension is scaled according to the following two criteria: i) the maximum extension size should not exceed a given fixed value; ii) the trace of the fracture advancement on the outer surface itself (in the case of a surface breaking crack) should be completely contained in one element, as in Figure 9. The number of load cycles corresponding to the computed amount of incremental extension of the crack can be evaluated using the formula of Paris law.

4.2 Remeshing Strategy

Let us consider Figure 9_a and suppose that: i) point A is an intersection of the old crack front with the outer surface ("surface crack tip"); ii) according to the propagation algorithm the new surface crack tip B lies at some given point of the shaded element ("propagation element"). The propagation element is partitioned into subregions as in Figure 9_b and depending on the subregion to which point B belongs, different remeshing strategies are adopted as evidenced in Figure 10.

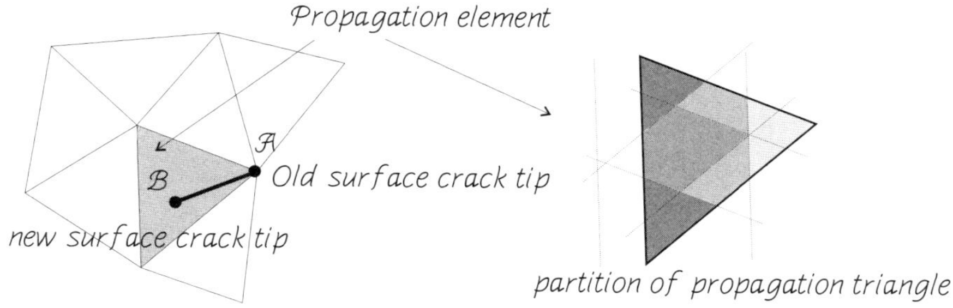

Figure 9. Old and new "surface crack tips" and partition of "propagation element" into subregions

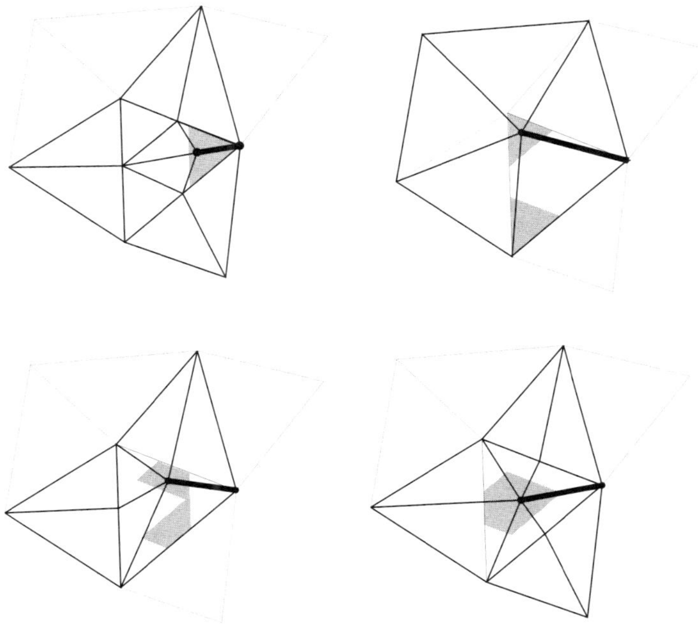

Figure 10. Different remeshing strategies according to the position of point B in the propagation element. The shaded portions of the propagation triangle correspond to the different subregions evidenced in Figure 9_b

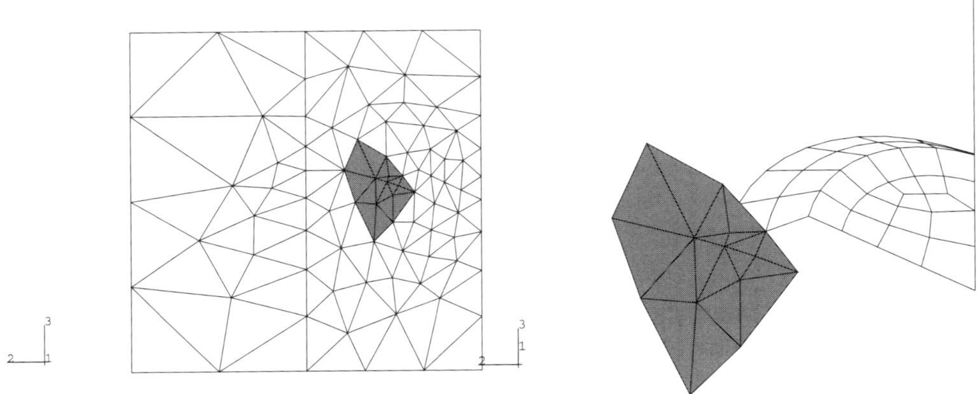

Figure 11. Elements around "surface crack tip" before the shape-regularization phase

Since the remeshing may create poorly shaped triangles, especially at later stages of the propagation process, it is necessary to introduce in the procedure a shape-regularizing step to improve the shape of the triangular elements generated on the outer surface (in the vicinity of the surface crack tip) by the remeshing algorithm represented in Figure 10. This shape-regularization phase should affect only few elements in order to reduce the cost of computing new entries in the BE matrices. Hence the adopted scheme provides for a rearrangement of only the nodes which share a triangle side with the new surface crack tip; this nodes will be named "near-tip nodes". The regularization is accomplished by minimizing an appropriate objective function which is the sum of terms of the type $(\pi/3 - \alpha_{\text{ver}})^2$, α_{ver} being the internal angle, at vertex ver, of one of the triangles to be reshaped. Since the triangular elements containing the new "surface crack tip" are the ones for which the improvement of the aspect ratio is more crucial, a weight factor λ_{el} associated to each element is introduced in the objective function: the choice adopted is to set $\lambda_{\text{el}} = 3$ if an element contains a surface crack tip node as vertex, $\lambda_{\text{el}} = 1$ if it contains at least one near-tip node, and $\lambda_{\text{el}} = 0$ otherwise.

The selected objective function eventually is:

$$\mathcal{F} = \sum_{\text{elements}} \lambda_{\text{el}} \sum_{\text{vertices}} \frac{(\pi/3 - \alpha_{\text{ver}})^2}{\sqrt{|\alpha_{\text{ver}}(\pi - \alpha_{\text{ver}})|}} \tag{33}$$

The presence of the term $\sqrt{|\alpha_{\text{ver}}(\pi - \alpha_{\text{ver}})|}$ at the denominator helps in preventing that some elements with a low value of λ_{el} might get too seriously distorted.

4.3 Elliptical Crack Embedded in a Cube.

The first example addressed is depicted in Figure 12 and basically consists of an elliptical crack with major semi-axis axis a and minor semi-axis b embedded in a homogeneous cube of side length $2L$ ($L >> a$) subjected to uniform tractions on the upper and lower surfaces. The elliptical crack forms angle α with the x_1, x_2 plane.

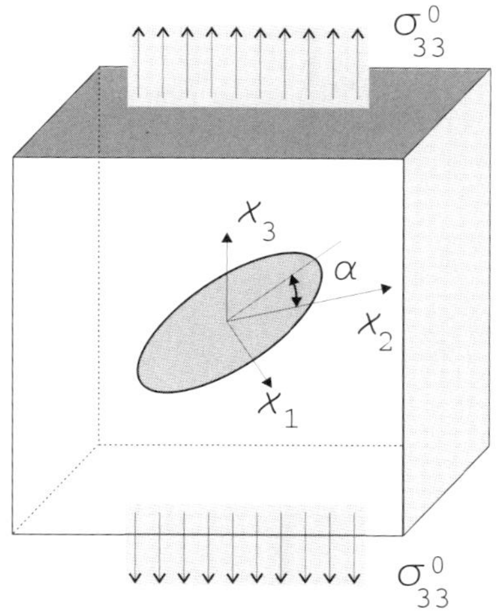

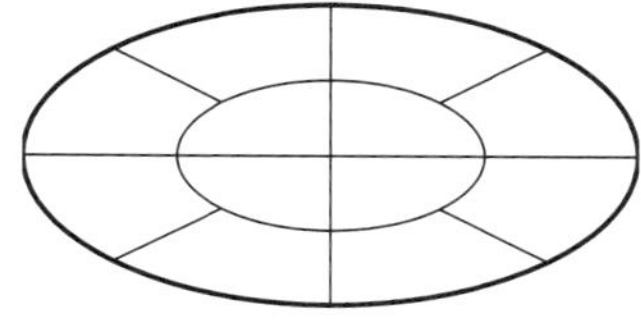

Figure 12. Elliptical crack embedded in a cube and initial mesh adopted

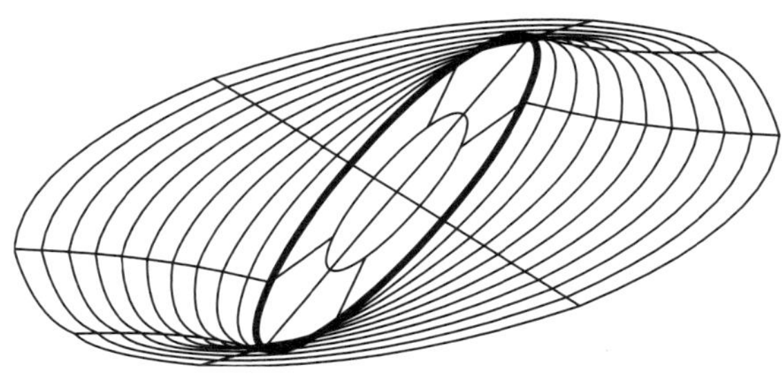

Figure 13. Propagation of an elliptical crack ($\alpha = 45°$): evolution of the crack surface for the first ten increments

The $\alpha = 45°$ configuration is analysed, adopting as material constants: $m = 3.88$, $C = 1.54\ 10^{-16}$, $B = 1$ (lengths in millimiters and forces in Newton). At each iteration, the maximum crack advancement Δa_{max} is set equal to $.25\, b$ and a set of eight new quarter-point 9-node BEs is added ahead of the crack front. For each BE the position of the three nodes lying on the new crack front is chosen according to the criterion specified above; when a node is shared by two elements an "averaged" position is adopted. The quarter-point nodes of the elements lying on the former crack front are moved to usual middle-element positions, and the relevant entries in the coefficient matrices are updated.

The crack front rapidly evolves towards a plane perpendicular to the applied tractions; K_{II} tends to zero, as a natural consequence of the propagation criterion adopted, while K_{III}, though decaying, is less sensitive to the propagation process, as evidenced in Figures 14.

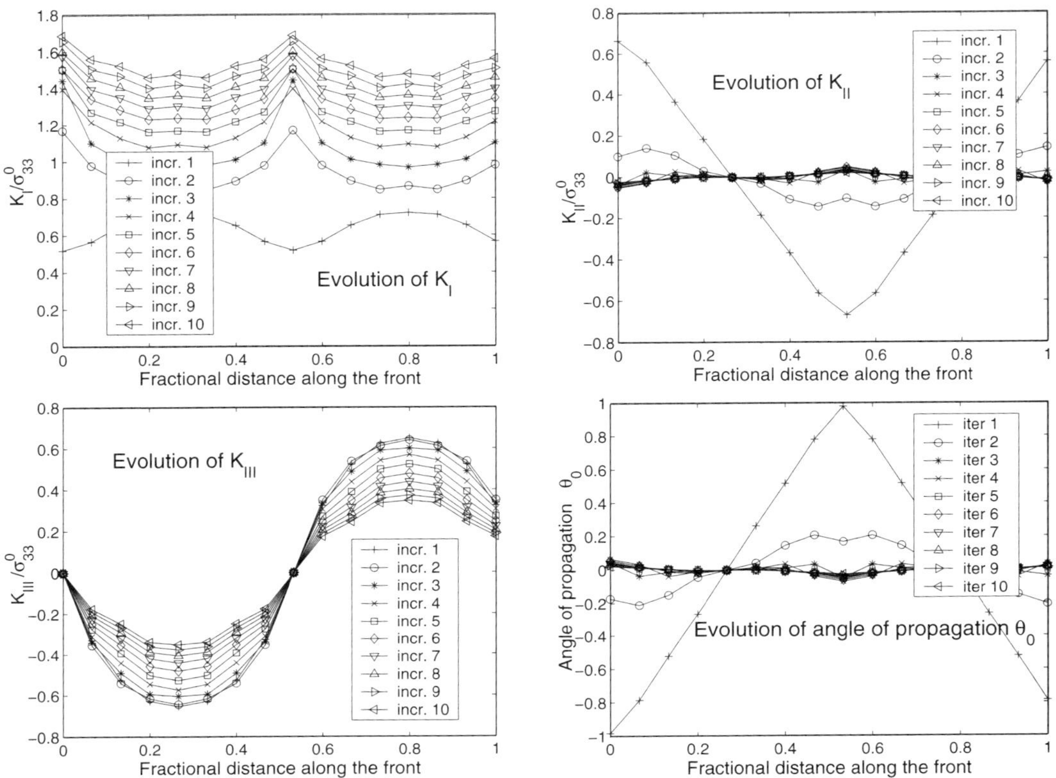

Figure 14. Propagation of an elliptical crack: K_I, K_{II}, K_{III} and θ_0 history at nodes along the crack front

4.4 Semicircular Surface Crack in a Cube

An elastic cube ($\nu = .3$) with semicircular surface crack (of radius $R = 1.5$ cm) is submitted to uniform tractions σ_0 (see Figures 15-16). The cube side length is $H = 10$ cm and parameters in Paris law have been chosen as $m = 2.5$, $C = 1.5\,10^{-17}$. The crack plane makes an angle $\alpha = 45°$ with the $x_1 - x_2$ plane.

In Figure 15_b the outer surface mesh is depicted, while Figure 16 presents a view of the cube interior and of the edge crack as seen from the viewpoint $(5, 10, 5)$ (the reference-system origin is placed in the corner node C).

Figures 17, 18 and 19, on the contrary, display the surface and crack meshes after the first six increments, along with the stress intensity factor K_I distribution along crack front. The fatigue life of the cracked bar is also displayed: the "equivalent" crack length, obtained as the sum of the maximum crack advancement at each iteration, is plotted versus the number of cycles, assuming that $\Delta\sigma = 10$ kN/cm^2.

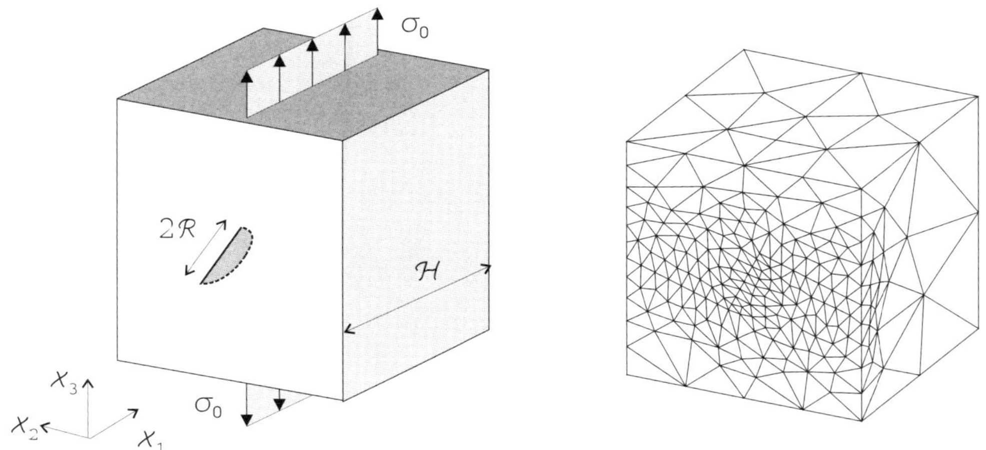

Figure 15. Elastic cube with a semi-circular surface crack and surface mesh adopted

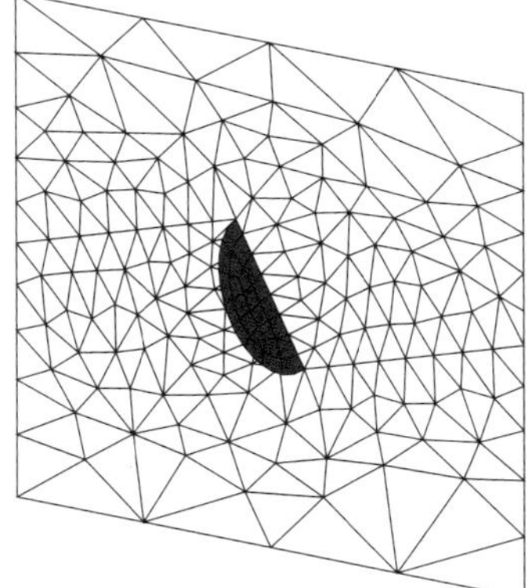

Figure 16. View of semicircular surface crack form the cube interior

5 Concluding Remarks

The 3D symmetric Galerkin BEM formulation implemented in this work using quadratic boundary elements has provided accurate results in terms of stress intensity factors evaluation. As for the simulation of crack propagation, the work carried out so far and reported in this contribution confirms the potential of the SGBEM in this context. However, further developments should take into account a number of issues linked both to the physics underlying fracture processes in 3D and to the numerical solution of large scale problems.

With reference to the first aspect, let us consider a node P where the crack front intersects the specimen surface. The asymptotic behaviour of the displacements near point P does not respect,

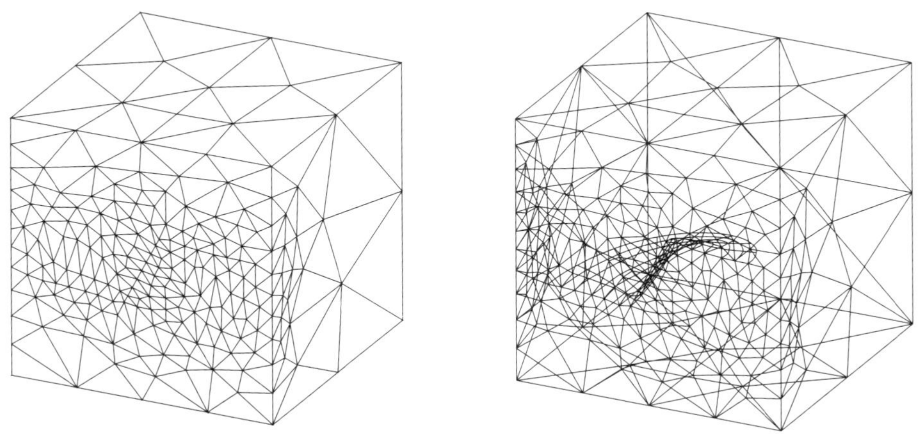

Figure 17. Semicircular surface breaking crack after six crack increments: "hidden line" plot and "wire-frame plot"

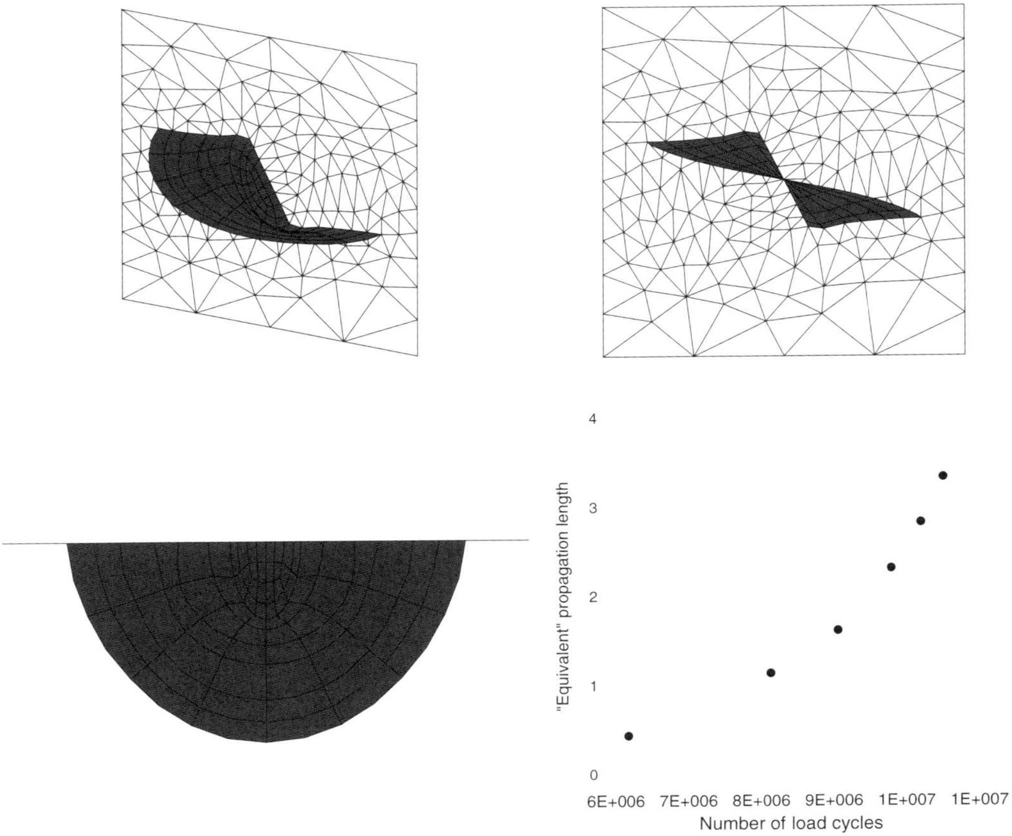

Figure 18. Semicircular surface breaking crack after six increments: different views of propagated crack from the cube interior and fatigue life of the bar

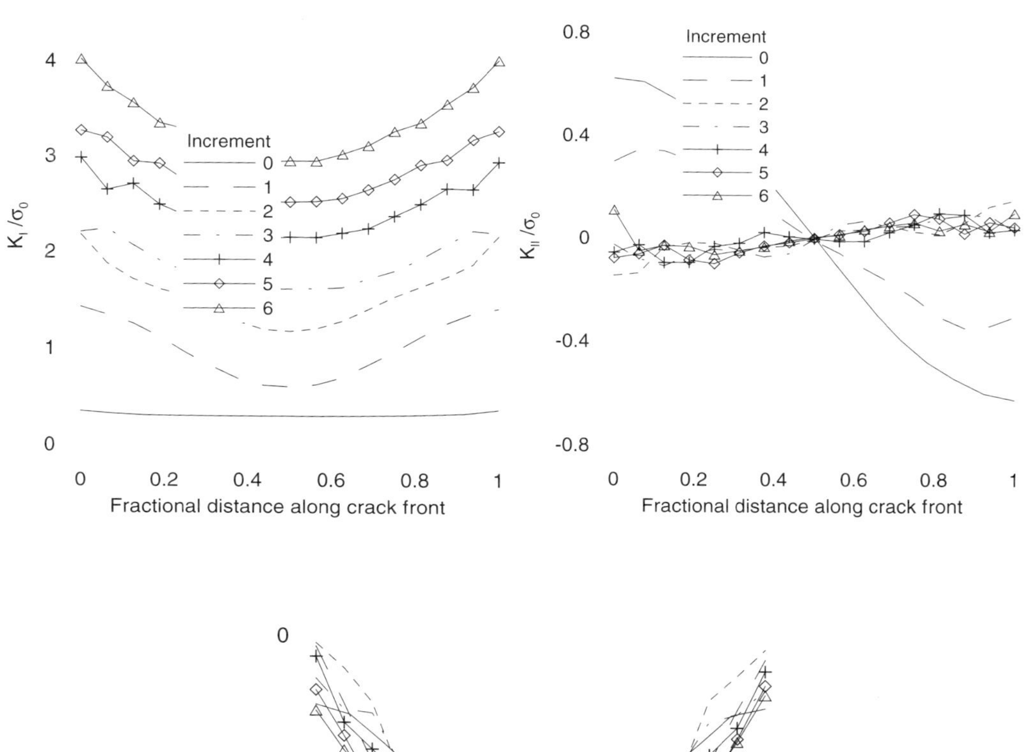

Figure 19. Stress intensity factor distributions along crack front for the first six crack extensions

in general, classical Williams' expansions, due to the discontinuity in the front geometry (see e.g. Pook, 1994). Indeed $u \simeq r^{\gamma}$, where γ depends on the Poisson's coefficient and the angle between the crack front line and the outer surface. The classical stress intensity factors, defined under the assumption of the usual $\sqrt{r}$ behaviour, tend either to zero or to infinity (according to whether γ is greater or smaller than $.5$) in a a boundary region near point P the thickness of which strongly depends on the problem geometry. In order to account for this and similar phenomena the material resistance to crack growth (as measured for instance by C in eqn. (31)) is often given in the literature (see e.g. Thomson and Sheppard, 1992) by a relation such as $C_P = Q^n C_A$, where C_A is the crack growth resistance at a crack front point inside the body and Q is a material parameter.

Another key factor, besides residual stress fields due to (thermal) treatments, is the combined presence of different fracture modes. Mixed $I - III$, $I - II$ modes have been the subject of a number of investigations (see e.g. Brown *et al.*, 1985) showing that the propagation direction is strongly dependent on the relative influence of the modes. Branching, "factory roof" cracks are alternative propagation possibilities that the numerical algorithm should be able to simulate, possibly following an energy-based approach. Work along these guidelines is in progress.

From the numerical standpoint, the SGBEM, like all BEM approaches, entails fully populated coefficient matrices. For the viability of the SGBEM in large-scale problems it is mandatory to reduce the computational cost and memory requirements by making recourse to recent algorithms such as fast multipole methods and panel clustering. Several contributions along this line have appeared recently with promising results (e.g. Lage and Schwab; Yoshida *et al.*, 2000; 2001) and are being envisaged for more efficient implementations.

Acknowledgements

A research grant (COFIN 2000) from the Italian Ministry of the University (MURST) is gratefully acknowledged.

References

Andrä, H and Schnack, E. (1997) Integration of singular Galerkin-type boundary element integrals for 3D elasticity, *Numerische Mathematik*, **76**, 143–165.

Balakrishna, C., Gray, L.J. and Kane, J.H. (1994) Efficient analytical integration of symmetric Galerkin boundary integrals over curved elements: elasticity, *Comp. Meth. Appl. Mech. Engng.*, **117**, 157–179.

Bonnet, M., Maier, G. and Polizzotto, C. (1998) Symmetric Galerkin boundary element method, *Appl. Mech. Rev.*, **51**, 669–704.

Bonnet, M. (1993) A regularized Galerkin symmetric BIE formulation for mixed 3D elastic boundary values problems, *Boundary Elements Abstracts & Newsletters*, **4**, 109–113.

Brown, M.W., Hay, E. and Miller, K.J. (1985) Fatigue at notches subjected to reversed torsion and static axial loads, *Fatigue Fract. Engng. Mater. Struct.*, **8**, 243–258.

Crouch, S.L. and Starfield, A.M. (1983) *Boundary Element Methods in Solid Mechanics*, George Allen and Unwin, .

Erichsen, S. and Sauter, S.A. (1998) Efficient automatic quadrature in 3-D Galerkin BEM, *Comp. Meth. Appl. Mech. Engng.*, **157**, 215–224.

Frangi, A. and Novati, G. (1996) Symmetric BE method in two dimensional elasticity: evaluation of double integrals for curved elements, *Computat. Mech.*, **19**, 58–68.

Frangi, A. (1998) Regularization of boundary element formulations by the derivative transfer method, in Sládek, V., Sládek, J. (eds.), *Singular Integrals in Boundary Element Methods, Advances in Boundary Elements, chap. 4*, Computational Mechanics Publications, 125–164.

Frangi, A., Novati, G., Springhetti, R. and Rovizzi, M. (2001) 3D fracture analysis by the symmetric Galerkin BEM, *Computat. Mech.*, accepted for publication.

Ganguly, S., Layton, J.B. and Balakrishna, C. (2000) Symmetric coupling of multi-zone curved Galerkin boundary elements with finite elements in elasticity, *Int. J. Num. Meth. Engng.*, **48**, 633–654.

Hartranft, R.J. and Sih, G.C. (1970) An approximate three-dimensional theory of plates with application to crack problems, *Int. J. Engng. Sci.*, **8**, 711–729.

Hills, D.A., Kelly, P.A. (1996) *Solution of Crack Problems*, Kluwer Academic Press, Dortrecht.

Kassir, M.K. and Sih, G.C. (1966) Three dimensional stress distribution around an elliptical crack under arbirary loadings, *J. Applied Mech.* , **33**, 602–615.

Lage, C. and Schwab, C. (2000) Advanced boundary element algorithms, in Whiteman, J.R. (eds.), *Mafelap 1999*, Elsevier, 283–306.

Li, X. and Keer, L.M. (1992) A direct method for solving crack growth problems - I, *Int. J. Solids Structures*, **29**, 2735–2747.

Li, X. and Keer, L.M. (1992) A direct method for solving crack growth problems - Shear meode problems II, *Int. J. Solids Structures*, **29**, 2749–2760.

Li, S., Mear, M.E. and Xiao, L. (1998) Symmetric weak-form integral equation method for three-dimensional fracture analysis, *Comp. Meth. Appl. Mech. Engng.*, **151**, 435–459.

Maier, G., Miccoli, S., Novati, G. and Sirtori, S. (1992) A Galerkin symmetric boundary element method in plasticity: formulation and implementation, in Kane, J.H., Maier, G. , Tosaka, N. and Atluri, S.N. (eds.), *Advances in Boundary Elements Techniques*, Springer Verlag, 288–328.

Nishimura, N. and Kobayashi, S. (1989) A regularized boundary integral equation method for elastodynamic crack problems, *Computat. Mech.*, **4**, 319–328.

Mi, Y. and Aliabadi, M.H. (1994) Three-dimensional crack growth simulation using BEM, *Computer & Structures*, **52**, 871–878.

Mi, Y. (1996) *Three-dimensional Analysis of Crack Growth*, Computational Mechanics Publications, Southampton.

Paulino, G.H. and Gray, L.J. (1999) Galerkin residuals for adaptive symmetric-Galerkin boundary element methods, *Journal of Engineering Mechanics (ASCE)*, **125**, 575–585.

Pook, L.P. (1994) Some implications of corner point singularities, *Engng. Fracture Mech.*, **48**, 367–378.

Raju, I.S. and Newman, J.C. (1977) Three dimensional finite-element analysis of finite-thickness fracture specimens, *NASA-TN*, D-8414.

Raju, I.S. and Newman, J.C. (1979) Stress-intensity factors for a wide range of semi-elliptical surface cracks in finite-thickness plates, *Engng. Fracture Mech.*, **11**, 817–829.

Sauter, S.A. and Schwab, C. (1997) Quadrature for hp-Galerkin BEM in 3-d, *Numerische Mathematik*, **78**, 211–258.

Sirtori, S. (1979) General stress analysis method by means of integral equations and boundary elements, *Meccanica*, **14**, 210–218.

Sirtori, S., Maier, G., Novati, G. and Miccoli, S. (1992) A Galerkin symmetric boundary element method in elasticity: formulation and implementation, *Int. J. Num. Meth. Engng.*, **35**, 255–282.

Tada, S., Paris, P. and Irwin, G. (1985) *The Stress Analysis of Cracks Handbook*, Dell Research Corporation, St. Louis.

Thomson, K.D. and Sheppard, S.D. (1992) Stress intesity factors in shafts subjected to torsion and axial loading, *Engng. Fracture Mech.*, **42**, 1019–1034.

Wawrzynek, P.A., Martha, L.F. and Ingraffea, A.R. (2000) A computational environment for the simulation of fracture processes in three dimensions, in Rosakis, A.J. (eds.), *Annal. Numer. Exper. Aspects of Three Dimensional Fracture Processes*, ASME, AND-91, 321–327.

Xu, G. and Ortiz, M. (1993) A variational boundary integral method for the analysis of 3-D cracks of arbitrary geometry modelled as continuous distributions of dislocation loops, *Int. J. Num. Meth. Engng.*, **36**, 3675–3701.

Yoshida, K., Nishimura, N. and Kobayashi, S. (2001) Application of fast multipole Galerkin boundary integral equation method to elastostatic crack problems, *Int. J. Num. Meth. Engng.*, **50**, 525–547.

A Transformations and equivalence of domains

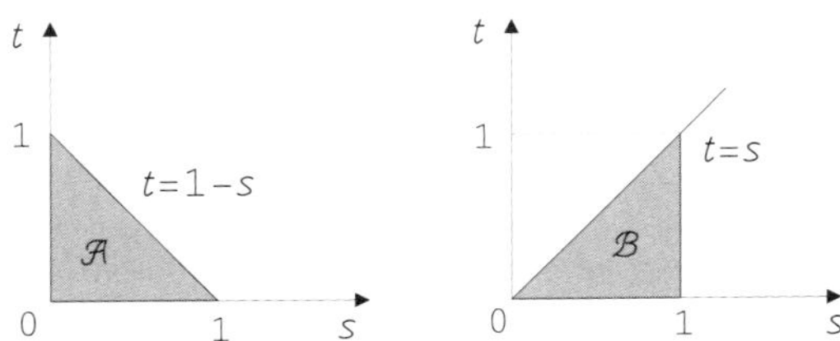

Figure 20. Geometrical representations of domains for Transformations A and B

Transformation A. Let us consider domain $\mathcal{A}$ in Figure (20) in the two dimensional space s, t. The following two sets of inequalities both define $\mathcal{A}$

$$\mathcal{A}(s,t) = \begin{cases} 0 \leq s \leq 1 \\ 0 \leq t \leq 1-s \end{cases} = \begin{cases} 0 \leq t \leq 1 \\ 0 \leq s \leq 1-t \end{cases} \tag{34}$$

Moreover the variable transformation $\tilde{s} = s + t, \; \tilde{t} = t$ allows to write

$$\mathcal{A}(\tilde{s} - \tilde{t}, \tilde{t}) = \begin{cases} 0 \leq \tilde{t} \leq 1 \\ \tilde{t} \leq \tilde{s} \leq 1 \end{cases} = \begin{cases} 0 \leq \tilde{s} \leq 1 \\ 0 \leq \tilde{t} \leq \tilde{s} \end{cases} = \mathcal{B}(\tilde{s}, \tilde{t}) \tag{35}$$

so mapping domain $\mathcal{A}$ onto domain $\mathcal{B}$ in Figure (20). This expedient is useful to exploit the Duffy coordinates transformations.

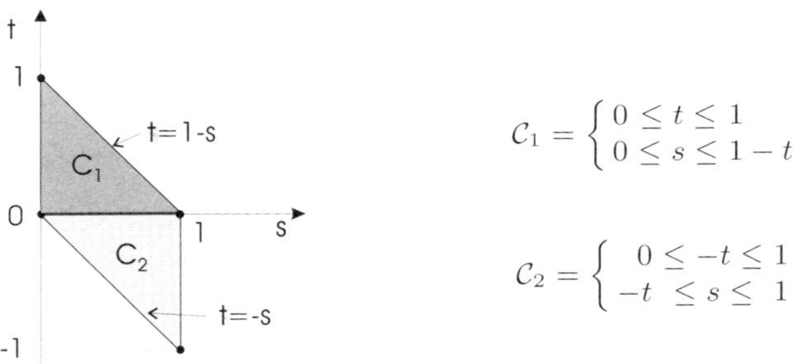

$$\mathcal{C}_1 = \begin{cases} 0 \leq t \leq 1 \\ 0 \leq s \leq 1-t \end{cases}$$

$$\mathcal{C}_2 = \begin{cases} 0 \leq -t \leq 1 \\ -t \leq s \leq 1 \end{cases}$$

Figure 21. Domain partition for Transformation C

Transformation B. Similar conclusions hold also for domain $\mathcal{B}$ in Figure (20):

$$\mathcal{B}(s,t) = \begin{cases} 0 \leq s \leq 1 \\ 0 \leq t \leq s \end{cases} = \begin{cases} 0 \leq t \leq 1 \\ t \leq s \leq 1 \end{cases} \tag{36}$$

Transformation C. Let us now focus on the domain $\mathcal{C}$ in Figure 21:

$$\mathcal{B}(s,t) = \begin{cases} 0 \le s \le 1 \\ -s \le t \le 1-s \end{cases}$$

The quadrangle $\mathcal{C}$ can be equivalently expressed as the union of two triangles $\mathcal{C}_1 \cup \mathcal{C}_2$.

B Equivalence of $\mathcal{D}_2^c(\tilde{\eta}_1, \tilde{\eta}_2, \eta_1, \eta_2)$ and $\mathcal{D}_1^c(\eta_1, \tilde{\eta}_2, \tilde{\eta}_1, \eta_2)$

Let us consider the two subdomains in eqn. (15) and focus on the first two inequalities in $\mathcal{D}_2^c$. Denoting $\tilde{u}_1 = \tilde{\eta}_1 - \eta_1 = -u_1$:

$$\begin{cases} 0 \le -u_1 \le 1 \\ -u_1 \ \le \tilde{\eta}_1 \le \ 1 \end{cases} \equiv \begin{cases} 0 \le \tilde{u}_1 \le 1 \\ \tilde{\eta}_1 - \eta_1 \le \tilde{\eta}_1 \le 1 \end{cases} \equiv \begin{cases} 0 \le \tilde{u}_1 \le 1 \\ 0 \le \eta_1 \le 1 - \tilde{u}_1 \end{cases} \tag{37}$$

If u_1 and $\tilde{u}_1$ are exchanged, eqn. (37) transforms into the first two inequalities defining $\mathcal{D}_1^c$ in eqn. (15), which completes the proof.

CHAPTER 14

Boundary contour method for plane problems in a dual formulation with quadratic shape functions

György Szeidl and Sándor Szirbik *

Department of Mechanics, University of Miskolc, Hungary
Gyorgy.SZEIDL@uni-miskolc.hu Sandor.SZIRBIK@uni-miskolc.hu

Abstract. The present section is devoted to the boundary contour method for plane problems in the dual system of elasticity. In this system the governing equations are given in terms of stress functions of order one. After clarifying the conditions of single valuedness we construct the fundamental solution for the dual basic equations. Then the integral equations of the direct method have been established. It has been shown that the integrals on the right side of the corresponding boundary integral equations are divergence free in the dual system provided that the unknown functions satisfy the field equations. Consequently these integrals can be given in closed form if appropriate shape functions have been chosen to approximate the unknown functions on the contour. Numerical examples prove the efficiency of this technique.

1 Introduction, Preliminaries

In spite of a great number of publications devoted to plane problems there are only a few dealing with these problems regarding the stress functions of order one as fundamental variables. As regards classical elasticity we refer to the paper by Jaswon et al. (1967) and the book [Jaswon and Symm, 1977] in which the unknown biharmonic function (stress function of order two) is given in terms of two harmonic functions as a single layer potential and the authors set up a pair of integral equations for the unknown source densities.

Application of stress functions of order one was initiated by de Veubeke (1975, 1976) in a new complementary energy based finite element procedure since the use of C^0 continuous stress functions of order one guarantees continuous surface tractions and makes possible to construct isoparametric elements. Further applications with an emphasis on three dimensional problems and laminated structures are due to Bertóti (1994, 1996).

A direct boundary element formulation in terms of stress functions of order one in the dual system of plane elasticity is due to Szeidl (1997).

It is proved in the article by Nagarjan et al. (1994) that the integrand of the direct boundary element method is divergence free in the primal system of the two and three-dimensional elasticity theory. Nagarjan et al. (1994) have come to the conclusion that the numerical solution of

* The support provided by the Hungarian National Research Foundation within the framework of the projects OTKA T022022 and T037118 is gratefully acknowledged.

three-dimensional problems requires the calculation of line integrals instead of surface integrals, while for plane problems evaluation of functions should be performed instead of calculating line integrals. This article supposes linear approximation. The accuracy is greatly increased if one uses quadratic elements (see Phan et al. (1997) for details). The method developed can also be employed for rewriting hypersingular integral equations into boundary contour equations. With this technique one can compute stresses and can solve shape optimization problems in two dimensions – Phan et al. (1998).

The boundary contour method in the dual system of plane elasticity has been formulated in the paper by Szirbik (2000). The approximation is, however, linear.

With regard to all that has been said above our aims are as follows:

- To clarify what are the conditions of single valuedness for a class of mixed boundary value problems assuming multiply connected regions.
- To derive the fundamental solutions for the stress functions of order one.
- To set up the dual Somigliana relations (both for inner regions and for exterior ones) from which the boundary integral equations of the direct method can be derived.
- To prove that the integrand of the direct boundary element method is divergence free in the dual system of plane elasticity and to determine the corresponding shape functions provided that the approximation is quadratic.
- To derive the discretized equations and to clarify how to compute stresses at internal points.
- To develop an algorithm for computations and to present some numerical examples.

The last Section is an Appendix in which the shape functions and some manipulations are presented.

2 The dual equation system of plane elasticity in terms of stress functions of order one

Throughout this section $x_1 = x$ and $x_2 = y$ are rectangular Cartesian coordinates, referred to an origin O. The totality of $x_1 = x$ and $x_2 = y$ is denoted by x. {Greek}[Latin subscripts] are assumed to have the range {(1,2)}[(1,2,3)], summation over repeated subscripts is implied. The triple connected region under consideration is denoted by A_i – inner region – and is bounded by the exterior contour

$$\mathcal{L}_0 = \mathcal{L}_{t1} \cup \mathcal{L}_{u2} \cup \mathcal{L}_{t3} \cup \mathcal{L}_{u4}$$

and two inner contours which – partly or wholly – consists of the arcs $\mathcal{L}_{t1}$, $\mathcal{L}_{t3}$, $\mathcal{L}_{t5}$ and $\mathcal{L}_{u2}$, $\mathcal{L}_{u4}$, $\mathcal{L}_{u6}$.

Further the inner contours $\mathcal{L}_1$ and $\mathcal{L}_2$ lie wholly in the interior of the exterior contour $\mathcal{L}_0$ and they have no points in common. We stipulate that each contour has a continuously turning unit tangent τ_κ and admits a nonsingular parametrization in terms of its arc length s. The outer normal is denoted by n_π. In accordance with the notations introduced $\delta_{\kappa\lambda}$ is the Kronecker symbol, ∂_α stands for the derivatives with respect to x_α and $\epsilon_{3\kappa\lambda}$ is the permutation symbol. The {symmetric}[skew] part of a tensor, say the tensor $t_{\kappa\lambda}$, is denoted by $\{t_{(\kappa\lambda)}\}[t_{[\kappa\lambda]}]$.

Assuming plane strain let u_κ, $e_{\kappa\lambda}$ and $t_{\kappa\lambda}$ be the displacement field and the in plane components of stress and strain, respectively. The stress functions of order one are denoted by $\mathcal{F}_\rho$.

For homogenous and isotropic material the plane strain problem of classical elasticity in the dual system of plane elasticity is governed by the dual kinematic equations

$$t_{\kappa\lambda} = \epsilon_{\kappa\rho 3}\mathcal{F}_\lambda \partial_\rho + \overset{o}{t}_{\kappa\lambda} \qquad x \in A_i \tag{1}$$

($\overset{o}{t}_{\kappa\lambda}$ is the particular solution in the presence of body forces), the inverse form of Hook's law

$$e_{\kappa\lambda} = \frac{1}{2\mu}\left(t_{(\kappa\lambda)} - \nu t_{\psi\psi}\delta_{\kappa\lambda}\right) \qquad x \in A_i \tag{2}$$

(μ is the shear modulus of elasticity, ν is the Poisson number), the dual balance equations

$$\epsilon_{\kappa\rho 3} e_{\lambda\kappa}\partial_\rho + \varphi_3\partial_\lambda = \epsilon_{\kappa\rho 3}\left(e_{\lambda\kappa} - \epsilon_{\lambda\kappa 3}\varphi_3\right)\partial_\rho = 0 \qquad x \in A_i \tag{3}$$

(equations of compatibility for a simply connected region; φ_3 is the rigid body rotation) and the symmetry condition

$$\epsilon_{3\kappa\lambda} t_{\kappa\lambda} = 0 \qquad x \in A_i \tag{4}$$

(equation of rotational equilibrium). If this equation is fulfilled then one of equations (2) can be omitted. In this way we have nine equations for the nine unknowns $\mathcal{F}_1$, $\mathcal{F}_2$, t_{11}, $t_{12} = t_{21}$, t_{22}, e_{11}, $e_{12} = e_{21}$, e_{22} and φ_3.

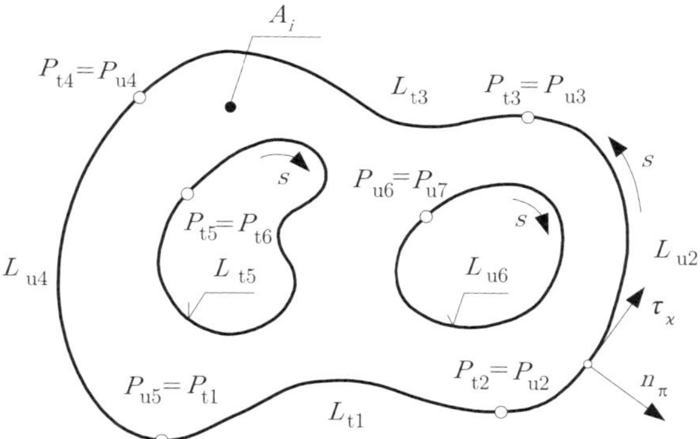

Figure 1. The inner region A_i

Field equations (1), (2), (3) and (4) should be associated with appropriate boundary conditions. If a contour is not divided into parts then either tractions or displacements are imposed on it. If a contour is divided then it is assumed to consists of arcs of even number on which displacements and tractions are imposed alternately. In the present case {tractions}[displacements] are given on the arc $\{\mathcal{L}_t = \mathcal{L}_{t1} \cup \mathcal{L}_{t3} \cup \mathcal{L}_{t5}\}[\mathcal{L}_u = \mathcal{L}_{u2} \cup \mathcal{L}_{u4} \cup \mathcal{L}_{u6}]$. We remark that hatted letters stand for the prescribed values.

Upon substitution of equation (1) into the traction boundary condition $n_\pi t_{\pi\rho} = \hat{t}_\rho$ we arrive at the differential equation

$$\hat{t}_\rho - \overset{o}{t}_\rho = n_\kappa \epsilon_{\kappa\nu 3}\mathcal{F}_\rho \partial_\nu = \frac{d\mathcal{F}_\rho}{ds}\ , \tag{5}$$

where $\overset{o}{t}_\rho = n_\pi \overset{o}{t}_{\pi\rho}$. One can readily check that the solution on the arcs of $\mathcal{L}_t$ assumes the form

$$\hat{\mathcal{F}}_\rho(s) = \int_{P_{ti}}^{s} \left[\hat{t}_\rho(\sigma) - \overset{o}{t}_\rho(\sigma) \right] d\sigma \qquad s \in \mathcal{L}_{ti} \qquad i = 1, 3, 5 \, .$$

Let $\underset{(ti)}{C}_\rho$ be constants of integration. Equation

$$\mathcal{F}_\rho(s) = \hat{\mathcal{F}}_\rho(s) + \underset{(ti)}{C}_\rho \qquad i = 1, 3, 5 \tag{6}$$

is equivalent to the boundary condition (5) and conversely.

REMARK 1.: Observe that the number of undetermined constants of integration is two times as much as the number of those arcs on which tractions are imposed.

Since the displacements do not belong to the unknowns of the dual system of plane elasticity one has to clarify what boundary conditions can be prescribed on the arcs constituting $\mathcal{L}_u$. Let

$$\mathcal{K} = \mathcal{K}(t_{\kappa\lambda}, \varphi_3) = -\frac{1}{2} \int_{A_i} t_{\kappa\lambda} e_{\kappa\lambda} \, dA + \int_{\mathcal{L}_u} n_\kappa t_{\kappa\lambda} \hat{u}_\lambda \, ds - \int_{A_i} t_{\kappa\lambda} \epsilon_{\kappa\lambda 3} \varphi_3 dA \tag{7}$$

be a modified form of the complementary energy functional. (The modification is a must in order to keep up the rotational equilibrium.) Solution to the problem posed can be sought by making use of the stationary condition

$$\delta \mathcal{K} = 0 \tag{8}$$

since the latter equation should ensure all the conditions the strains $e_{\kappa\lambda}$ and the rigid body rotation φ_3 are to be met in order to be kinematically admissible. In the functional (7) $e_{\kappa\lambda}$ is given in terms of the stresses $t_{\kappa\lambda}$ via Hook's law while the stresses $t_{\kappa\lambda}$ should satisfy the equilibrium equation and the traction boundary condition though it is not necessary for them to be symmetric. Consequently the variations of stresses can not be arbitrary but should meet the conditions

$$\delta t_{\kappa\lambda} \partial_\kappa = 0 \qquad x \in A \qquad \text{and} \qquad n_\kappa \delta t_{\kappa\lambda} = 0 \qquad x \in \mathcal{L}_t \, . \tag{9}$$

Both conditions are satisfied if $\delta t_{\kappa\lambda}$ is given in terms of the variations of stress functions

$$\delta t_{\kappa\lambda} = \epsilon_{\kappa\rho 3} \delta \mathcal{F}_\lambda \partial_\rho \, , \tag{10}$$

where $\delta \mathcal{F}_\lambda$ is arbitrary on A_i. However, with regard to (6) it follows that on $\mathcal{L}_t$

$$\delta \mathcal{F}_\rho(s) = \delta \underset{(ti)}{C}_\rho \qquad i = 1, 3, 5 \, . \tag{11}$$

Derivation of the conditions the strains $e_{\kappa\lambda}$ and the rigid body rotation should meet in order to be kinematically admissible requires the transformation of the stationary condition

$$\delta \mathcal{K} = -\int_{A_i} e_{\kappa\lambda} \, \delta t_{\kappa\lambda} \, dA + \int_{\mathcal{L}_u} n_\kappa \delta t_{\kappa\lambda} \hat{u}_\lambda \, ds - \int_{A_i} \delta t_{\kappa\lambda} \epsilon_{\kappa\lambda 3} \varphi^3 dA - \int_{A_i} t_{\kappa\lambda} \epsilon_{\kappa\lambda 3} \delta \varphi^3 dA = 0 \tag{12}$$

The main steps of the transformations are as follows:

1. Substitution of the condition (10) into the first and second surface integrals and substitution cf (1) into the third surface integral.
2. Substitution of the relation
$$n_\kappa \epsilon_{\kappa\rho 3} \delta \mathcal{F}_\lambda \partial_\rho = \frac{d\delta\mathcal{F}_\lambda}{ds}$$
into the line integral taken on $\mathcal{L}_u$.
3. Application of the Green-Gauss theorem [Eringen (1951)] to the first and second surface integrals.
4. Performation of partial integrations on the arcs constituting $\mathcal{L}_u$ taking into account the validity of (11) at the extremities of the arcs.
5. Division of the line integrals obtained by the application of the Green-Gauss theorem by using the relation
$$\int_{\mathcal{L}} .... = \int_{\mathcal{L}_u} .... + \int_{\mathcal{L}_t} ....$$
then substitution of equation (11) into the line integrals taken on $\mathcal{L}_t$.

Transformation of the result making use of the equation
$$n_\rho \epsilon_{\kappa\rho 3} = -\tau_\kappa \tag{13}$$

After performing the steps listed above
$$\begin{aligned}\delta K = &\int_{A_i} (\epsilon_{\kappa\rho 3} e_{\kappa\lambda} \partial_\rho + \varphi_3 \partial_\lambda)\, \delta\mathcal{F}_\lambda dA - \int_{A_i} (\mathcal{F}_\psi \partial_\psi)\, \delta\varphi_3 dA \\ &+ \sum_{i=2,4,6} \int_{\mathcal{L}_{ui}} \left\{ n_\pi [\epsilon_{\pi\kappa 3} e_{\kappa\lambda} - \delta_{\pi\lambda}\varphi_3] - \frac{d\hat{u}_\lambda}{ds} \right\} \delta\mathcal{F}_\lambda ds \\ + &\sum_{i=1,3,5} \left\{ \int_{\mathcal{L}_{ti}} n_\pi [\epsilon_{\pi\kappa 3} e_{\kappa\lambda} - \delta_{\pi\lambda}\varphi_3] ds - \hat{u}_\lambda|_{P_{ti}}^{P_{t,i+1}} \right\} \delta \underset{(ti)}{C}_\lambda = 0\end{aligned}$$

is the stationary condition. Since the variations are arbitrary this condition implies the compatibility condition (3), the symmetry condition (4) – in the latter $t_{\kappa\lambda}$ is given in terms of $\mathcal{F}_\psi$ –, the strain boundary condition
$$\frac{d\hat{u}_\lambda}{ds} = n_\pi [\epsilon_{\pi\kappa 3} e_{\kappa\lambda} - \delta_{\pi\lambda}\varphi_3]\,, \tag{14}$$
the compatibility condition in the large
$$\int_{\mathcal{L}_{t5}} n_\pi [\epsilon_{\pi\kappa 3} e_{\kappa\lambda} - \delta_{\pi\lambda}\varphi_3] ds = 0 \tag{15}$$
and the supplementary condition of single valuedness
$$\int_{\mathcal{L}_{ti}} n_\pi [\epsilon_{\pi\kappa 3} e_{\kappa\lambda} - \delta_{\pi\lambda}\varphi_3] ds - \hat{u}_\lambda|_{P_{ti}}^{P_{t,i+1}} = 0 \qquad i = 1, 3\,. \tag{16}$$

REMARK 2.: The strain boundary condition can also be obtained if one regards the primal kinematic equation
$$e_{\kappa\lambda} = \frac{1}{2}(u_\kappa \partial_\lambda + u_\lambda \partial_\kappa)$$

on the contour, multiplies it by $n_\pi \epsilon_{\pi\kappa 3} = \tau_\kappa$ taking into account that $u_{[\kappa}\partial_{\lambda]} = -\epsilon_{\kappa\lambda 3}\varphi_3$.

REMARK 3.: Both the compatibility condition in the large (15) and the supplementary condition of single valuedness (16) can be set up by integrating the strain boundary condition (14) appropriately.

REMARK 4.: It can be shown that only two of the three conditions (as a matter of fact three times two conditions) (15) and (16) are independent of each other. In accordance with this one can set one times two of the three times two undetermined constants of integration $\underset{(ti)}{C}_\rho$, say $\underset{(t1)}{C}_\rho$, to zero since there belong no stresses to the stress function $\mathcal{F}_\rho = \underset{(t1)}{C}_\rho =$ constant. In other words we have as many independent macro conditions of single valuedness as there are undetermined constants of integration.

3 Fundamental solutions and the dual Somigliana formulae

3.1 Basic equations and fundamental solutions

Here and in the sequel we shall assume that there are no body forces. Substituting the dual kinematic equation (1) into Hook's law (2) and the result into the compatibility equations (3) we have

$$\frac{1}{2\mu}(1-\nu)\Delta\mathcal{F}_1 - \frac{1}{2\mu}(\frac{1}{2}-\nu)(\mathcal{F}_1\partial_1 + \mathcal{F}_2\partial_2)\partial_1 + \varphi_3\partial_1 = 0 \tag{17a}$$

$$\frac{1}{2\mu}(1-\nu)\Delta\mathcal{F}_2 - \frac{1}{2\mu}(\frac{1}{2}-\nu)(\mathcal{F}_1\partial_1 + \mathcal{F}_2\partial_2)\partial_2 + \varphi_3\partial_2 = 0 \tag{17b}$$

These equations are associated with the symmetry condition in terms of $\mathcal{F}_\rho$:

$$\mathcal{F}_1\partial_1 + \mathcal{F}_2\partial_2 = 0\ . \tag{17c}$$

Upon substitution of (17c) into (17a,b) the latter two equations become much simpler. In spite of that and for the sake of a comparison with the plane orthotropic case, the work on that problem is in progress, we do not change the above equations. Introducing the notations

$$[\mathfrak{D}_{lk}] = \begin{bmatrix} \frac{1}{2\mu}(1-\nu)\Delta - \frac{1}{2\mu}(\frac{1}{2}-\nu)\partial_1\partial_1 & -\frac{1}{2\mu}(\frac{1}{2}-\nu)\partial_1\partial_2 & -\partial_1 \\ -\frac{1}{2\mu}(\frac{1}{2}-\nu)\partial_2\partial_1 & \frac{1}{2\mu}(1-\nu)\Delta - \frac{1}{2\mu}(\frac{1}{2}-\nu)\partial_2\partial_2 & -\partial_2 \\ -\partial_1 & -\partial_2 & 0 \end{bmatrix} \tag{18a}$$

and

$$\mathfrak{u}_k = (\mathcal{F}_1, \mathcal{F}_2, -\varphi_3) \tag{18b}$$

the basic equation takes the form

$$\mathfrak{D}_{ik}\mathfrak{u}_k = 0 \tag{19}$$

Let $Q(\xi_1, \xi_2)$ and $M\ (x_1, x_2)$ be two points in the plane of strain (the source point and the point of effect). Further let $\mathbf{e}$ with components e_i be a unit vector at Q. We shall assume temporarily that the point Q is fixed. The distance between Q and M is R, the position vector of M relative to Q is r_κ. Solution to the differential equation

$$\mathfrak{D}_{ik}\mathfrak{u}_k + \delta(M-Q)e_i = 0$$

can be cast – see [Szeidl (1997)] for details – into the form

$$\mathfrak{u}_k = \mathfrak{U}_{kl}(M,Q)e_l(Q)\,, \tag{20}$$

where

$$[\mathfrak{U}_{kl}(M,Q)] = \frac{\mu}{4\pi(1-\nu)}\begin{bmatrix} -2\ln R - 3 - 2\dfrac{r_2r_2}{R^2} & 2\dfrac{r_1r_2}{R^2} & \dfrac{2}{\mu}(1-\nu)\dfrac{r_1}{R^2} \\ 2\dfrac{r_2r_1}{R^2} & -2\ln R - 3 - 2\dfrac{r_1r_1}{R^2} & \dfrac{2}{\mu}(1-\nu)\dfrac{r_2}{R^2} \\ \dfrac{2}{\mu}(1-\nu)\dfrac{r_1}{R^2} & \dfrac{2}{\mu}(1-\nu)\dfrac{r_2}{R^2} & 0 \end{bmatrix} \tag{21}$$

is referred to as fundamental solution.

REMARK 5.: The fundamental solution $\mathfrak{U}_{kl}(M,Q)$ satisfies the symmetry conditions

$$\mathfrak{U}_{kl}(M,Q) = \mathfrak{U}_{lk}(M,Q) = \mathfrak{U}_{kl}(Q,M) = \mathfrak{U}_{lk}(Q,M)\,. \tag{22}$$

Consequently

$$\mathfrak{u}_k = \mathfrak{U}_{kl}(M,Q)e_l(Q) = e_l(Q)\mathfrak{U}_{lk}(M,Q)\,. \tag{23}$$

REMARK 6.: Each row and column of $\mathfrak{U}_{kl}(M,Q)$ as a three dimensional vector satisfies the basic equation (19) both in M and in Q.

For our later considerations we shall introduce the notation

$$\mathfrak{t}_\lambda = -\frac{du_\lambda}{ds}\,. \tag{24}$$

where the vector $\mathfrak{t}_\lambda$ is referred to as displacement derivative. Making use of equation (14) but omitting the details, for the vector $\mathfrak{t}_\lambda$ derived from the fundamental solution we obtain

$$\mathfrak{t}_\lambda(\overset{o}{M}) = e_l(Q)\mathfrak{T}_{l\lambda}(\overset{o}{M},Q)\,, \tag{25a}$$

where

$$\mathfrak{T}_{l\lambda}(\overset{o}{M},Q) = \frac{\hat{C}}{R^2}\begin{bmatrix} \begin{matrix} n_1r_1\left(4\dfrac{r_2^2}{R^2} - 2(3-2v)\right) \\ +n_2r_2\left(4\dfrac{r_2^2}{R^2} - 2(3-2v)\right)\end{matrix} & \begin{matrix} -n_2r_1\left(4\dfrac{r_2^2}{R^2} + 2(1-2v)\right) \\ -n_1r_2\left(4\dfrac{r_1^2}{R^2} - 2(1-2v)\right)\end{matrix} \\ \begin{matrix} -n_1r_2\left(4\dfrac{r_1^2}{R^2} + 2(1-2v)\right) \\ -n_2r_1\left(4\dfrac{r_2^2}{R^2} - 2(1-2v)\right)\end{matrix} & \begin{matrix} n_2r_2\left(4\dfrac{r_1^2}{R^2} - 2(3-2v)\right) \\ +n_1r_1\left(4\dfrac{r_1^2}{R^2} - 2(3-2v)\right)\end{matrix} \\ \begin{matrix} -n_1\dfrac{2}{\mu}(1-\nu)\dfrac{r_1^2-r_2^2}{R^2} \\ -n_2\dfrac{4}{\mu}(1-\nu)\dfrac{r_1r_2}{R^2}\end{matrix} & \begin{matrix} -n_1\dfrac{4}{\mu}(1-\nu)\dfrac{r_1r_2}{R^2} \\ +n_2\dfrac{2}{\mu}(1-\nu)\dfrac{r_1^2-r_2^2}{R^2}\end{matrix} \end{bmatrix} \tag{25b}$$

and

$$\hat{C} = \frac{1}{8\pi(1-\nu)}\,. \tag{25c}$$

Here and in the sequel the small circle over the letters M and/or Q has the meaning that the corresponding point is located on the contour. The normal n_λ is taken at the point $\overset{o}{M}$.

REMARK 7.: Recalling that in the circle of the boundary value problems considered either the stress functions or the derivative of the displacements with respect to the arc coordinate can be prescribed at a point on the contour for our later consideration it is worth giving the value of the stress functions from the fundamental solution on the boundary:

$$\mathfrak{u}_\lambda = e_l(Q)\mathfrak{U}_{l\lambda}(\overset{o}{M}, Q) \tag{26}$$

The displacement derivative obtained from the fundamental solution is given by (25a,b,c).

3.2 The dual Somigliana relations for inner regions

In the sequel it is assumed that the region A_i under consideration is simply connected and is bounded. The contour $\mathcal{L}_0$ is divided into arcs of even number on which displacements (or their derivatives with respect to s) and tractions (or stress functions) can be imposed alternately. In Figure 2 the region A_i is divided into four arcs though this fact does not play any role in the transformations.

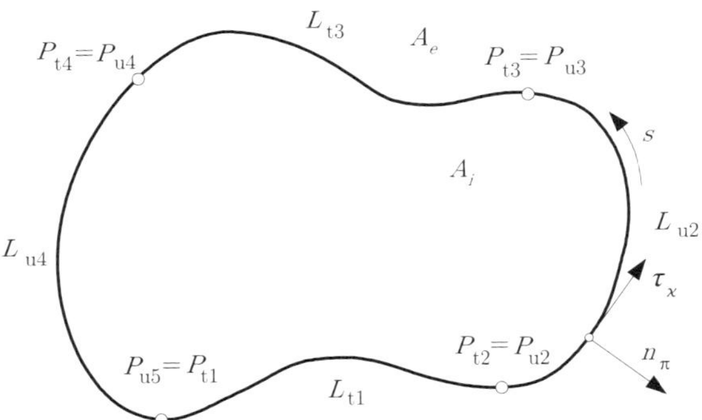

Figure 2. The simply connected inner region A_i and exterior regin A_e

The functions $\mathcal{F}_\psi$, $t_{\kappa\lambda}$, $e_{\kappa\lambda}$ and φ_3 are referred to as an elastic state of the region A_i provided that they satisfy the field equations (1), (2), (3) and (4). Let

$$\mathcal{F}_\psi,\ t_{\kappa\lambda},\ e_{\kappa\lambda},\ \varphi_3 \qquad \text{and} \qquad \overset{*}{\mathcal{F}}_\psi,\ \overset{*}{t}_{\kappa\lambda},\ \overset{*}{e}_{\kappa\lambda},\ \overset{*}{\varphi}_3$$

be two elastic states of the region A_i. Applying the Green–Gauss theorem and taking (1) into account (since there are no body forces $\overset{o}{t}_{\kappa\lambda} = 0$) one can write

$$\int_{A_i} [\epsilon_{\kappa\rho 3} e_{\kappa\lambda}\partial_\rho + \varphi_3\partial_\lambda]\overset{*}{\mathcal{F}}_\lambda dA - \int_{A_i} (\mathcal{F}_\psi\partial_\psi)\,\overset{*}{\varphi}_3 dA = \oint_{\mathcal{L}_o} n_\pi[\epsilon_{\pi\kappa 3}e_{\kappa\lambda} - \delta_{\pi\lambda}\varphi_3]\overset{*}{\mathcal{F}}_\lambda ds$$
$$- \int_{A_i}\left(\overset{*}{\mathcal{F}}_\psi\partial_\psi\right)\varphi_3 dA - \int_{A_i} (\mathcal{F}_\psi\partial_\psi)\,\overset{*}{\varphi}_3 dA\,. \tag{27}$$

REMARK 8.: Observe that the first surface integral and the sum of the last two surface integrals on the right side do not depend on the placement of the asterisk which can be put over the first or the second factor of the corresponding products.

If we replace the asterisk over the letters denoting the first elastic state and subtract (27) from the resulting equation then we get the dual Somigliana identity for plane problems

$$\int_{A_i} [\epsilon_{\kappa\rho 3}\overset{*}{e}_{\kappa\lambda}\partial_\rho + \overset{*}{\varphi}_3\partial_\lambda]\mathcal{F}_\lambda dA - \int_{A_i} \left(\overset{*}{\mathcal{F}}_\psi\partial_\psi\right)\varphi_3 dA$$
$$- \int_{A_i} [\epsilon_{\kappa\rho 3}e_{\kappa\lambda}\partial_\rho + \varphi_3\partial_\lambda]\overset{*}{\mathcal{F}}_\lambda dA - \int_{A_i} (\mathcal{F}_\psi\partial_\psi)\overset{*}{\varphi}_3 dA =$$
$$= \oint_{\mathcal{L}_o} n_\pi[\epsilon_{\pi\kappa 3}\overset{*}{e}_{\kappa\lambda} - \delta_{\pi\lambda}\overset{*}{\varphi}_3]\mathcal{F}_\lambda ds - \oint_{\mathcal{L}_o} n_\pi[\epsilon_{\pi\kappa 3}e_{\kappa\lambda} - \delta_{\pi\lambda}\varphi_3]\overset{*}{\mathcal{F}}_\lambda ds. \tag{28}$$

On the left side we have the integrals of the basic equations. As regards the right side we have the integrals of those quantities one can prescribe on the boundary. Recalling the relations (17a,b,c), (18a,b), (19) giving the basic equations and the notation (24) in which the derivative is given by (13) one can cast the Somigliana identity into a form similar to the Green identity [Jaswon and Symm (1977)]

$$\int_{A_i} \left[\mathfrak{u}_k\left(\mathfrak{D}_{kl}\overset{*}{\mathfrak{u}}_l\right) - \overset{*}{\mathfrak{u}}_k\left(\mathfrak{D}_{kl}\mathfrak{u}_l\right)\right] dA = \oint_{\mathcal{L}_o} [\mathfrak{u}_\lambda\overset{*}{\mathfrak{t}}_\lambda - \overset{*}{\mathfrak{u}}_\lambda\mathfrak{t}_\lambda]\, ds\,. \tag{29}$$

REMARK 9.: When deriving (29) we had never taken into consideration that $\overset{*}{\mathfrak{u}}_k$ and $\mathfrak{u}_k$ are compatible, i.e., fulfill the basic equation (19). Consequently, equation (29) is really an identity which is always valid if $\overset{*}{\mathfrak{u}}_k$ and $\mathfrak{u}_k$ are differentiable as many times as required – in other respects both functions can be arbitrary.

In order to establish the dual Somigliana relations it is assumed that $\overset{*}{\mathfrak{u}}_k$ is an elastic state given by (20) and (21). Quantities in the line integrals are defined by (26) and (25a,b).

In what follows we shall utilize that $\mathfrak{u}_k$ is also an elastic state.

Since the state identified by the asterisk is singular at the point Q (at the source point) we distinguish three cases depending on the location of Q with respect to the region A_i.

1. If $Q \in A_i$, then the neighborhood of Q with radius R_ε, which is denoted by A_ε and is assumed to lie in A_i, is removed from A_i and we apply the dual Somigliana identity to the double connected domain $A' = A_i \setminus A_\varepsilon$. We remark that the contour $\mathcal{L}_\varepsilon$ of A_ε and the arc $\mathcal{L}_\varepsilon^{'}$ of the contour $\mathcal{L}_\varepsilon$ within A_i coincide with each other.
2. If $Q = \overset{o}{Q} \in \partial A_i = \mathcal{L}_o$, then the part $A_i \cap A_\varepsilon$ of the neighborhood A_ε of Q is removed from A_i and we apply the dual Somigliana identity to the simply connected region $A' = A_i \setminus (A_i \cap A_\varepsilon)$. If this is the case the contour of the simply connected region consists of two arcs, the arc $\mathcal{L}_o^{'}$ left from $\mathcal{L}_o$ after the removal of A_ε and the arc $\mathcal{L}_\varepsilon^{'}$, i.e., the part of $\mathcal{L}_\varepsilon$ that lies within A_i.
3. If $Q \notin (A_i \cup \mathcal{L}_o)$ we apply the Somigliana identity to the original region A_i.

Since both $\overset{*}{\mathfrak{u}}_k$ and $\mathfrak{u}_k$ are elastic states the surface integrals in (29) are identically equal to zero. Following the well known procedure but omitting the details for the three cases we obtain:

1. the first dual Somigliana relation:

$$\mathfrak{u}_k(Q) = \oint_{\mathcal{L}_o} \mathfrak{U}_{k\lambda}(\overset{o}{M}, Q)\mathfrak{t}_\lambda(\overset{o}{M})\, ds_{\overset{o}{M}} - \oint_{\mathcal{L}_o} \mathfrak{T}_{k\lambda}(\overset{o}{M}, Q)\mathfrak{u}_\lambda(\overset{o}{M})\, ds_{\overset{o}{M}} \,. \tag{30}$$

   which is in fact a representation formula,
2. the second dual Somigliana relation

$$c_{\kappa\lambda}(\overset{o}{Q})\mathfrak{u}_\lambda(\overset{o}{Q}) = \oint_{\mathcal{L}_o} \mathfrak{U}_{\kappa\lambda}(\overset{o}{M}, \overset{o}{Q})\mathfrak{t}_\lambda(\overset{o}{M})\, ds_{\overset{o}{M}} - \oint_{\mathcal{L}_o} \mathfrak{T}_{\kappa\lambda}(\overset{o}{M}, \overset{o}{Q})\mathfrak{u}_\lambda(\overset{o}{M})\, ds_{\overset{o}{M}} \,. \tag{31}$$

   which is the integral equation of the direct method with unknowns $\mathfrak{u}_\lambda(\overset{o}{M})$ on $\mathcal{L}_u$ and $\mathfrak{t}_\lambda(\overset{o}{M})$ on $\mathcal{L}_t$. (The two line integrals in (31) should be taken in principal value.)
3. and the third dual SOMIGLIANA formula:

$$0 = \oint_{\mathcal{L}_o} \mathfrak{U}_{k\lambda}(\overset{o}{M}, Q)\mathfrak{t}_\lambda(\overset{o}{M})\, ds_{\overset{o}{M}} - \oint_{\mathcal{L}_o} \mathfrak{T}_{k\lambda}(\overset{o}{M}, Q)\mathfrak{u}_\lambda(\overset{o}{M})\, ds_{\overset{o}{M}} \,. \tag{32}$$

Making use of the first dual Somigliana formula (30) and the dual kinematic equation (1) (in the latter case one has to recall that there are no body forces consequently the particular solution is zero) one obtains a formula for the stresses $\mathfrak{s}_k = (t_{11}, t_{12}, t_{22})$:

$$\mathfrak{s}_k(Q) = \oint_{\mathcal{L}_o} \mathrm{D}_{k\lambda}(\overset{o}{M}, Q)\mathfrak{t}_\lambda(\overset{o}{M})\, ds_{\overset{o}{M}} - \oint_{\mathcal{L}_o} \mathrm{S}_{k\lambda}(\overset{o}{M}, Q)\mathfrak{u}_\lambda(\overset{o}{M})\, ds_{\overset{o}{M}} \,, \tag{33}$$

where the elements of $\mathrm{D}_{k\lambda}(\overset{o}{M}, Q)$ and $\mathrm{S}_{k\lambda}(\overset{o}{M}, Q)$ are given by

$$\mathrm{D}_{k\lambda}(\overset{o}{M}, Q) = \frac{\mu}{4\pi(1-\nu)R^2}\hat{\mathrm{D}}_{k\lambda}(\overset{o}{M}, Q) \tag{34a}$$

$$\hat{\mathrm{D}}_{k\lambda}(\overset{o}{M}, Q) = \begin{bmatrix} 6r_2 - 4r_2^3/R^2 & -2r_1 + 4r_1r_2^2/R^2 \\ -2r_1 + 4r_1r_2^2/R^2 & 2r_2 - 4r_1^2r_2/R^2 \\ 2r_2 - 4r_1^2r_2/R^2 & -6r_1 + 4r_1^3/R^2 \end{bmatrix} \tag{34b}$$

and

$$\mathrm{S}_{k\lambda}(\overset{o}{M}, Q) = \frac{1}{8\pi(1-\nu)R^2}\hat{\mathrm{S}}_{k\lambda}(\overset{o}{M}, Q) \tag{35a}$$

$$\hat{\mathrm{S}}_{11} = \frac{1}{R^2}(n_1r_1 + n_2r_2)\left[16\frac{r_2^3}{R^2} - 4(5-2\nu)r_2\right] - n_2\left[4\frac{r_2^2}{R^2} - 2(3-2\nu)\right] \tag{35b}$$

$$\hat{\mathrm{S}}_{12} = n_1\left[4\frac{r_1^2}{R^2} - 2(1-2\nu)\right] - \frac{1}{R^2}n_2r_1\left[16\frac{r_2^3}{R^2} - 4(1+2\nu)r_2\right] - \\ - \frac{1}{R^2}n_1r_2\left[16\frac{r_1^2r_2}{R^2} - 4(1-2\nu)r_2\right] \tag{35c}$$

$$\hat{\mathrm{S}}_{22} = \frac{1}{R^2} n_1 r_2 \left[16 \frac{r_1^3}{R^2} - 4(3-2\nu) r_1 \right] + \frac{1}{R^2} n_2 r_1 \left[16 \frac{r_1 r_2^2}{R^2} - 4(1-2\nu) r_2 \right] - \\ - n_2 \left[4 \frac{r_2^2}{R^2} + 2(1-2\nu) \right] \quad (35d)$$

$$\hat{\mathrm{S}}_{32} = \frac{1}{R^2}(n_1 r_1 + n_2 r_2) \left[16 \frac{r_1^3}{R^2} - 4(5-2\nu) r_1 \right] - n_1 \left[4 \frac{r_1^2}{R^2} - 2(3-2\nu) \right] \quad (35e)$$

$$\hat{\mathrm{S}}_{31} = \hat{\mathrm{S}}_{22} , \qquad \hat{\mathrm{S}}_{21} = \hat{\mathrm{S}}_{12} \quad (35f)$$

We remark that the normal is taken at $\overset{o}{M}$.

3.3 The dual Somigliana relations for exterior regions

By the exterior region A_e we mean the region outside the contour $\mathcal{L}_0$. We shall assume that the stresses are constants at infinity. These are denoted by

$$t_{11}(\infty),\ t_{12}(\infty) = t_{21}(\infty) \ \text{ and } \ t_{22}(\infty).$$

We shall also assume that there is no rigid body rotation at infinity, that is,

$$\varphi_3(\infty) = 0 . \quad (36)$$

Observe that the strains obtainable from the stresses at infinity via Hook's law are compatible for they satisfy the compatibility condition (3). The corresponding stress functions are of the form

$$\tilde{\mathrm{u}}_\lambda(Q) = \epsilon_{\alpha 3 \rho} \xi_\alpha t_{\lambda\rho}(\infty) + c_\lambda(\infty) \quad (37)$$

where $c_\lambda(\infty)$ is a constant vector to which there belong no stresses. Further let

$$\tilde{\mathrm{u}}_3(Q) = -\varphi_3(\infty) = 0 . \quad (38)$$

When deriving the dual Somigliana formula for the exterior region A_e we shall follow the line of thought of the previous section with an emphasis placed on the difference. It is assumed again that $\overset{*}{\mathrm{u}}_k$ is the elastic state described by the fundamental solutions (20) and (21). Further u_k is also an elastic state arbitrary at finite but it is to meet the conditions

$$\mathrm{u}_\kappa = \tilde{\mathrm{u}}_\lambda \ \text{ and } \ \mathrm{u}_3 = \tilde{\mathrm{u}}_3 = 0$$

at infinity. Depending on the location of the point Q we distinguish three cases in the same way as we did for the inner region A_i. It is assumed that the origin O is within the region A_i.

As regards the details we shall omit the lengthy formal transformations which, of course, need some care though the technique to be applied is well known – Szeidl (1997, 1999).

1. If $Q \in A_e$ we shall consider the triple connected region A'_e bounded by the contours $\mathcal{L}_0$, $\mathcal{L}_\varepsilon$ and the circle $\mathcal{L}_R$ with radius ${}_eR$ and center at O. Here $\mathcal{L}_\varepsilon$ is the contour of the neighborhood A_ε of Q with radius R_ε while ${}_eR$ is sufficiently large to involve both $\mathcal{L}_0$, and $\mathcal{L}_\varepsilon$. In addition A_ε is to lie wholly in A'_e. Now we apply the dual Somigliana identity to the region A'_e and take the limit of the resulting equation:

$$\mathrm{u}_k(Q) = \tilde{\mathrm{u}}_k(Q) + \oint_{\mathcal{L}_o} \mathfrak{U}_{k\lambda}(\overset{o}{M}, Q) \mathrm{t}_\lambda(\overset{o}{M})\, ds_{\overset{o}{M}} - \oint_{\mathcal{L}_o} \mathfrak{T}_{k\lambda}(\overset{o}{M}, Q) \mathrm{u}_\lambda(\overset{o}{M})\, ds_{\overset{o}{M}} . \quad (39)$$

2. If $Q = \overset{o}{Q} \in \partial A = \mathcal{L}_o$ we shall consider the double connected region A'_e bounded by $\mathcal{L}'_o$, $\mathcal{L}'_\varepsilon$ and $\mathcal{L}_R$ where $\mathcal{L}'_o$ is the part of $\mathcal{L}_o$ that is left after the removal of A_ε and $\mathcal{L}'_\varepsilon$ is the part of $\mathcal{L}_\varepsilon$ that lies within A_e. Applying again the dual Somigliana identity to A'_e and taking the limit as as $R_\varepsilon \longrightarrow 0$ and ${}_eR \longrightarrow \infty$ we get the second dual Somigliana relation for the exterior region A_e:

$$c_{\kappa\lambda}(\overset{o}{Q})\mathfrak{u}_\lambda(\overset{o}{Q}) = \tilde{\mathfrak{u}}_\kappa(Q) + \oint_{\mathcal{L}_o} \mathfrak{U}_{\kappa\lambda}(\overset{o}{M},\overset{o}{Q})\mathfrak{t}_\lambda(\overset{o}{M})\, ds_{\overset{o}{M}} - \oint_{\mathcal{L}_o} \mathfrak{T}_{\kappa\lambda}(\overset{o}{M},\overset{o}{Q})\mathfrak{u}_\lambda(\overset{o}{M})\, ds_{\overset{o}{M}} \ . \tag{40}$$

REMARK 10.: The integral equation (40) with unknowns $\mathfrak{u}_\lambda(\overset{o}{M})$ on $\mathcal{L}_u$ and $\mathfrak{t}_\lambda(\overset{o}{M})$ on $\mathcal{L}_t$ is that of the direct method for exterior regions.

3. It is obvious on the basis of all that had been said above that for $Q \in A_i$ the third dual Somigliana relation for the exterior region is of the form

$$0 = \tilde{\mathfrak{u}}_k(Q) + \oint_{\mathcal{L}_o} \mathfrak{U}_{\kappa\lambda}(\overset{o}{M},Q)\mathfrak{t}_\lambda(\overset{o}{M})\, ds_{\overset{o}{M}} - \oint_{\mathcal{L}_o} \mathfrak{T}_{\kappa\lambda}(\overset{o}{M},Q)\mathfrak{u}_\lambda(\overset{o}{M})\, ds_{\overset{o}{M}} \ . \tag{41}$$

Let

$$\tilde{\mathfrak{s}}_k = (t_{11}(\infty), t_{12}(\infty), t_{22}(\infty)) \ . \tag{42}$$

By repeating the line of thought leading to (33) we obtain a formula for the stresses at the internal points of the exterior region A_e:

$$\mathfrak{s}_k(Q) = \tilde{\mathfrak{s}}_k(Q) + \oint_{\mathcal{L}_o} \mathrm{D}_{k\lambda}(\overset{o}{M},Q)\mathfrak{t}_\lambda(\overset{o}{M})\, ds_{\overset{o}{M}} - \oint_{\mathcal{L}_o} \mathrm{S}_{k\lambda}(\overset{o}{M},Q)\mathfrak{u}_\lambda(\overset{o}{M})\, ds_{\overset{o}{M}} \tag{43}$$

where $\mathrm{D}_{k\lambda}(\overset{o}{M},Q)$ and $\mathrm{S}_{k\lambda}(\overset{o}{M},Q)$ are given by (34a,b) and (35a,...,f).

4 Basic equations for quadratic shape functions

4.1 Freedom from divergence

First we confine ourselves to the inner region A_i. We shall assume that the vector $\mathfrak{u}_l$ fulfills the basic equation (19) and $\mathfrak{t}_\lambda$ is calculated from $\mathfrak{u}_l$ by utilizing equation (14):

$$\mathfrak{t}_\lambda(\overset{o}{M}) = -n_\rho(\overset{o}{M})\left(\epsilon_{\rho\pi 3}e_{\pi\lambda}(\overset{o}{M}) - \delta_{\rho\lambda}\varphi_3(\overset{o}{M})\right) \tag{44}$$

where $\varphi_3 = \mathfrak{u}_l$ and $e_{\pi\lambda}$ is obtained from $\mathfrak{u}_\lambda$ via the dual kinematic equation (1) and the Hook law (2) in which we write $\mathfrak{u}_\lambda$ for $\mathcal{F}_\lambda$. The quantities derived from the fundamental solution will again be denoted by asterisks.

Let $\overset{*}{e}_{k\pi\lambda}(\overset{o}{M},\overset{o}{Q})$ be the strain tensor obtained from the stress function vector $\overset{*}{\mathfrak{u}}_\lambda(\overset{o}{M}) = \mathfrak{U}_{k\lambda}(\overset{o}{M},\overset{o}{Q})$. It is also clear that the corresponding rotation $\overset{*}{\varphi}_3(\overset{o}{M})$ is $\mathfrak{U}_{k3}(\overset{o}{M},Q)$. Making use of these notations and equation (44) we can write

$$\overset{*}{\mathfrak{t}}_\lambda(\overset{o}{M},Q) = \mathfrak{T}_{k\lambda}(\overset{o}{M},\overset{o}{Q}) = -n_\rho(\overset{o}{M})\left(\epsilon_{\rho\pi 3}\overset{*}{e}_{k\pi\lambda}(\overset{o}{M},\overset{o}{Q}) - \delta_{\rho\lambda}\mathfrak{U}_{k3}(\overset{o}{M},\overset{o}{Q})\right) \tag{45}$$

We can avoid computation of strongly singular integrals if we assume that $\mathfrak{u}_\lambda(\overset{o}{Q}) = \mathfrak{u}_\lambda(\overset{o}{M}) =$ constant. Under this condition $\mathfrak{t}_\lambda(\overset{o}{M}) = 0$ and equation (31) yields

$$c_{\kappa\lambda}(\overset{o}{Q})\mathfrak{u}_\lambda(\overset{o}{Q}) = \oint_{\mathcal{L}_o} n_\rho(\overset{o}{M}) \left(\epsilon_{\rho\pi 3} \overset{*}{e}_{k\pi\lambda}(\overset{o}{M},\overset{o}{Q}) - \delta_{\pi\lambda}\mathfrak{U}_{k3}(\overset{o}{M},\overset{o}{Q}) \right) \mathfrak{u}_\lambda(\overset{o}{Q})\, \mathrm{d}s_{\overset{o}{M}} \,. \tag{46}$$

Subtracting now equation (46) from (31) we have

$$I(\overset{o}{Q}) = \oint_{\mathcal{L}_o} n_\rho(\overset{o}{M}) \left[\left(\epsilon_{\rho\pi 3} \overset{*}{e}_{\kappa\pi\lambda}(\overset{o}{M},\overset{o}{Q}) - \delta_{\rho\lambda}\overset{*}{\mathfrak{U}}_{\kappa 3}(\overset{o}{M},\overset{o}{Q}) \right) \left(\mathfrak{u}_\lambda(\overset{o}{M}) - \mathfrak{u}_\lambda(\overset{o}{Q}) \right) - \right.$$
$$\left. - \overset{*}{\mathfrak{U}}_{\kappa\lambda}(\overset{o}{M},\overset{o}{Q}) \left(\epsilon_{\rho\pi 3} e_{\pi\lambda}(\overset{o}{M}) - \delta_{\rho\lambda}\varphi_3(\overset{o}{M}) \right) \right] \mathrm{d}s_{\overset{o}{M}} = 0 \,. \tag{47}$$

Let the coefficient of $n_\rho(\overset{o}{M})$ in (47) be denoted by $P_{\kappa\rho}(\overset{o}{M}) = P_{\kappa\rho}(\overset{o}{M},\overset{o}{Q})$. If we apply Gauss's theorem integral $I_\kappa(\overset{o}{Q})$ can be transformed into a surface integral:

$$I_\kappa(\overset{o}{Q}) = \oint_{\mathcal{L}_o} P_{\kappa\rho}(\overset{o}{M},\overset{o}{Q}) n_\rho(\overset{o}{M}) \mathrm{d}s_{\overset{o}{M}} = \int_{A_i} P_{\kappa\rho}(M,\overset{o}{Q}) \overset{M}{\partial}_\rho \mathrm{d}A_M \,, \tag{48}$$

where, as can be seen after some hand-made calculations – see Section A.1. of the Appendix for details – it holds

$$P_{\kappa\rho}(M,\overset{o}{Q}) \overset{M}{\partial}_\rho = 0 \,, \tag{49}$$

that is, there exists a function $\phi_\kappa(M,\overset{o}{Q})$ such that

$$P_{\kappa 1} = \frac{\partial \phi_\kappa(M,\overset{o}{Q})}{\partial x_2} \qquad \text{and} \qquad P_{\kappa 2} = -\frac{\partial \phi_\kappa(M,\overset{o}{Q})}{\partial x_1} \,. \tag{50}$$

In other words the integrand $P_{\kappa\rho}(M,\overset{o}{Q})$ is divergence free.

Let $\overset{o}{M}_1$ and $\overset{o}{M}_3$be two points on the contour for which $s_1 < s_3$. Taking now the line integral between the points $\overset{o}{M}_1$ and $\overset{o}{M}_3$ and making use of the above results we have

$$\int_{\overset{o}{M}_1}^{\overset{o}{M}_3} P_{\kappa\rho}(\overset{o}{M},\overset{o}{Q}) n_\rho(\overset{o}{M}) \mathrm{d}s_{\overset{o}{M}} = \int_{\overset{o}{M}_1}^{\overset{o}{M}_3} \tau_\pi(\overset{o}{M}) \phi_\kappa(\overset{o}{M},\overset{o}{Q}) \overset{M}{\partial}_\pi \mathrm{d}s_{\overset{o}{M}} = \phi_\kappa(\overset{o}{M}_3,\overset{o}{Q}) - \phi_\kappa(\overset{o}{M}_1,\overset{o}{Q}) \,. \tag{51}$$

REMARK 11.: It can also be proved [Szirbik (2000)] that the sum of the integrals on the right side of the dual Somigliana formulae are also divergence free independently of the position of Q relative to the region A_i (or A_e).

Assume that the contour is divided into n_{be} boundary elements. The extremities of the elements are locally denoted by M_1 and M_3. (Here and in the sequel for simplicity we have omitted the zero standing over the letters M and Q.) Then integrating element by element we have

$$I_\kappa(Q) = \sum_{e=1}^{n_{be}} \left[\phi_\kappa^e(M_3,Q) - \phi_\kappa^e(M_1,Q) \right] \,, \tag{52}$$

where the upper index e shows that ϕ_κ is taken on the e-th element.

If the region under consideration is the exterior one A_e the regularized integral equation assumes the form

$$0 = \tilde{\mathfrak{u}}_\lambda(Q) + I_\kappa(Q)\,. \tag{53}$$

4.2 Shape functions

It is clear from Figure 3 that an element has five nodal points. The first, third, and fifth nodal points are denoted by M_1, M_2 and M_3 the second and fourth by K_1 and K_2. At the points M_1, M_2 and M_3 the stress functions $\mathcal{F}_\lambda$ (or which is the same the components $\mathfrak{u}_\lambda$) are regarded as unknowns or prescribed physical quantities. At the points K_1 and K_2 the displacement derivative $\mathfrak{t}_\lambda$ is taken as an unknown or a prescribed physical quantity.

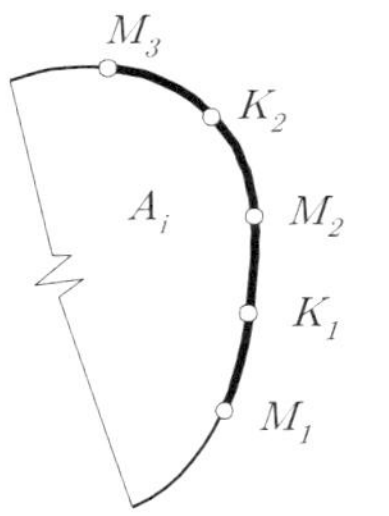

Figure 3. Nodal points on a boundary element

Over an element and its neighborhood we shall approximate the unknown vector $\mathfrak{u}_k$ by quadratic functions for the stress functions, and by linear functions for the rigid body rotation, i.e.,

$$\begin{bmatrix} \mathcal{F}_1 \\ \mathcal{F}_2 \\ -\varphi_3 \end{bmatrix}^e = \begin{bmatrix} a_1 + a_2x_1 + a_3x_2 - 2a_7x_1x_2 + a_8x_1^2 + a_9x_2^2 \\ a_4 + a_5x_1 - a_2x_2 + a_7x_2^2 - 2a_8x_1x_2 + a_{10}x_1^2 \\ a_6 + ka_7x_2 + ka_8x_1 + ka_9x_1 + ka_{10}x_2 \end{bmatrix}, \tag{54}$$

where $k = (1-\nu)/\mu$. The constants

$$(\mathbf{a}^e)^T = \begin{bmatrix} a_1 \; a_2 \; a_3 \; a_4 \; a_5 \; a_6 \; a_7 \; a_8 \; a_9 \; a_{10} \end{bmatrix} \tag{55}$$

in (54) are related to the ten physical quantities

$$(\mathbf{p}^e)^T = \left[\, \mathcal{F}_1^{M_1} \mid \mathcal{F}_2^{M_1} \mid \mathfrak{t}_1^{K_1} \mid \mathfrak{t}_2^{K_1} \mid \mathcal{F}_1^{M_2} \mid \mathcal{F}_2^{M_2} \mid \mathfrak{t}_1^{K_2} \mid \mathfrak{t}_2^{K_2} \mid \mathcal{F}_1^{M_3} \mid \mathcal{F}_2^{M_3} \,\right] \tag{56}$$

taken on the element e via the equation

$$\underbrace{\mathbf{T}^e}_{(10\times10)}\underbrace{\mathbf{a}^e}_{(10\times1)} = \underbrace{\mathbf{p}^e}_{(10\times1)}, \tag{57}$$

where the transformation matrix $\mathbf{T}^e$ depends only on the nodal coordinates and the outward unit normal at K. It can be proved that the transformation is one to one if the nodal points are different. For our latter considerations a new local coordinate system (η_1, η_2), centered at the first nodal point $M_1(x_1, x_2)$ is introduced in order to make the shape function variables conform to those of the fundamental solutions $\overset{*}{\mathfrak{U}}_{\kappa\lambda}$ and $\overset{*}{e}_{\kappa\pi\lambda}$. The new axes are parallel to the global ones. In this coordinate system

$$\begin{bmatrix} \mathcal{F}_1 \\ \mathcal{F}_2 \\ -\varphi_3 \end{bmatrix}^e = \begin{bmatrix} \hat{a}_1 + a_2\eta_1 + a_3\eta_2 - 2a_7\eta_1\eta_2 + a_8\eta_1^2 + a_9\eta_2^2 \\ \hat{a}_4 + a_5\eta_1 - a_2\eta_2 + a_7\eta_2^2 - 2a_8\eta_1\eta_2 + a_{10}\eta_1^2 \\ \hat{a}_6 + ka_7\eta_2 + ka_8\eta_1 + ka_9\eta_1 + ka_{10}\eta_2 \end{bmatrix}, \tag{58}$$

where

$$\hat{a}_1 = a_1 + a_2 x_1 + a_3 x_2 - 2a_7 x_1 x_2 + a_8 x_1^2 + a_9 x_2^2 \,, \tag{59a}$$

$$\hat{a}_4 = a_4 + a_5 x_1 - a_2 x_2 + a_7 x_2^2 - 2a_8 x_1 x_2 + a_{10} x_1^2 \,, \tag{59b}$$

$$\hat{a}_6 = k a_7 x_2 + k a_8 x_1 + k a_9 x_1 + k a_{10} x_2 \,. \tag{59c}$$

By

$$(\hat{\mathbf{a}}^e)^T = \left[\hat{a}_1 \; a_2 \; a_3 \; \hat{a}_4 \; a_5 \; \hat{a}_6 \; a_7 \; a_8 \; a_9 \; a_{10}\right] \tag{60}$$

we denote the vector of constants in the local system. For the relation between $\mathbf{a}$ and $\hat{\mathbf{a}}$ one can write

$$\underbrace{\hat{\mathbf{a}}^e}_{(10\times1)} = \underbrace{\mathbf{B}}_{(10\times10)} \underbrace{\mathbf{a}^e}_{(10\times1)} \,, \tag{61}$$

where the transformation matrix $\mathbf{B}$ depends only on the coordinates x_1 and x_2 of the point M_1. Relation (58) is a linear combination of the linearly independent state vectors

$$\begin{aligned}
\mathfrak{u}_1^T &= \left[\,1 \;\; 0 \;\; 0\,\right] , & \mathfrak{u}_6^T &= \left[0 \;\; 0 \;\; 1\right] , \\
\mathfrak{u}_2^T &= \left[\eta_1 \; -\eta_2 \; 0\right] , & \mathfrak{u}_7^T &= \left[-2\eta_1\eta_2 \; \eta_2^2 \; k\eta_2\right] , \\
\mathfrak{u}_3^T &= \left[\eta_2 \; 0 \;\; 0\right] , & \mathfrak{u}_8^T &= \left[\eta_1^2 \; -2\eta_1\eta_2 \; k\eta_1\right] , \\
\mathfrak{u}_4^T &= \left[\,0 \;\; 1 \;\; 0\,\right] , & \mathfrak{u}_9^T &= \left[\eta_2^2 \;\; 0 \;\; k\eta_1\right] , \\
\mathfrak{u}_5^T &= \left[\,0 \;\; \eta_1 \; 0\,\right] , & \mathfrak{u}_{10}^T &= \left[0 \;\; \eta_1^2 \; k\eta_2\right] .
\end{aligned} \tag{62}$$

which satisfy the fundamental equation (19).

5 Discretized equations

Fulfillment of the boundary element equations is enforced only at the points M_1, M_2 and M_3 of the elements. Assume that the collocation point Q coincides with the first nodal point of the h-th element. Further for the stress functions at this point we introduce the notations $\mathfrak{u}_1(Q) = \hat{a}_1^h$ and $\mathfrak{u}_2(Q) = \hat{a}_2^h$ in the local coordinate system. With these notations we can write

$$\left(\mathfrak{u}_\lambda(M) - \mathfrak{u}_\lambda(Q)\right)^e = \mathbf{U}^e\left(\eta_1, \eta_2\right) \tilde{\mathbf{a}}^e \,, \tag{63}$$

where the columns of the matrix $\mathbf{U}^e\left(\eta_1, \eta_2\right)$ are formed by the vectors $\mathfrak{u}_i$ $(i = 1, \ldots, 10)$. The vector of constants $\tilde{\mathbf{a}}^e$ can now be rewritten as:

$$\begin{gathered}
(\tilde{\mathbf{a}}^e)^T = \left[\beta_1^e \; a_2^e \; a_3^e \; \beta_4^e \; a_5^e \; \hat{a}_6^e \; a_7^e \; a_8^e \; a_9^e \; a_{10}^e\right] , \\
\beta_1^e = \hat{a}_1^e - \hat{a}_1^h \,, \qquad \beta_4^e = \hat{a}_4^e - \hat{a}_4^h \,.
\end{gathered} \tag{64}$$

It is clear – see Appendix A for details – that ϕ_{11}, ϕ_{21}, ϕ_{14} and ϕ_{24} are singular when the point of effect approaches the source point. In this case $\hat{a}_1^e = \hat{a}_1^h$ and $\hat{a}_2^e = \hat{a}_2^h$, therefore we can avoid the evaluation of the corresponding potential functions. Let n_{bn} be the number of nodal points. Boundary integral equation (47) can be manipulated into the form

$$\mathbf{0} = \sum_{e=1}^{n_{be}} \mathbf{\Phi}^{je} \, \tilde{\mathbf{a}}^e ; \qquad j = 1, \ldots, n_{bn} \tag{65}$$

where

$$\mathbf{\Phi}^{je} = \begin{bmatrix} \phi_{11}^{je}(M_3) - \phi_{11}^{je}(M_1) \; \phi_{12}^{je}(M_3) - \phi_{12}^{je}(M_1) \; ... \; \phi_{1,10}^{je}(M_3) - \phi_{1,10}^{je}(M_1) \\ \phi_{21}^{je}(M_3) - \phi_{21}^{je}(M_1) \; \phi_{22}^{je}(M_3) - \phi_{22}^{je}(M_1) \; ... \; \phi_{2,10}^{je}(M_3) - \phi_{2,10}^{je}(M_1) \end{bmatrix} \tag{66}$$

in which j is the number of the source point Q. With the previous notations we get

$$\mathbf{0} = \sum_{e=1}^{n_{be}} \mathbf{\Psi}^{je}\,\hat{\mathbf{a}}^e = \sum_{e=1}^{n_{be}} \mathbf{\Psi}^{je}\,\mathbf{B}^j\,\left(\mathbf{T}^{-1}\right)^e\,\mathbf{p}^e; \qquad j = 1, \ldots, n_{bn}\,, \tag{67}$$

where it has been taken into account tacitly that the difference (63) vanishes if $e = h$. For this reason we have written $\mathbf{\Psi}^{je}$ for $\mathbf{\Phi}^{je}$.

If the region under consideration is an exterior one the right side is not zero but is equal to the stress function representation $\tilde{\mathbf{u}}(\infty)$ of the stresses at infinity. Consequently, the equation system to be solved can be written as

$$\sum_{e=1}^{n_{be}} \mathbf{N}^{je}\mathbf{p}^e = \begin{cases} \mathbf{0} & \text{inner region} \\ \tilde{\mathbf{u}}(\infty) & \text{outer region} \end{cases} \qquad j = 1, \ldots, n_{bn} \tag{68}$$

where,

$$\mathbf{N}^{je} = \mathbf{\Psi}^{je}\,\mathbf{B}^j\,\left(\mathbf{T}^e\right)^{-1} \tag{69}$$

Since the stress functions are continuous at the extremities of the elements

$$\mathcal{F}_\lambda^{n_{be}M_3} = \mathcal{F}_\lambda^{1M_1}, \quad \mathcal{F}_\lambda^{eM_3} = \mathcal{F}_\lambda^{(e+1)M_1} \qquad e = 1, \ldots, n_{be} - 1\,. \tag{70}$$

Accordingly, the corresponding columns in matrices $\mathbf{N}^{je}$ can be added to each other. The continuity conditions the elements of the matrices $\mathbf{p}^e$ should meet are also to be taken into account. Under these conditions we get

$$\mathbf{K}\,\mathbf{f} = \begin{cases} \mathbf{0} & \text{inner region} \\ \tilde{\mathbf{u}}(\infty) & \text{outer region}\,. \end{cases} \tag{71}$$

Here $\mathbf{K}$ is a matrix with size $(4n_{be} * 8n_{be})$ and $\mathbf{f}$ denotes the vector of physical quantities. The columns of the matrix $\mathbf{K}$ that are multiplied by the prescribed quantities should be grouped on the right side of the equation in order to obtain the equation system to be solved. We should know at least $4n_{be}$ physical quantities from the boundary conditions to get a solvable linear equation system.

If the vectors of physical quantities $\mathbf{p}^e$ are known, then

$$\mathbf{a}^e = \left[\mathbf{T}^e\right]^{-1}\,\mathbf{p}^e\,. \tag{72}$$

is the vector of constants. With the knowledge of the vector $\mathbf{a}^e$ we can compute the stress functions at an arbitrary point. Making use of the equation (43) and applying the notations we have introduced, equation (43) can be manipulated into the form

$$\mathbf{u}(Q) = \sum_{e=1}^{n_{be}} \mathbf{\Phi}^{Qe}\,\mathbf{B}^Q\,\mathbf{a}^e \tag{73}$$

where $\mathbf{u}^T(Q) = [\mathrm{u}_1(Q) \mid \mathrm{u}_2(Q)]$ is the vector of stress functions and $\mathbf{B}^Q$ is the transformation matrix corresponding to the internal source point Q. Recalling equation (1) for the stress components at Q we can write

$$\begin{aligned} \sigma_{11}(Q) &= \left.\frac{\partial \mathrm{u}_1}{\partial x_2}\right|_Q, \qquad \sigma_{22}(Q) = -\left.\frac{\partial \mathrm{u}_2}{\partial x_1}\right|_Q, \\ \tau_{12}(Q) = \tau_{21}(Q) &= \left.\frac{\partial \mathrm{u}_1}{\partial x_1}\right|_Q = -\left.\frac{\partial \mathrm{u}_2}{\partial x_2}\right|_Q. \end{aligned} \tag{74}$$

Derivatives of the stress functions are obtained from equation (73)

$$\frac{\partial}{\partial x_\kappa}\mathbf{u}(Q) = \sum_{e=1}^{n_{be}} \left[\left(\frac{\partial}{\partial x_\kappa}\mathbf{\Phi}^{Qe}\right)\mathbf{B}^Q + \mathbf{\Phi}^{Qe}\left(\frac{\partial}{\partial x_\kappa}\mathbf{B}^Q\right)\right] \mathbf{a}^e. \tag{75}$$

Exploiting the relationship

$$\frac{\partial}{\partial x_\kappa(Q)}\mathbf{\Phi}(M,Q) = -\frac{\partial}{\partial \eta_\kappa}\mathbf{\Phi}(M,Q) \tag{76}$$

the above equation (75) can be rewritten as

$$\frac{\partial}{\partial x_\kappa}\mathbf{u}(Q) = \sum_{e=1}^{n_{be}} \left[\mathbf{\Phi}^{Qe}\left(\frac{\partial}{\partial x_\kappa}\mathbf{B}^Q\right) - \left(\frac{\partial}{\partial \eta_\kappa}\mathbf{\Phi}^{Qe}\right)\mathbf{B}^Q\right] \mathbf{a}^e. \tag{77}$$

REMARK 12.: There is no need to calculate the derivatives of known shape functions in the local coordinate system because they are given by equation (50) as $P_{\kappa 1}(M,Q)$ and $-P_{\kappa 2}(M,Q)$.

REMARK 13.: For exterior regions the above formulae should be modified to include $\tilde{\mathrm{u}}_\lambda(Q)$.

6 Examples

Three examples are presented. The region under consideration including its matter, is the same for the first two cases. $r_0 = 10[\mathrm{mm}]$, $\mu = 8 \cdot 10^4[\mathrm{MPa}]$, $\nu = 0.3$

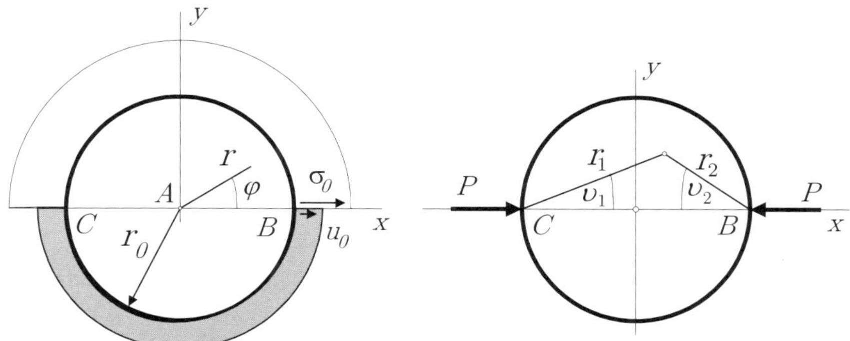

Figure 4. Circular regions

Problem 1. On arc BC the radial stress (normal stress) is $\sigma_o = 100$ [MPa] (there are no shear stresses). On arc CB the radial displacement is $u_o = (1-2\nu)\sigma_o r_o/2\mu$ (there is no tangential displacement). One can check with ease that these values determine a homogenous state of stress of the region. The exact solutions are given by the equations

$$\mathcal{F}_1 = \mathcal{F}_x = \sigma_o y = \sigma_o r \sin\varphi\,, \qquad \mathcal{F}_2 = \mathcal{F}_y = -\sigma_o x = -\sigma_o r\cos\varphi\,,$$

$$\sigma_{xx} = \sigma_{yy} = \sigma_o\,, \qquad \tau_{xy} = 0\,,$$

$$u_x = \frac{1-2\nu}{2\mu}\sigma_o x = \frac{1-2\nu}{2\mu}\sigma_o r\cos\varphi\,,$$

$$u_y = \frac{1-2\nu}{2\mu}\sigma_o y = \frac{1-2\nu}{2\mu}\sigma_o r\sin\varphi\,,$$

where r and φ are polar coordinates. On arc BC and CB

$$\mathfrak{u}_x = \mathcal{F}_x = \sigma_o r_o \sin\varphi\,, \qquad \mathfrak{u}_y = \mathcal{F}_y = -\sigma_o r_o\cos\varphi$$

and

$$-\mathfrak{t}_x = \frac{d\mathfrak{u}_x}{ds} = -\frac{1-2\nu}{2\mu}\sigma_o\sin\varphi\,, \qquad -\mathfrak{t}_y = \frac{d\mathfrak{u}_y}{ds} = \frac{1-2\nu}{2\mu}\sigma_o\cos\varphi$$

are the boundary conditions. The contour was divided into 16 equidistant elements. The table below contains the numerical results for the stresses

x [mm]	y [mm]	σ_{xx} [MPa]	τ_{xy} [MPa]	σ_{yy} [MPa]
-7.50	0.00	100.0000	0.0000000	100.0000
-5.00	0.00	100.0000	0.0000002	100.0000
-2.50	0.00	100.0000	0.0000003	100.0000
0.00	0.00	100.0000	0.0000005	100.0000
2.50	0.00	100.0000	0.0000005	100.0000
5.00	0.00	100.0000	0.0000004	100.0000
7.50	0.00	100.0000	0.0000001	100.0000
7.50	5.00	100.0000	0.0000002	100.0000
5.00	7.50	100.0000	0.0000004	100.0000
9.00	1.00	100.0000	0.0000002	100.0000

Problem 2. The region is subjected to a pair of compressive forces with magnitude 100.0 N/mm Consequently the boundary conditions on the arcs AB and BC are

$$\mathcal{F}_x = P = -100\,, \qquad \mathcal{F}_y = 0\,,$$

and

$$\mathcal{F}_x = 0\,, \qquad \mathcal{F}_y = 0\,,$$

respectively. With the notations of Figure 4

$$\sigma_{xx} = \frac{2P}{\pi}\left[\frac{\cos^3\vartheta_1}{r_1} + \frac{\cos^3\vartheta_2}{r_2}\right] - \frac{P}{\pi r_o}\,, \qquad \tau_{xy} = -\frac{2P}{\pi}\left[\frac{\sin\vartheta_1\cos^2\vartheta_1}{r_1} - \frac{\sin\vartheta_2\cos^2\vartheta_2}{r_2}\right]\,,$$

$$\sigma_{yy} = \frac{2P}{\pi}\left[\frac{\sin^2\vartheta_1\cos\vartheta_1}{r_1} + \frac{\sin^2\vartheta_2\cos\vartheta_2}{r_2}\right] - \frac{P}{\pi r_o}$$

are the exact solutions [Muskhelisvili (1966)]. Figures 5 to 8 represent the exact and the numerical solutions. The latter is denoted by diamonds. In this case the contour was divided into 40 equidistant elements. The pairs of elements that meet at A and B are partially discontinuous.

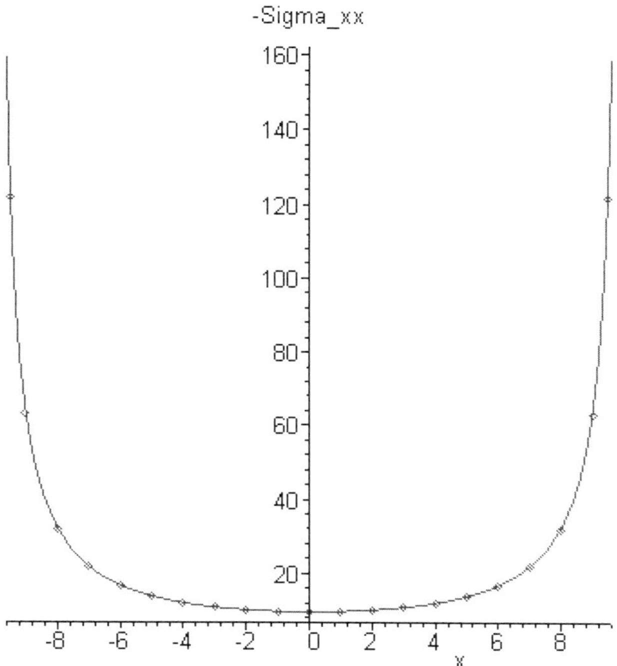

Figure 5. Exact and numerical solution: $-\sigma_{xx}$ along the horizontal diameter

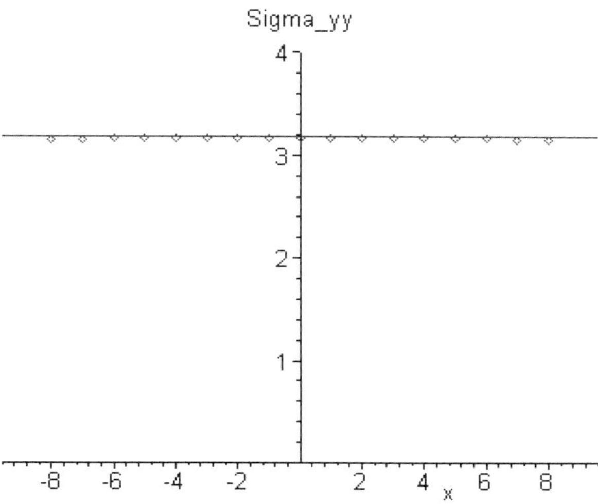

Figure 6. Exact and numerical solution: σ_{yy} along the horizontal diameter

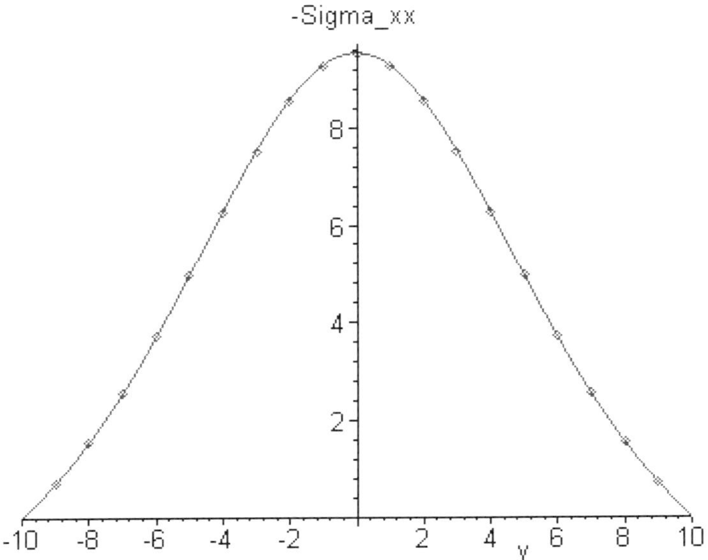

Figure 7. Exact and numerical solution: $-\sigma_{xx}$ along the vertical diameter

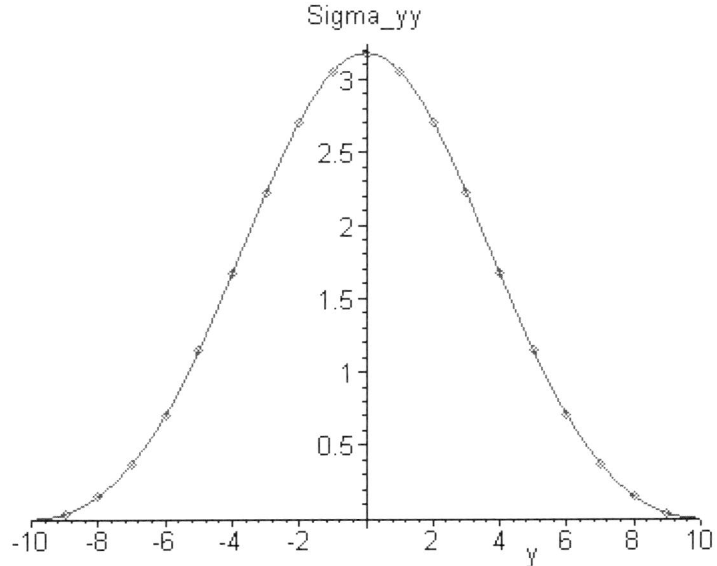

Figure 8. Exact and numerical solution: σ_{yy} along the vertical diameter

Problem 3. Though the contour $\mathcal{L}_o$ and the material are the same as in the previous examples the region under consideration is the outer one for which a constant stress state $\sigma_{xx}(\infty) = 100[\text{MPa}]$, $\sigma_{xy}(\infty) = \sigma_{yx}(\infty) = \sigma_{yy}(\infty) = 0$ is prescribed at infinity.

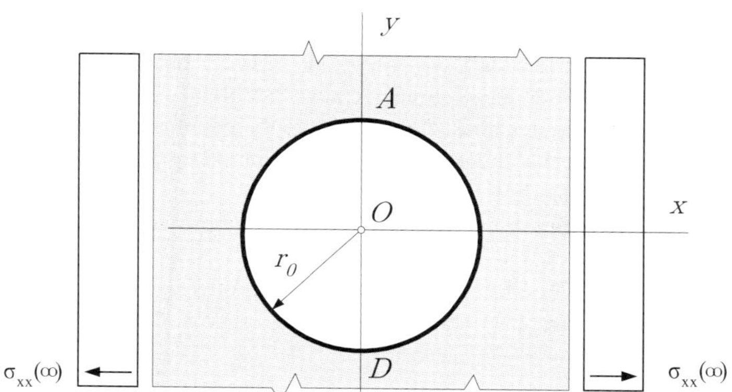

Figure 9. Outer region bounded by a circle with radius $r_o = 10\,[\text{mm}]$ and centered at O

It is well known that the formulae

$$\sigma_{rr} = \frac{\sigma_{xx}(\infty)}{2}\left[\left(1-\frac{r_o^2}{r^2}\right)+\left(1+\frac{3r_o^4}{r^4}-\frac{4r_o^2}{r^2}\right)\cos 2\varphi\right],$$

$$\sigma_{\varphi\varphi} = \frac{\sigma_{xx}(\infty)}{2}\left[\left(1+\frac{r_o^2}{r^2}\right)-\left(1+\frac{3r_o^4}{r^4}\right)\cos 2\varphi\right],$$

$$\sigma_{r\varphi} = \frac{\sigma_{xx}(\infty)}{2}\left[\left(1-\frac{3r_o^4}{r^4}+\frac{2r_o^2}{r^2}\right)\sin 2\varphi\right]$$

written in polar coordinates give the exact solution to this problem – see [Timoshenko and Goodier (1951)] or [Muskhelisvili (1966)]. The table below show both the stresses we computed and the exact solution on the y axis. The contour was divided into 16 equidistant elements.

x [mm]	y [mm]	σ_{xx} [MPa]	τ_{xy} [MPa]	σ_{yy} [MPa]
0.00	10.00	300.5371	0.0000000	0.000000
		300.0000	0.0000000	0.000000
0.00	11.00	243.8070	0.0000000	21.72130
		243.7743	0.0000000	21.51494
0.00	12.00	206.9581	0.0000000	31.98909
		207.0602	0.0000000	31.82870
0.00	13.00	182.0110	0.0000000	36.32535
		182.1049	0.0000000	36.23823
0.00	14.00	164.4815	0.0000000	37.53069
		164.5564	0.0000000	37.48438
0.00	15.00	151.7924	0.0000000	37.06093
		151.8518	0.0000000	37.03704

7 Concluding remarks

In accordance with our aims we have clarified what the supplementary conditions of single valuedness are for a class of mixed boundary value problems in dual system assuming multiply connected domains.

The fundamental solutions for the stress functions of order one have also been constructed. In the knowledge of the fundamental solutions we have established the dual Somigliana relations both for inner regions and for outer ones. These involve the equations of the direct method.

It has also been shown that the integrand of the direct boundary element method is divergence free in the dual system of plane elasticity.

The corresponding shape functions have been determined provided that the approximation is quadratic.

A program has been developed in Fortran 90 for the numerical solution by using quadratic shape functions. The three examples illustrate the applicability of the algorithm.

Appendix A

A.1. Transformation of the integrand in the surface integral (48). If we recall that $\overset{*}{\mathrm{u}}_\lambda = \mathfrak{U}_{\lambda\kappa}$ and $\overset{*}{\varphi} = \mathfrak{U}_{3\kappa}$ then the corresponding compatibility equation and the symmetry conditions for the stresses are of the form

$$\left[\epsilon_{\rho\pi 3}\overset{*}{e}_{\kappa\pi\lambda}(M,\overset{o}{Q}) - \delta_{\rho\lambda}\mathfrak{U}_{3\kappa}(M,\overset{o}{Q})\right]\overset{M}{\partial}_\rho = 0 \tag{78}$$

and

$$\epsilon_{\lambda\pi 3}\overset{*}{\mathrm{t}}_{\kappa\pi\lambda} = 0\,. \tag{79}$$

Since $\overset{*}{\mathrm{t}}_{\kappa\pi\lambda}$ and $e_{\kappa\lambda}$ are elastic states we can write

$$\overset{*}{\mathrm{t}}_{\kappa\pi\lambda}e_{\pi\lambda} = \mathrm{t}_{\pi\lambda}\overset{*}{e}_{\kappa\pi\lambda} \tag{80}$$

With regard to (3), (4), (78), (79) and (80) the left side of equation (49) can be transformed as follows

$$\begin{aligned}
P_{\kappa\rho}(M,\overset{o}{Q})\overset{M}{\partial}_\rho = &-\left(\mathfrak{U}_{\kappa\lambda}(M,\overset{o}{Q})\overset{M}{\partial}_\rho\right)\left[\epsilon_{\rho\pi 3}e_{\pi\lambda}(M) - \delta_{\rho\lambda}\varphi_3(M)\right] - \\
&-\mathfrak{U}_{\kappa\lambda}(M,\overset{o}{Q})\left(\left[\epsilon_{\rho\pi 3}e_{\pi\lambda}(M) - \delta_{\rho\lambda}\varphi_3(M)\right]\overset{M}{\partial}_\rho\right) + \\
&+\left(\left[\epsilon_{\rho\pi 3}\overset{*}{e}_{\kappa\pi\lambda}(M,\overset{o}{Q}) - \delta_{\rho\lambda}\mathfrak{U}_{\kappa 3}(M,\overset{o}{Q})\right]\overset{M}{\partial}_\rho\right)\mathrm{u}_\lambda(M) + \\
&+\left(\epsilon_{\rho\pi 3}\overset{*}{e}_{\kappa\pi\lambda}(M,\overset{o}{Q}) - \delta_{\rho\lambda}\mathfrak{U}_{\kappa 3}(M,\overset{o}{Q})\right)\left(\mathrm{u}_\lambda(M)\overset{M}{\partial}_\rho\right) = \\
= &-\left(\mathfrak{U}_{\kappa\lambda}(M,\overset{o}{Q})\overset{M}{\partial}_\rho\right)\epsilon_{\rho\pi 3}e_{\pi\lambda}(M) + \epsilon_{\rho\pi 3}\overset{*}{e}_{\kappa\pi\lambda}(M,\overset{o}{Q})\left(\mathrm{u}_\lambda(M)\overset{M}{\partial}_\rho\right) + \\
&+\left(\mathfrak{U}_{\kappa\lambda}(M,\overset{o}{Q})\overset{M}{\partial}_\rho\epsilon_{\rho\pi 3}\right)\epsilon_{\lambda\pi 3}\varphi_3(M) - \left(\mathrm{u}_\lambda(M)\overset{M}{\partial}_\rho\epsilon_{\rho\pi 3}\right)\epsilon_{\lambda\pi 3}\mathfrak{U}_{\kappa 3}(M,\overset{o}{Q}) = \\
= &-\overset{*}{\mathrm{t}}_{\kappa\pi\lambda}e_{\pi\lambda} + \mathrm{t}_{\pi\lambda}\overset{*}{e}_{\kappa\pi\lambda} + \epsilon_{\lambda\pi 3}\overset{*}{\mathrm{t}}_{\kappa\pi\lambda}\varphi_3(M) - \epsilon_{\lambda\pi 3}\mathrm{t}_{\kappa\pi}\overset{*}{\varphi}_3(M) = 0
\end{aligned}$$

In other words $P_{\kappa\rho}$ is divergence free.

A.2. Quadratic shape functions. If the approximation is quadratic the shape functions are as follows:

$$\phi_{11} = \frac{1}{2\pi}\arctan\frac{\eta_2}{\eta_1} + \frac{1}{4\pi(1-\nu)}\frac{\eta_1\eta_2}{\eta_1^2+\eta_2^2},$$

$$\phi_{12} = \frac{-\eta_2}{4\pi(1-\nu)}\left(\ln\sqrt{\eta_1^2+\eta_2^2} + \frac{4\nu-3}{2} + \frac{2\eta_2^2}{\eta_1^2+\eta_2^2}\right),$$

$$\phi_{13} = \frac{\eta_1}{4\pi(1-\nu)}\left((1-\nu)\ln\sqrt{\eta_1^2+\eta_2^2} + \frac{3-\nu}{2} - \frac{\eta_1^2}{\eta_1^2+\eta_2^2}\right),$$

$$\phi_{14} = \frac{-1}{4\pi(1-\nu)}\left((1-2\nu)\ln\sqrt{\eta_1^2+\eta_2^2} + \frac{\eta_1^2}{\eta_1^2+\eta_2^2}\right),$$

$$\phi_{15} = \frac{\eta_1}{4\pi(1-\nu)}\left(\nu\ln\sqrt{\eta_1^2+\eta_2^2} + \frac{5\nu}{2} - \frac{\eta_1^2}{\eta_1^2+\eta_2^2}\right), \quad \phi_{16} = \frac{-\mu\eta_2}{4\pi(1-\nu)}\left(2\ln\sqrt{\eta_1^2+\eta_2^2} + 3\right),$$

$$\phi_{17} = \frac{1}{8\pi(1-\nu)}\left\{\left[-(1-\nu)\eta_1^2 - (\nu-2)\eta_2^2\right]\ln\left(\eta_1^2+\eta_2^2\right) - 2(1-\nu)\left(\eta_1^2+\eta_2^2\right) + \frac{6\eta_2^4}{\eta_1^2+\eta_2^2}\right\},$$

$$\phi_{18} = \frac{1}{4\pi(1-\nu)}\left\{\left[3-5\nu-\nu\ln\left(\eta_1^2+\eta_2^2\right)\right]\eta_1\eta_2 - \frac{3\eta_1\eta_2^3}{\eta_1^2+\eta_2^2}\right\},$$

$$\phi_{19} = \frac{1}{4\pi(1-\nu)}\left\{\left[4-3\nu+(1-\nu)\ln\left(\eta_1^2+\eta_2^2\right)\right]\eta_1\eta_2 - \frac{\eta_1^3\eta_2}{\eta_1^2+\eta_2^2}\right\},$$

$$\varphi_{1,10} = \frac{1}{8\pi(1-\nu)}\left\{\eta_1^2\left[(\nu-1)+\nu\ln\left(\eta_1^2+\eta_2^2\right)\right] + \eta_2^2(1-\nu)\left[4+\ln\left(\eta_1^2+\eta_2^2\right)\right] - \frac{2\eta_1^4}{\eta_1^2+\eta_2^2}\right\},$$

$$\phi_{21} = \frac{1}{4\pi(1-\nu)}\left((1-2\nu)\ln\sqrt{\eta_1^2+\eta_2^2} + \frac{\eta_2^2}{\eta_1^2+\eta_2^2}\right),$$

$$\phi_{22} = \frac{-\eta_1}{4\pi(1-\nu)}\left(\ln\sqrt{\eta_1^2+\eta_2^2} + \frac{4\nu-3}{2} + \frac{2\eta_1^2}{\eta_1^2+\eta_2^2}\right),$$

$$\phi_{23} = \frac{-\eta_2}{4\pi(1-\nu)}\left(\nu\ln\sqrt{\eta_1^2+\eta_2^2} + \frac{5\nu}{2} - \frac{\eta_2^2}{\eta_1^2+\eta_2^2}\right),$$

$$\phi_{24} = -\frac{1}{2\pi}\arctan\frac{\eta_1}{\eta_2} - \frac{1}{4\pi(1-\nu)}\frac{\eta_1\eta_2}{\eta_1^2+\eta_2^2},$$

$$\phi_{25} = \frac{-\eta_2}{4\pi(1-\nu)}\left((1-\nu)\ln\sqrt{\eta_1^2+\eta_2^2} + \frac{3-\nu}{2} - \frac{\eta_2^2}{\eta_1^2+\eta_2^2}\right),$$

$$\phi_{26} = \frac{\eta_1}{4\pi(1-\nu)}\left(2\ln\sqrt{\eta_1^2+\eta_2^2} + 3\right),$$

$$\phi_{27} = \frac{1}{4\pi(1-\nu)}\left\{\eta_1\eta_2\left[5\nu-3+\nu\ln\left(\eta_1^2+\eta_2^2\right)\right] + \frac{3\eta_1^3\eta_2}{\eta_1^2+\eta_2^2}\right\},$$

$$\phi_{28} = \frac{1}{8\pi(1-\nu)}\left\{\left[(1-\nu)\eta_2^2 + (\nu-2)\eta_1^2\right]\ln\left(\eta_1^2+\eta_2^2\right) + 2(1-\nu)\left(\eta_1^2+\eta_2^2\right) - \frac{6\eta_1^4}{\eta_1^2+\eta_2^2}\right\},$$

$$\phi_{29} = \frac{1}{8\pi(1-\nu)}\left\{\eta_2^2\left[(1-\nu)-\nu\ln\left(\eta_1^2+\eta_2^2\right)\right] - \eta_1^2(1-\nu)\left[4+\ln\left(\eta_1^2+\eta_2^2\right)\right] + \frac{2\eta_2^4}{\eta_1^2+\eta_2^2}\right\},$$

$$\phi_{2,10} = \frac{1}{4\pi(1-\nu)}\left\{\left[3\nu-4-(1-\nu)\ln\left(\eta_1^2+\eta_2^2\right)\right]\eta_1\eta_2 + \frac{\eta_1\eta_2^3}{\eta_1^2+\eta_2^2}\right\}.$$

References

Bertóti, E. (1994). Indeterminacy of first order stress functions and the stress and rotation based formulation of linear elasticity. *Computational Mechanics* 14:249–265.

Bertóti, E. (1996). Stress and rotation-based hierarchic models for laminated composites. *International Journal for Numnerical Methods in Engineering* 39:2647–2671.

de Veubeke, B. M. F., and Millard, A. (1976). Discretization of stress fields in finite element method. *J. Franklin Inst.* 302:389–412.

de Veubeke, B. M. F. (1975). Stress function approach. In *Proc. World Cong. on Finite Element Methods in Structural Mechanics, Bournemouth, U.K.*, J1–J51.

Eringen, A. C. (1951). *Mechanics of Continua*. New York London Sydney: John Wiley & Sons. Inc.

Jaswon, M. A., and Symm, G. T. (1977). *Integral Equation Methods in Potential Theory and Elastostatics*. London – NewYork – San Francisco: Academic Press.

Jaswon, M. A., Maiti, M., and Symm, G. T. (1967). Numerical biharmonic analysis and some applications. *Int. J. Solids Structures* 3:309–332.

Muskhelisvili, I. I. (1966). *Some Fundamental Problems of Mathematical Theory of Elasticity*. Moscow: Publisher NAUKA, sixth edition.

Nagarjan, A., Lutz, E., and Mukherjee, S. (1994). A novel boundary element method for linear elasticity with no numerical integration for two dimensional and line integrals for three-dimensional problems. *Journal of Applied Mechanics* 264(61):264–269.

Phan, A. V., Mukherjee, S., and Mayer, J. R. R. (1997). The boundary contour method for two-dimensional linear elasticity with quadratic boundary elements. *Computational Mechanics* 20:310–319.

Phan, A. V., Mukherjee, S., and Mayer, J. R. R. (1998). Stresses, stress sensitivities and shape optimization in two dimensional linear elasticity by the boundary contour method. *International Journal for Numerical Methods in Engineering* 42:1391–1407.

Szeidl, G. (1997). *Dual Problems of Continuum Mechanics (Derivation of Defining Equations, Single Valuedness of Mixed Boundary Value Problems, Boundary Element Method for Plane problems)*. Habilitation Booklets of the Miskolc University, Faculty of Mechanical Engineering, University of Miskolc, Department of Mechanics. 52-63, (In Hungarian).

Szeidl, G. (1999). Boundary integral equations for plane problems – remark to the formulation for exterior regions. *Publications of the University of Miskolc, Series D, Natural Sciences, Mathematics* 40:79–88.

Szirbik, S. (2000). Boundary contour method for plane problems in a dual formulation with linear elements. *Journal of Computational and Applied Mechanics* 1(2):205–222.

Timoshenko, S., and Goodier, J. N. (1951). *Theory of Elasticity*. New York Toronto London: McGraw-Hill.